EXERCISES

SECOND EDITION

for the General, Organic, & Biochemistry LABORATORY

William G. O'Neal

University of Richmond

MORTON PUBLISHING

925 W. Kenyon Avenue, Unit 12
Englewood, CO 80110

morton-pub.com

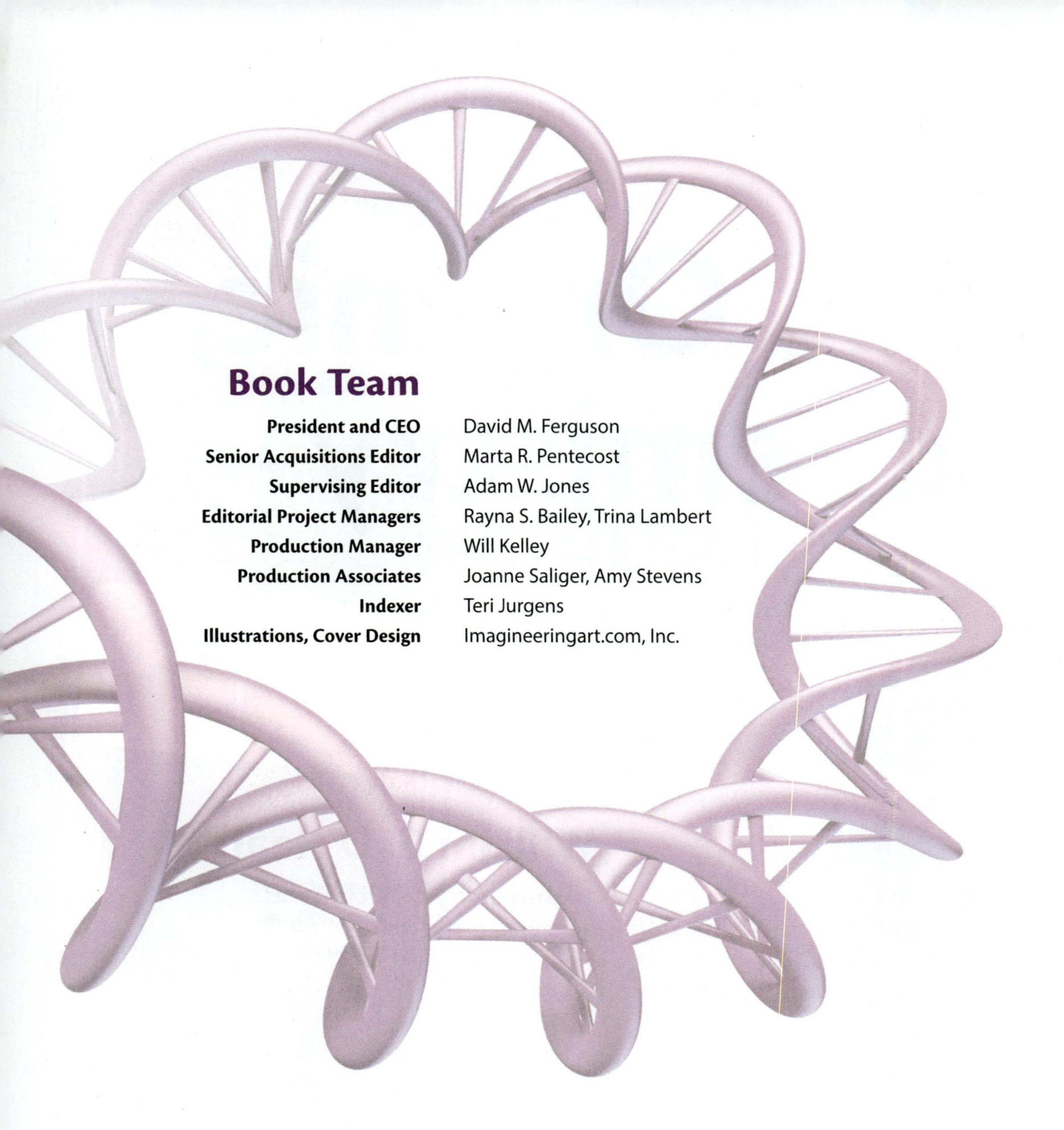

Book Team

President and CEO	David M. Ferguson
Senior Acquisitions Editor	Marta R. Pentecost
Supervising Editor	Adam W. Jones
Editorial Project Managers	Rayna S. Bailey, Trina Lambert
Production Manager	Will Kelley
Production Associates	Joanne Saliger, Amy Stevens
Indexer	Teri Jurgens
Illustrations, Cover Design	Imagineeringart.com, Inc.

Printed in the United States of America

10 9 8 7 6 5 4 3 2 1

ISBN-10: 1-64043-044-X
ISBN-13: 978-1-64043-044-0

Library of Congress Control Number: 2019943024

Preface

This text grew from my desire to bring chemistry to life in a new way for students in the General, Organic, and Biochemistry and Introductory Chemistry courses I taught for many years at Coastal Carolina Community College. While the curricula in these types of courses can be very different, they share a common challenge: For most students, this is the first and last chemistry course they will ever take. It also may be one of the most difficult academic obstacles they will face. Like many instructors, I believe students are inspired to achieve greater success when they are given context and relevance for their coursework. This laboratory manual was written with that guiding principle in mind. Many of the labs are derivations of classic teaching experiments, but my hope is that they have been reimagined here in a way that is easily approachable by today's students and enhances the academic experience.

Like the first edition, the experiments and organization of the second edition were deliberately designed to support an integrated approach to the one-semester, terminal General, Organic, and Biochemistry course. The early labs examine certain fundamental aspects of chemistry, and subsequent labs are intended to explore concepts of general and organic chemistry in a biological context as much as possible. As a result, the sequence of labs is somewhat atypical, and the manual is not divisible into three clearly defined sections. For example, the functional group concept is introduced in the context of an aspirin synthesis lab positioned much earlier than in traditional texts. However, each lab is written as a stand-alone exercise in order for instructors to maintain maximum flexibility in planning their courses.

The experiments and post-laboratory exercises in this edition also are designed to illustrate concepts as clearly as possible and to teach the skills of scientific thinking. The emphasis is not on teaching rigorous experimental technique because that task is more fitting for a major's curriculum. Where trade-offs between experimental design and conceptual illustration were necessary, you will find that I invariably favored a clearer illustration of the concept. Color photographs and high-quality art were a high priority because I believe they both engage students and significantly improve pre- and post-laboratory comprehension and analysis. To help students develop proficiency in skillful scientific argumentation, this edition explicitly emphasizes the "claim, evidence, rationale" (CER) paradigm. Labs that require students to determine something "unknown" from their data incorporate a "CER table" designed so that students will learn to include and connect these three pillars of effective scientific reasoning.

This edition includes three new laboratory exercises, which were added in response to the valuable feedback from instructors who had experience with the first edition.

- **Lab 1** (**Be Methodical: Think Like a Chemist**) serves as an introduction to the scientific process. Through simple tasks, students develop hypotheses, make careful observations, interpret those observations, recognize the limits of their data, and become familiar with the structure of a scientific argument.

- **Lab 18** (**Pretty in Pink: Titration of Vinegar**) provides an opportunity to perform a traditional titration experiment in a real-world context.
- **Lab 21** (**The Blue-Green Protein: Denaturation of Phycocyanin**) allows students to explore protein structure through visible changes in a sample derived from a natural source.

Several labs from the first edition have been revised to improve the student experience and/or enhance safety.

- **Lab 1** (**Measure Up: Laboratory Measurements**) in the first edition introduced students to laboratory measurements and then applied those skills to a density exercise. The length of this lab could be challenging, especially since many instructors wanted to use it during the first day of lab. In this second edition, the content has been split into separate labs. Lab 2 (It's Significant: Laboratory Measurements and Units) focuses on tools and measurements. It also includes new exercises on unit conversion. Lab 3 (Be Precise: Density and Measurement Precision) introduces students to the concept of density and encourages them to think about the value of taking multiple measurements of the same quantity in order to formulate a claim.
- The procedure in **Lab 7** (**Eat Your Wheaties: Nutritional Minerals in Cereal**) was restructured in response to feedback that it was too long and students were sometimes confused by the number of samples they were asked to test. The streamlined procedure reduces and consolidates the sample testing process.
- The procedure in **Lab 12** (**Skeletons and Suffixes: Hydrocarbon Isomerism and Bond Reactivity**) was revised to provide more reliable results and to eliminate the use of molecular bromine. Students now generate the reactive species in solution from the much-less hazardous mixture of sodium bromide, acid, and bleach.
- **Lab 15** (**A Lot of Hot Air: Simple Gas Laws in Action**) was revised to incorporate a more reliable apparatus and more realistic consideration of the set of gas laws underlying the exercise.
- **Lab 23** (**Blueprint of a Strawberry: DNA Extraction**) was revised to incorporate additional chemical testing to provide more structural evidence for the identity of the extracted DNA.

Throughout the lab manual, much of the introductory text has been revised for clarity and many of the illustrations have been enhanced. Nearly all the data sheets have been reformatted so that most of the data is collected in well-labeled tables. Students reported that this format streamlined the in-lab recording of data and the post-lab analysis. Instructors may also find that is simplifies the grading process.

It is my sincere hope that the second edition of *Exercises for the General, Organic, & Biochemistry Laboratory* will provide you with a productive, enlightening, and fun laboratory experience. I welcome all comments and suggestions for future editions of this book. Please feel free to contact me at **chemistry@morton-pub.com**.

Acknowledgments

I am so grateful for all the wonderful people who helped make this book a reality. First and foremost, thank you to my family who has continued to support my aspirations while keeping me grounded in everything that is truly important. Thank you for making me want to come home early.

I am so proud to be associated with the mission and work of the talented individuals at Morton Publishing, as established by founder Doug Morton and continued by President David Ferguson. Students deserve high quality, affordable educational materials, and Morton's dedication to that vision makes it a leader in the field. Marta Pentecost, I still trace all of this back to your not hanging up the phone when I first rambled through a wish list for a GOB lab manual. I appreciate your willingness to take a chance on that new author and for your support ever since. Your enthusiasm for this edition was just as contagious as for the first. Rayna Bailey, Trina Lambert, Adam Jones, Will Kelley, Joanne Saliger, Amy Stevens, and the art team at Imagineering, you are (as always) wizards who can make words shine, turn pages into landscapes, and breathe life into sketches. Thanks to the whole team for your experience and expertise.

I'd like to thank the following reviewers, who gave their input for the first edition of *Exercises for the General, Organic, & Biochemistry Laboratory:* Rebecca Hailey, Florida State College at Jacksonville; Dr. Stephanie A. Havemann, Alvin Community College; L. Jaye Hopkins, Spokane Community College; and Cheryl Lavoie, Simmons College. The following adopters provided valuable feedback for the second edition: Mark Erickson, Hartwick College; Olga Fryszman, MiraCosta College; Louis Fadel, Ivy Tech Community College; Elizabeth Bolen, Muskegon Community College; Aaron Hutchinson, Cedarville University; Bridget Loftus, Spoon River College; Jeff Vargason, George Fox University; Cynthia Gilley, San Diego Miramar College; Rosana Pedroza, Cuyamaca College; Jean Dupon, Coastline College; Aaron Goodpaster, Ivy Tech Community College; Julia Keller, Florida State College at Jacksonville; and Arun Datta, Muskegon Community College. Their comments, suggestions, and critiques were invaluable to development of this edition.

Finally, I continue to be indebted to my students from Coastal Carolina Community College, who were the original inspiration for this project. They slogged through typo-laden draft versions for the first edition with a sense of humor and a willingness to accept and analyze failures right along with me. When we finally had the first edition to test drive, they taught me about everything I had right and—more importantly—everything I had wrong. Thank you.

Biography

Will O'Neal is a faculty member in the Chemistry Department at the University of Richmond, where he directs the organic chemistry laboratory curriculum. He was previously a chemistry instructor and physical sciences department head at Coastal Carolina Community College in Jacksonville, North Carolina. He received a Ph.D. in chemistry from Dartmouth College, where his research focused on the development of new synthetic routes to macrocyclic tetrapyrroles of the chlorin and bacteriochlorin classes. Will served as an American Chemical Society Congressional Fellow from 2008 to 2009 and subsequently worked as a congressional staffer. He and his family reside in Richmond, Virginia.

Contents

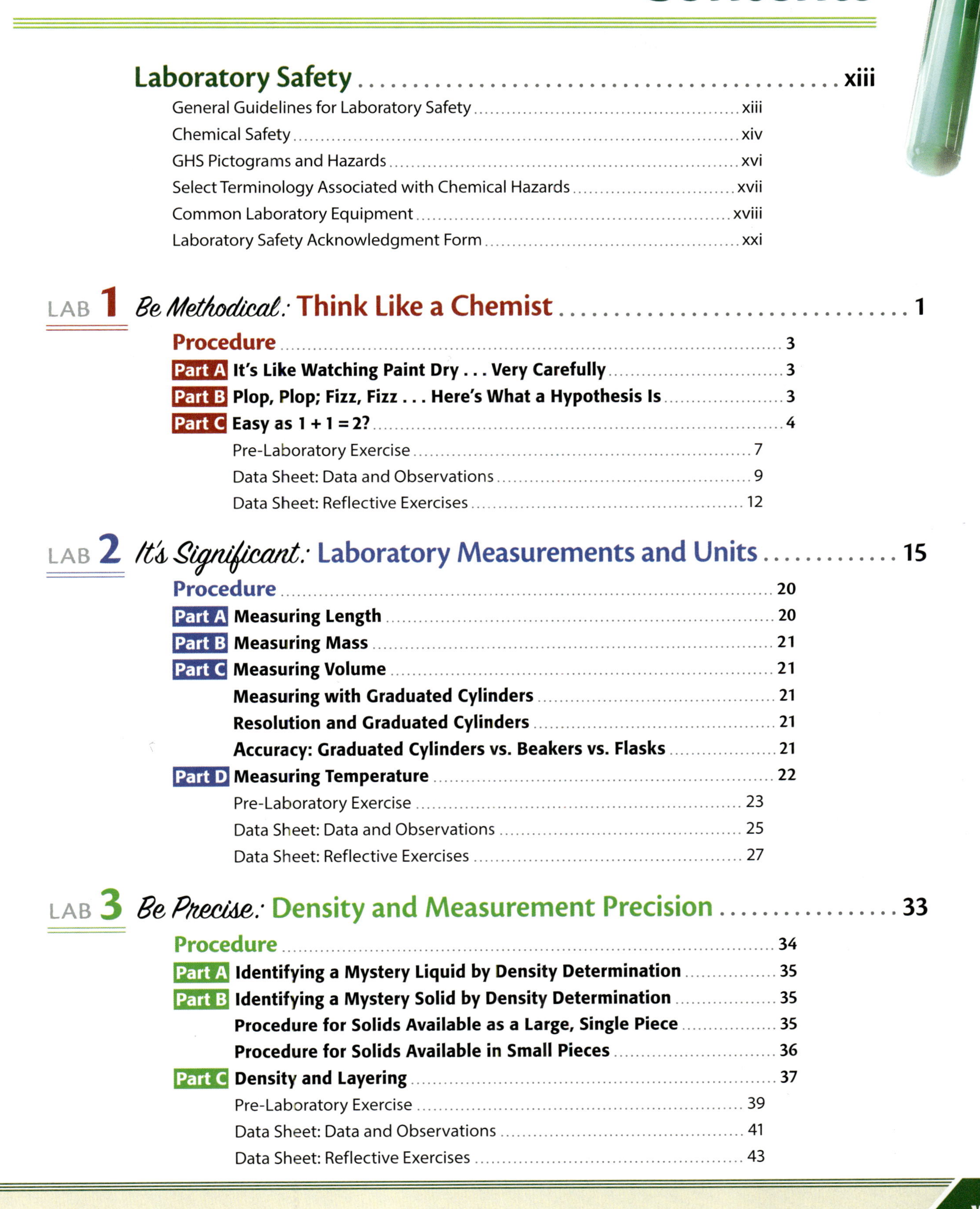

Laboratory Safety

Working in a chemistry lab is a hands-on experience similar to cooking in your kitchen at home. You will use different instruments to measure, transfer, and mix things. You may experience a pleasant smell, but an offensive odor would not be out of the question either. You may spill something and need to clean it up. You often will heat or chill a substance. You may need to use a sharp tool or accidentally break a piece of glassware.

All of these actions are accompanied by hazards. You can help minimize, control, and/or manage the hazards of working in a laboratory by adhering to the safety practices described in the guidelines below. Of course, the major difference between lab work and cooking is the use of chemicals, broadly defined as substances that are not approved for consumption. Safe handling of chemicals is discussed after the general guidelines for laboratory behavior.

General Guidelines for Laboratory Safety

Personal Conduct and Preparation

1. Always conduct yourself in a professional manner. Horseplay and pranks are strictly prohibited in the laboratory.
2. Carefully follow all written and verbal instructions. If you are uncertain about how to proceed, ask your instructor.
3. Prepare for the experiment by carefully reading all assigned background material and procedures before entering the laboratory.
4. Never work alone in a laboratory. Should an accident occur, it is imperative that help is nearby.
5. Perform only the experiments and procedures described by your instructor. Do not attempt to perform any unauthorized operation in the laboratory space.
6. Never eat or drink anything in the laboratory.
7. Never apply cosmetics, handle contact lenses, or use a cell phone in the laboratory.
8. Practice good housekeeping while performing a lab. Do not bring unnecessary materials (book bags, notebooks, purses, cell phones, etc.) into the work-space, and keep your assigned area tidy. Clean up dirty glassware and material as you go along.

Attire and Personal Protection

9. Always wear shoes that completely cover your feet.
10. Do not wear loose or baggy clothing.
11. Long hair and any dangling jewelry should be secured before working in the laboratory.
12. Always wear chemical splash goggles while in the laboratory.
13. Wear appropriate chemical resistant gloves when handling hazardous substances and when advised by your instructor. If chemicals are spilled on your gloves, remove them immediately and wash your hands. Always remove gloves and wash your hands before leaving the laboratory.

Laboratory Operations

14. When using a sharp instrument, always carry it with the sharp end pointed down and away from you and others. Use caution with sharp instruments, and always cut or puncture items in a direction away from yourself and others.
15. Carry glassware with two hands and in a vertical position to prevent inadvertently bumping or breaking the glass.
16. Examine glassware before each use. Never use dirty, cracked, or chipped glassware.
17. Exercise caution with fire and other heat sources. Never leave an open flame or heated material unattended. Ensure that the area is free of flammable materials before lighting a flame.
18. If heating a test tube, always point the open end upward and away from yourself and others.
19. Hot liquids and steam can cause severe burns. Use caution when handling all heated substances.

20. Remember that objects usually look the same whether they are hot or cold. Check the temperature of glassware, hot plates, and other materials before picking them up. Use caution if it is necessary to move a hot object.

Accidents and Emergencies

21. Know the location and operation of all emergency exits and equipment, including: first aid kits, eyewash stations, safety showers, fire alarms, fire extinguishers, and fire blankets.
22. All spills and accidents should be immediately reported to the instructor.
23. All broken glass should be disposed of in a properly labeled container, not the trash can. Do not pick up broken glassware with your hands. Your instructor will use a broom and dustpan to retrieve the pieces of glass.
24. If a chemical splashes in your eyes, immediately flush your eyes with water from the eyewash station for at least 15 minutes. Be sure to hold your eyes open while flushing with water.
25. If a small amount of a hazardous chemical splashes on your skin, hold the exposed area under running water for at least 15 minutes.
26. If a large amount of chemical splashes onto your skin or clothing, proceed immediately to the safety shower and wash with water for at least 20 minutes. Remove contaminated clothing as quickly as possible while standing under the shower.

Chemical Safety

There is always some level of risk associated with the use of any chemical. In general, we may think about two major categories of chemical hazards: physical and health. Physical hazards are potential dangers to your physical safety posed by a chemical. For example, many organic chemicals are flammable and therefore increase the risk of fire in the laboratory. Health hazards are associated with acute or chronic biological conditions that potentially result from exposure to a chemical. Ingestion of lead, for example, is known to cause developmental delays and memory loss.

The experimental procedures in this text were designed to minimize the use of and potential exposure to chemicals known to present serious hazards to health and physical safety. When possible, traditional solvents and reagents have been replaced with less hazardous alternatives. Nevertheless, you should minimize your risk of exposure to all laboratory chemicals by observing the following general guidelines at all times.

General Guidelines for Handling Chemicals

1. Always double check the label to ensure that you are using the appropriate chemical.
2. Remove only the quantity of chemical necessary to achieve the task from the original container. Securely replace the cap as soon as you have removed the chemical from the container.
3. Always use a scoop or microspatula to remove solids from a container. Never handle chemicals with your hands.
4. Clean all scoops, spatulas, and glassware as soon as possible after you are finished using them.
5. Dispose of all chemicals in the appropriate waste container, as directed by your instructor.
6. Never return unused chemicals to the original container. Unused portions should be disposed of in the appropriate waste container.
7. Keep chemicals under the fume hood or within another properly ventilated space as instructed. When working in a fume hood, the materials should be placed at least six inches from the front sash of the hood.
8. When transporting a chemical, always hold the container securely with two hands, and proceed slowly and carefully around the laboratory.
9. While working with chemicals, do not touch your face, eyes, mouth, nose, hair, other body parts, or personal items such as a cell phone. Wash your hands after completing your work and before leaving the laboratory.
10. You will be instructed in the safe handling of acids. If you are required to dilute an acid sample, always pour the acid slowly into the water. Do not add water to a sample of concentrated acid.
11. Never taste or smell a chemical by holding it directly under your nose to inhale the vapors. If necessary, you will be instructed on the proper technique to smelling a chemical.
12. Always wear proper personal protective equipment, including proper attire, chemical splash goggles, and disposable gloves. Never assume that personal

protective equipment will safeguard you from every type of chemical exposure. Follow proper emergency procedures in the event of an accident or spill. Always remove gloves and wash your hands before exiting the laboratory.

13. Be aware of the hazards associated with each chemical you use, and follow any recommended special handling practices.

The final guideline raises an important question, especially if you have little or no experience in a chemistry lab. You have a right and a responsibility to understand the hazards associated with the chemicals you use in the laboratory. After all, you must first be aware of any hazards if you are going to be able to take action to minimize or control the risk. But how do you find this information?

There are a number of text-based and online reference materials that provide information about the physical properties and hazards associated with many common substances. Your instructor may require that you regularly refer to one or more of these resources. By far, the most common and uniform source of information on a chemical's known characteristics is the **Safety Data Sheet (SDS)** (formerly known as the Material Safety Data Sheet or MSDS), which each manufacturer of a hazardous chemical is required by law to produce and make available to users. Your institution should maintain a copy of the SDS for each chemical in the laboratory. Manufacturers also generally make their SDSs available on their website. Throughout this text, each laboratory procedure highlights the most important safety information from the SDSs of chemicals used during the lab. It is therefore critical that you are familiar with certain information and terminology found on an SDS.

The United States has adopted the Globally Harmonized System of Classification and Labeling of Chemicals (GHS) as the standard format for communicating chemical hazard information on an SDS. Under this system, the SDS is divided into 16 sections, each of which includes a different category of information about the chemical. Hazards associated with the chemical are identified in Section 2 and communicated in the following ways:

- **Pictograms:** There are nine graphic symbols used to communicate major categories of hazards. A pictogram will be present if a chemical presents one or more hazards found within the associated category. Some hazards fall into more than one category and therefore may be represented by more than one pictogram. In such cases, the pictogram that carries the higher warning is used. The nine pictograms and associated hazards are summarized on page xvi. Definitions of select terms used to communicate chemical hazards are provided on page xvii.
- **Signal Words:** There are two signal words used to communicate the overall level of hazard associated with a chemical. The word *Danger* signifies a more severe hazard level, and the word *Warning* is used when the hazards are less severe.
- **Hazard Statements:** Hazard statements summarize specific hazards associated with the chemical. Some examples are: "Causes skin irritation"; "Harmful if swallowed or inhaled"; "Extremely flammable liquid and vapor."
- **Precautionary Statements:** Precautionary statements summarize actions to prevent or minimize adverse effects of handling a chemical. First aid information is also provided in the precautionary statements. Some examples are: "Keep away from heat/sparks/open flames/hot surfaces"; "Wear protective gloves/eye protection/face protection"; and "IF INHALED: Remove victim to fresh air and keep at rest in a position comfortable for breathing."

This text uses the GHS pictograms to visually summarize hazards associated any chemicals used in the laboratory procedures. In addition, "Safety Notes" boxes inserted in the text highlight specific hazards and suggest appropriate precautions.

There are several others sections on an SDS that students using the manual may also find useful. Section 3 describes the composition of a substance, including the molecular weight. Sections 4 and 5 provide detailed information on first aid and firefighting measures, respectively. Section 9 summarizes the known physical and chemical properties of the substance, including appearance, melting point, boiling point, density, etc. The remaining sections include additional handling, storage, disposal, accident response, and other regulatory information.

Safety is always the top priority in any laboratory. By following the guidelines discussed in this section and any guidance from your instructor, you can avoid accidents and minimize the risk of lab work. If you are ever in doubt about the safest way to proceed, do not hesitate to ask! Work hard, have fun, and always stay safe!

GHS Pictograms and Hazards

Health Hazard

- Carcinogen
- Mutagenicity
- Reproductive toxicity
- Respiratory sensitizer
- Target organ toxicity
- Aspiration toxicity

Flame

- Flammables
- Pyrophorics
- Self-heating
- Emits flammable gas
- Self-reactives
- Organic peroxides

Exclamation Mark

- Irritant (skin and eye)
- Skin sensitizer
- Acute toxicity (harmful)
- Narcotic effects
- Respiratory tract irritant
- Hazardous to ozone layer (non-mandatory)

Gas Cylinder

- Gases under pressure

Corrosion

- Skin corrosion/burns
- Eye damage
- Corrosive to metals

Exploding Bomb

- Explosives
- Self-reactives
- Organic peroxides

Flame over Circle

- Oxidizers

Environment (non-mandatory)

- Aquatic toxicity

Skull and Crossbones

- Acute toxicity (fatal or toxic)

Select Terminology Associated with Chemical Hazards

Term	Definition
Acute toxicity	A severe adverse effect caused by single or short-term exposure to a chemical
Aspiration toxicity	A severe effect caused by the entry of a liquid or solid into the trachea or lower respiratory system either directly through the oral or nasal cavity or indirectly through vomiting
Carcinogen	A chemical or substance that induces or increases the incidence of cancer
Corrosive	A chemical that causes destruction of living tissue at the site of contact
Corrosive to metals	A chemical that will materially damage or destroy metals
Explosive	A chemical or substance that may produce a damaging release of energy and gas when subjected to ignition, shock, or heat
Flammable	A chemical or substance that is easily ignited and burns readily
Irritant	A chemical or substance that causes reversible damage, such as redness, swelling, itching, or rash, to the site of contact
Lachrymator	A chemical or substance that causes the production of tears upon exposure
Mutagen	A chemical or substance known to cause an increase in the occurrence of mutations within populations of cells and/or organisms
Oxidizer	A chemical or substance that may cause or enhance the combustion of other substances
Organic peroxide	A chemical that contains the R-O-O-R bonding pattern; they are often explosion hazards and may react violently with other chemicals
Poison	A chemical or substance that is extremely toxic
Pyrophoric	A chemical or substance that may spontaneously ignite within five minutes after coming into contact with air
Reproductive toxicity	A chemical or substance that has adverse effects on sexual function and fertility in adults and/or developmental effects in offspring
Self-heating	A substance other than a pyrophoric chemical that will heat itself by reaction with air and without an additional energy supply
Self-reactive	A chemical that is thermally unstable and even in the absence of air may decompose in a fashion that gives off significant heat
Sensitizer	A chemical or substance that may cause an allergic reaction upon repeated exposure
Target organ toxicity	An adverse effect that manifests in a specific organ or system in the body
Teratogen	A chemical or substance that causes physical defects in a developing fetus or embryo
Toxic	A chemical or substance that causes adverse health effects

Common Laboratory Equipment

Common Laboratory Equipment

Beaker	1	Mortar and pestle	19
Büchner funnel	2	Pasteur pipette	20
Bunsen burner	3	Powder funnel	21
Burette	4	Ring stand	22
Burette clamp	5	Rubber stoppers	23
Clay triangle	6	Scoopula	24
Cork stoppers	7	Stirring rod	25
Crucible and cover	8	Test tubes	26
Dropper	9	Test-tube brush	27
Erlenmeyer flask	10	Test-tube holder	28
Evaporating dish	11	Test-tube rack	29
Filter flask	12	Thermometer	30
Forceps	13	Three-pronged clamp	31
Goggles	14	Tongs	32
Graduated cylinder	15	Utility clamp	33
Iron ring	16	Volumetric pipette	34
Long-stem funnel	17	Watch glass	35
Microspatula	18	Wire gauze	36

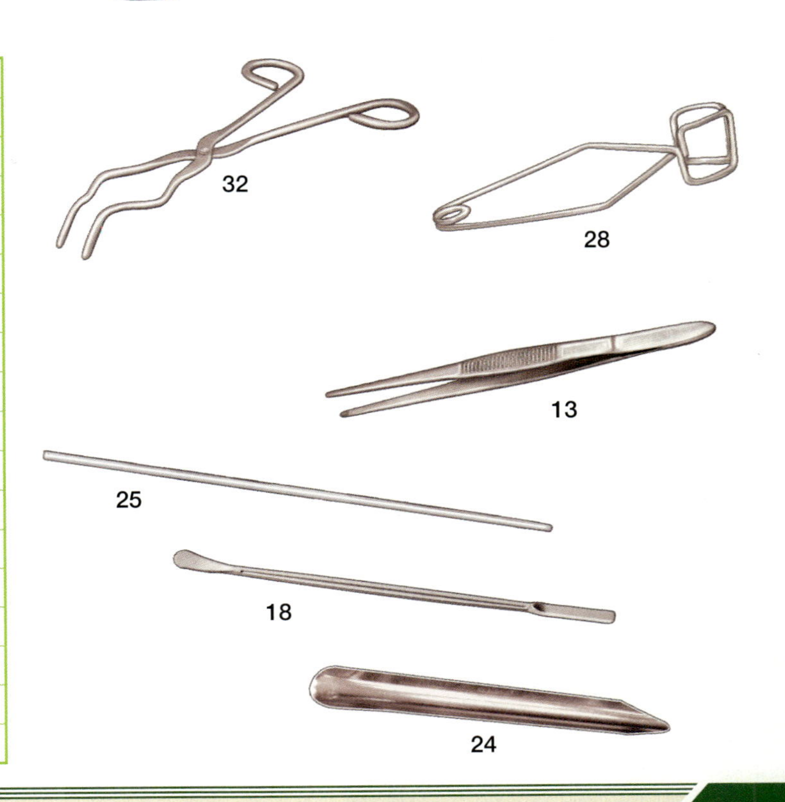

Laboratory Safety Acknowledgment Form

I have read and agree to abide by the aforementioned safety rules and guidelines for handling chemicals. I understand that willful violation of these practices may result in my expulsion from the laboratory.

Name ____

Date ______________________

Lab Section ____________________

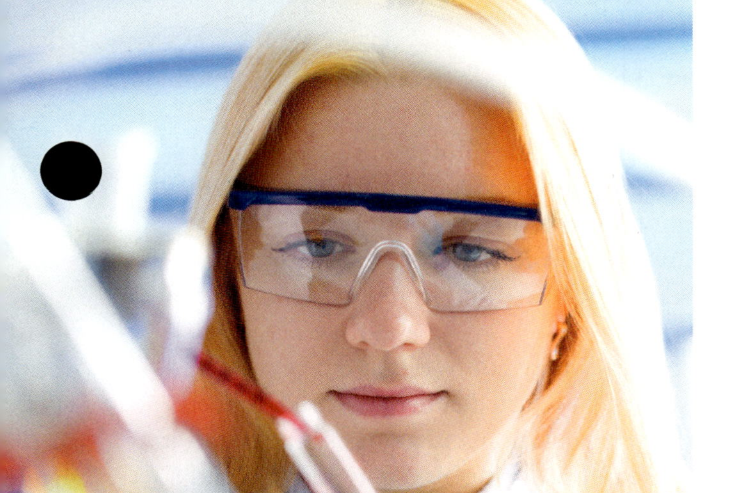

▲ *Scientist pipetting liquid into a test tube.*

Lab 1

Be Methodical

Think Like a Chemist

Scientists attempt to understand phenomena by employing a set of general principles known as the **scientific method.** This term often is misinterpreted to imply that there is a single, "correct" process through which science is conducted. On the contrary, there are an infinite number of pathways through which scientific knowledge is discovered, acquired, and developed.

The process is often messy, and mistakes are common. The scientific method is more accurately understood as a set of guidelines used to direct the process of scientific investigation and correct any mistakes or unjustified conclusions.

Figure 1.1 offers a schematic representation of how the scientific method might be applied. Any scientific investigation begins with an observation about something interesting or unexpected in the universe. A scientist then asks a question about such an observation.

For example: If you cooked beets in a glass dish and then scrubbed it with soap and a little water, you might find that the deep red residue changes to a pale yellow color. A scientist might ask: What causes this color change?

After identifying a question to explore, the scientific method calls for the formulation of a **hypothesis,** a testable explanation of the observation or answer to the question. The most important feature of a hypothesis is that it is a *testable* idea. Scientists are therefore able to use the

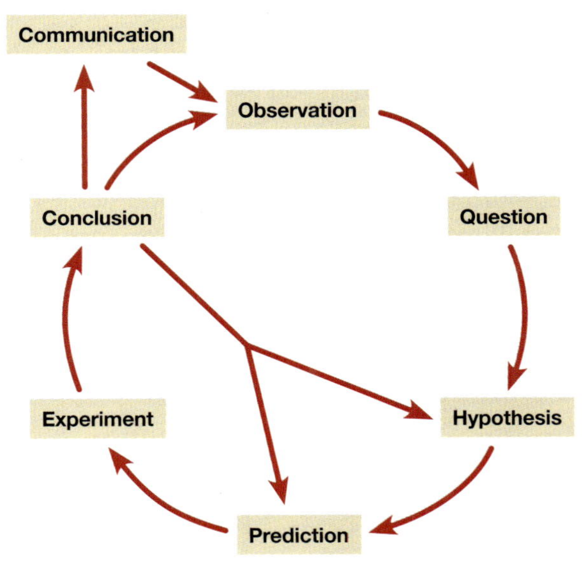

1.1 Schematic flow diagram of the principles and application of the scientific method.

Objectives

After completing this lab you should be able to:

- Explain and generally apply the principles of the scientific method.
- Exhibit proficiency in making careful observations during an experiment.
- Exhibit proficiency in making a scientific argument.
- Demonstrate an understanding of how experimental results may be analyzed in order to draw reasonable conclusions based on the data.
- Demonstrate the ability to identify and control for potential sources of experimental error.

hypothesis to make a prediction about the phenomenon during an **experiment**, a controlled set of conditions designed to examine the hypothesis.

In our example, the scientist might hypothesize that the dark beet juice residue became diluted in the water. A reasonable prediction based on this hypothesis is that adding water *without* soap to another pan of beet residue would produce a similar color change. A scientist would carry out this experiment in order to test the hypothesis.

After performing an experiment, the results are analyzed in order to draw a conclusion about the validity of the hypothesis. The results may match the prediction and support the hypothesis, in which case additional experimentation would provide stronger evidence that the hypothesis is valid.

On the other hand, the results may not match the prediction completely, or they may contradict it entirely. Such results are evidence that the hypothesis is partially or wholly invalid, and it may then be rejected entirely, revised, refined, or otherwise modified. The new hypothesis is then subject to the same cycle of experimentation and analysis. Of course, it is also possible that brand new observations are encountered during an experiment, producing a new hypothesis and a new cycle of scientific inquiry.

Thinking again of our beet juice example, the scientist may find that adding water alone to the residue does not cause the interesting color change. The scientist should recognize the "dilution hypothesis" as invalid, propose a different hypothesis to explain the change in color, and test the new idea through experimentation.

The final principle of the scientific method is that experimental results and conclusions must be communicated among scientists. When scientists report their findings, they must make a clear argument that their conclusions are supported by experimental data. A well-communicated scientific argument starts with a **claim**, which is some form of answer to a research question (e.g., an explanation, descriptive statement, conclusion, etc.). The scientist next summarizes the **evidence** (i.e., relevant data). Finally, a **rationale** is presented to explain how and why the evidence supports the claim. Figure 1.2 illustrates the components of a well-composed scientific argument.

Whether the results validate or invalidate a scientist's hypothesis should not be a factor in the decision to communicate the findings. It is the *sharing* of scientific evidence that expands our collective understanding of phenomena and provides a self-correcting mechanism for science. The independent evaluation of experimental results that occurs when results are reported publicly helps eliminate bias and mistakes from scientific investigations.

Furthermore, scientific understanding should never be viewed as static. Scientists maintain a healthy skepticism about their own understanding of the universe. A hypothesis is always subject to repetitive and additional experimentation by different scientists.

The accumulated body of experimental evidence either for or against a particular hypothesis defines its validity. Similarly, the sum total of the principles embodied in the scientific method is what allows scientists to have confidence in our current, albeit always evolving, understanding of the universe.

Chemists are especially interested in understanding phenomena related to physical substances (**matter**) and their transformations. For centuries chemists have described matter as being composed of submicroscopic particles. As a result, it is often necessary for chemists to apply the principles of the scientific method to processes that cannot be seen directly.

In this laboratory exercise, you will perform various tasks and make careful observations about what you actually *observe* happening. You will then attempt to formulate reasonable hypotheses and/or experiments related to your observations. Of course, you also will analyze your results, formulate conclusions, perform additional experimentation, and report your findings. The primary goal of this exercise is for you to gain confidence in applying the scientific method.

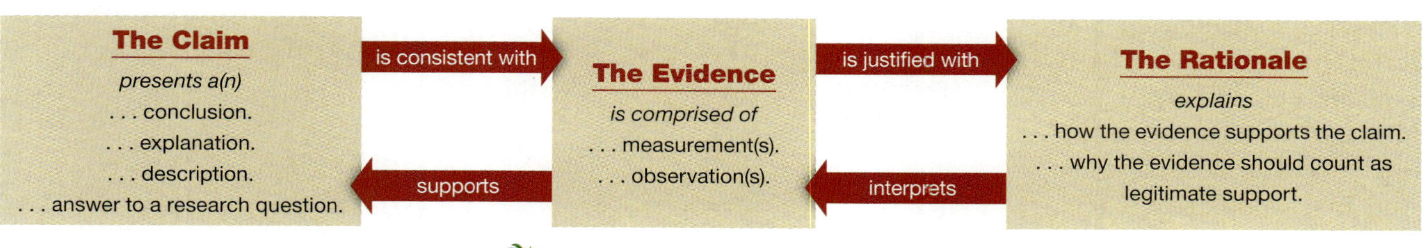

1.2 Diagram of a scientific argument.

Procedure

Materials

- ❑ Watch glass
- ❑ Dropper
- ❑ Test tubes (6)
- ❑ 50 mL graduated cylinders (2)
- ❑ Glass stirring rod
- ❑ Scoopula or forceps

Part A: Observing the changes in a thin layer of liquid (ethyl acetate) on a watch glass and formulating reasonable conclusions based on your data.

Part B: Mixing pieces of solid magnesium with various liquids, observing any changes that take place, and formulating reasonable hypotheses based on your observations.

Part C: Mixing equal volumes of two different liquids, predicting the volume of the mixture, testing your prediction, and analyzing the validity of the experiment and its outcome.

Part A It's Like Watching Paint Dry . . . Very Carefully

1. Use a dropper to add 5 drops of ethyl acetate to a watch glass.
2. Swirl the watch glass to spread the liquid out in a thin layer over its surface.
3. Watch the liquid for several minutes, and record your observations of any changes on the data sheet, page 9. Carefully describe exactly what you *observe* without including any conclusions, inferences, or assumptions about the changes that take place.
4. Answer Questions 1 and 2 on the data sheet, pages 9–10, before moving on to Part B.

Part B Plop, Plop; Fizz, Fizz . . . Here's What a Hypothesis Is

1. Label three test tubes 1–3.
2. To each test tube, add 1 dropper full of the appropriate liquid as indicated below:
   - Test Tube 1: 1 M hydrochloric acid (HCl)
   - Test Tube 2: 1 M sodium hydroxide (NaOH)
   - Test Tube 3: methanol
3. Drop one chip (or "turning") of magnesium into each test tube. Watch the test tubes for several minutes and record your observations on the data sheet, page 10.

SAFETY NOTE

Hydrochloric acid (HCl) is a strong acid. Handle with caution! Sodium hydroxide (NaOH) is caustic. Wear eye protection and handle it with care!

4. Chemicals often are categorized based on similarities in their properties. Magnesium belongs to a class of substances called metals, and the three liquids used in Step 2 belong to the classes presented in Table 1.1. Use this information and your observations to formulate hypotheses about the observable changes in magnesium when it is exposed to each category of liquid. Record your answers on the data sheet, page 10.
5. Label six test tubes 1–6.
6. To each test tube, add 1 dropper full of the appropriate liquid as indicated below:
   - Test Tube 1: 1 M sulfuric acid (H_2SO_4)
   - Test Tube 2: 1 M acetic acid (CH_3CO_2H)
   - Test Tube 3: 1 M potassium hydroxide (KOH)
   - Test Tube 4: 1 M sodium bicarbonate ($NaHCO_3$)
   - Test Tube 5: ethanol
   - Test Tube 6: isopropanol
7. Predict the outcome of adding a turning of magnesium to each test tube. Record your predictions on the data sheet, page 11.
8. Drop one turning of magnesium into each test tube. Observe the test tubes for several minutes and record your observations on the data sheet.

TABLE 1.1 ■ Some Common Classes and Examples of Substances Used in the Chemistry Laboratory

Acids (dissolved in water)	Bases (dissolved in water)	Alcohols
Hydrochloric acid (HCl)	Sodium hydroxide (NaOH)	Methanol
Sulfuric acid (H_2SO_4)	Potassium hydroxide (KOH)	Ethanol
Nitric acid (HNO_3)	Magnesium hydroxide [$Mg(OH)_2$]	Isopropanol
Hydrobromic acid (HBr)	Calcium hydroxide [$Ca(OH)_2$]	Amyl alcohol
Acetic acid (CH_3CO_2H)	Sodium bicarbonate ($NaHCO_3$)	Butanol
Perchloric acid ($HClO_4$)	Ammonium hydroxide (NH_4OH)	Cyclohexanol

Part C Easy as 1 + 1 = 2?

1. Use a 50 mL graduated cylinder to obtain approximately 20 mL of distilled water. You do not need to spend time obtaining *exactly* 20 mL, but you do need to accurately record the exact volume of water you obtain. Technique Tip 1.1 describes the correct way to measure volume with a graduated cylinder. Record the volume on the data sheet, page 11.
2. Repeat Step 1 using a separate 50 mL graduated cylinder. Record the actual volume of water you obtain on the data sheet.

3. On the data sheet, record your prediction for the total volume of water you would obtain by mixing the two samples.
4. Pour the contents of one graduated cylinder into the other.
5. Stir the mixture with a glass stirring rod.
6. Record the volume of the mixed sample on the data sheet.
7. Repeat Step 1. Record the actual (measured) volume of water you obtain on the data sheet.
8. Rinse one of the graduated cylinders well with ethanol to remove any residual water droplets. Obtain approximately 20 mL of ethanol. Record the actual volume of ethanol you obtain on the data sheet.
9. On the data sheet, record your prediction for the total volume of liquid you would obtain by mixing the water and ethanol samples.
10. Pour the ethanol into the graduated cylinder containing the water.
11. Stir the mixture with a glass stirring rod.
12. Record the volume of the ethanol/water mixture on the data sheet.
13. If the result from Step 12 does not match your prediction, there are two possible explanations: Either your prediction was incorrect or there was an error in your experiment. Let's assume your prediction was correct and that any discrepancy is due to an experimental error. Identify two possible error sources that might account for the discrepancy. Describe these potential errors on the data sheet, page 12.
14. On the data sheet, describe an experiment you could perform to detect each of the possible errors you described.
15. Describe the expected result of each experiment if the corresponding error is detected.
16. Perform each experiment and record your results on the data sheet.

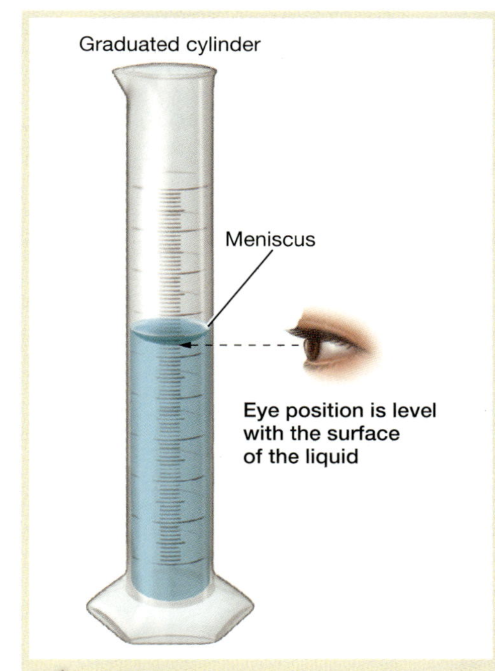

Technique Tip 1.1

Fill a graduated cylinder until it is within about 0.5 mL of the desired volume and "top-off" the volume with a pipette or dropper. Measure the volume by positioning your eye level with the liquid surface and reading the scale at the bottom of the meniscus (curve).

Name ______

Lab Partner ______

Lab Section ______ Date ______

Lab 1 Pre-Laboratory Exercise

1 Provide a term that matches each description below.

a Testable idea proposed as an explanation of an observation. ______

b Guidelines that direct the process of scientific investigation. ______

c An explanation for why experimental data support a claim. ______

d A controlled set of conditions used to examine a hypothesis. ______

e Refers to physical substances that make up the universe. ______

f A statement presenting a scientific conclusion or answer to a research question. ______

2 Imagine a pot of boiling water. After a while, the level of water in the pot decreases. Where does the water go? What have you observed about a pot of boiling water that supports your conclusion?

3 Some ways that scientists communicate their experimental results and conclusions with each other include publishing articles and making presentations at conferences. Other scientists are then able to provide critiques of their methods or help verify their conclusions by repeating the experiment. How might you simulate this type of communication and feedback in an instructional laboratory setting?

4 When using a graduated cylinder, the liquid often forms a curve (meniscus) at the surface. Where along the curve should you measure the volume?

Name

Lab Partner

Lab Section Date

Lab 1 DATA SHEET

Data and Observations

Part A It's Like Watching Paint Dry . . . Very Carefully

Event	Observations
Ethyl acetate on watch glass	

1 Using only your observations, are you able to make a justifiable claim about the location of the ethyl acetate at the end of the observation period? In the table below, check the "Yes" or "No" box to indicate your answer. If you choose "Yes," in the corresponding boxes write your claim and provide a rationale that explains how your observations support the claim. If you choose "No," explain why your observations do not provide evidence that allows you to make a claim.

Yes ☐		No ☐
Claim:		Explanation:
Rationale:		

2 Suppose a lab mate suggested that the ethyl acetate was absorbed into the watch glass.

a Do your observations provide any evidence about the validity of this claim? Explain.

b Describe a simple experiment you could perform that would provide evidence about whether or not the ethyl acetate was absorbed into the watch glass.

c What result would you expect if the ethyl acetate were absorbed during your experiment?

d If approved by your instructor, perform the experiment you described in Question 2b. Use your results to make a claim, then describe your evidence and provide a rationale that explains how your evidence supports your claim.

Claim	
Evidence	**Rationale**

Part B Plop, Plop; Fizz, Fizz . . . Here's What a Hypothesis Is

Test Tube	Sample	Observations
1	Magnesium + HCl	
2	Magnesium + NaOH	
3	Magnesium + methanol	

Combination	Hypothesis
Magnesium + acid	
Magnesium + base	
Magnesium + alcohol	

Name ______________________________

Lab Partner ______________________________

Lab Section ____________________ Date ____________________

Lab 1 DATA SHEET

(continued)

Test Tube	Sample	Predictions	Observations
1	Magnesium + H_2SO_4		
2	Magnesium + CH_3CO_2H		
3	Magnesium + KOH		
4	Magnesium + $NaHCO_3$		
5	Magnesium + ethanol		
6	Magnesium + isopropanol		

Part C Easy as 1 + 1 = 2?

Procedure Reference	Quantity	Data
Step 1	Measured volume: first water sample (mL)	
Step 2	Measured volume: second water sample (mL)	
Step 3	Predicted volume: mixed water sample (mL)	
Step 6	Measured volume: mixed water sample (mL)	
Step 7	Measured volume: water (mL)	
Step 8	Measured volume: ethanol (mL)	
Step 9	Predicted volume: water + ethanol (mL)	
Step 12	Measured volume: water + ethanol (mL)	

Procedure Reference	Description	Error Source 1	Error Source 2
Step 13	Possible error source		
Step 14	Experiment to detect error source		
Step 15	Prediction if error detected		
Step 16	Experiment data		

Reflective Exercises

1 In ancient Greece, philosophers held two competing views of matter: Aristotle and many others proposed that substances are continuous, or composed of a single strand of material; Democritus proposed that all substances are composed of small particles.

a Did the change you observed in the ethyl acetate take place all at once or bit by bit? Use your data to justify your answer.

b Does your answer to 1a support the description of matter proposed by Aristotle or Democritus? Explain.

2 Did your data from Part B support the hypotheses you developed? Justify your answer. If you your answer is "no," what should you do to the hypotheses?

Name ______________________

Lab Partner ______________________

Lab Section ______________ Date ______________

Lab 1
DATA SHEET
(continued)

3 Like magnesium, the substances sodium, aluminum, and copper are categorized as metals.

a Using the hypotheses you developed in Part B, how would you predict that these metals might interact with each category of liquid you examined?

Combination	Prediction
Metal + acid	
Metal + base	
Metal + alcohol	

b Consider the data below, which were obtained by mixing a small amount of each metal in a test tube with the appropriate liquid. Do these data support the hypotheses you developed? Explain.

Combination	Observation
Copper + HNO_3	Solid dissolves and a brown gas is produced. The mixture eventually takes on a greenish-blue color.
Aluminum + NaOH	Solid dissolves and a vigorous bubbling is observed.
Sodium + ethanol	A steady stream of bubbles is observed and the solid dissolves.

c What does this information suggest about the reliability of hypotheses based on small numbers of experiments, small amounts of data, and/or experiments that involve examination of a single substance?

4 In Part C, were the possible errors you described on the data sheet avoidable or unavoidable when you mixed the water and ethanol? That is, were they actions you could control and correct, or were they a component of the experimental design?

5 Based on your data from Part C, did the possible errors you identified account for the discrepancy between your prediction and the measured volume of ethanol + water? Explain.

6 If experimental errors were not responsible for the discrepancy encountered in Part C, your prediction must have been incorrect. We often encounter unexpected observations because we are not used to thinking about matter as being composed of tiny particles. Use this idea to offer a hypothesis that explains the discrepancy between your prediction and the measured volume of ethanol + water. Keep in mind that volume is a measure of the space occupied by matter. This analogy might help: If you mixed a bucket of softballs and a bucket of marbles, would the combination take up two buckets worth of space?

▲ *Measurement is necessary in a chemistry lab.*

Lab 2

It's Significant

Laboratory Measurements and Units

In order to investigate or test their ideas and hypotheses, chemists often need to measure the physical attributes of a substance. You probably have more experience *counting* objects than *measuring* them. Scientists use many tools to take measurements, and their overall confidence in these measured values should reflect a combination of factors such as **resolution** (detail) and **accuracy** (correctness).

It is important that you learn to properly use certain laboratory tools and to report and interpret the uncertainty associated with measured quantities. This skill allows scientists to gauge their confidence in the evidence they gather and any claims based on them.

When a quantity is *counted*, the result is an *exact number*. For example, you may count the exact number of students in your classroom. As long as you know how to count, there is zero uncertainty associated with the number of students present. By contrast, measurement involves the *inexact* process of using a tool to compare a physical quantity to some reference value. As a result, measured values can never be exact numbers and therefore never have zero uncertainty.

For example, you could use a ruler to measure the length of this piece of paper as 10.9 inches. This number indicates that the paper length is 10.9 times larger than the distance defined as *one inch*. However, you should not interpret the measurement to indicate that the paper length is *exactly* 10.9 inches. Each instrument and each practitioner have some inherent limitations that introduce uncertainty into the measured quantity. For example, the markings on a ruler determine the quality of any measured value obtained by using that specific ruler.

Consider the two rulers illustrated in Figure 2.1. The markings on Ruler 1 give you information about length using a scale based on 1 centimeter (cm) increments. The markings on Ruler 2 give you information based on 0.1 cm increments. Smaller incremental markings on the scale of a measuring device enable the user to distinguish between smaller differences among measured objects.

Objectives

After completing this lab you should be able to:

- Properly measure and report values of length, volume, mass, and temperature using common laboratory instruments and SI units.
- Describe the relationship between the digits in a measured number and the certainty of the measurement.
- Describe how an instrument's scale determines the certainty of values measured using the instrument.
- Explain the meaning of and differences between the terms *resolution* and *accuracy*.
- Express an appropriate level of confidence in measured values.
- Write conversion factors and use them to carry out unit conversions.

This concept is referred to as the **resolution** of a measuring device, which is the smallest difference between measured objects that can be faithfully detected with the device. Ruler 2 has a higher resolution than Ruler 1, and measurements obtained from Ruler 2 will always give you more detail about the length of an object. In general, higher resolution in a measuring device corresponds to greater certainty in measurements obtained with the device.

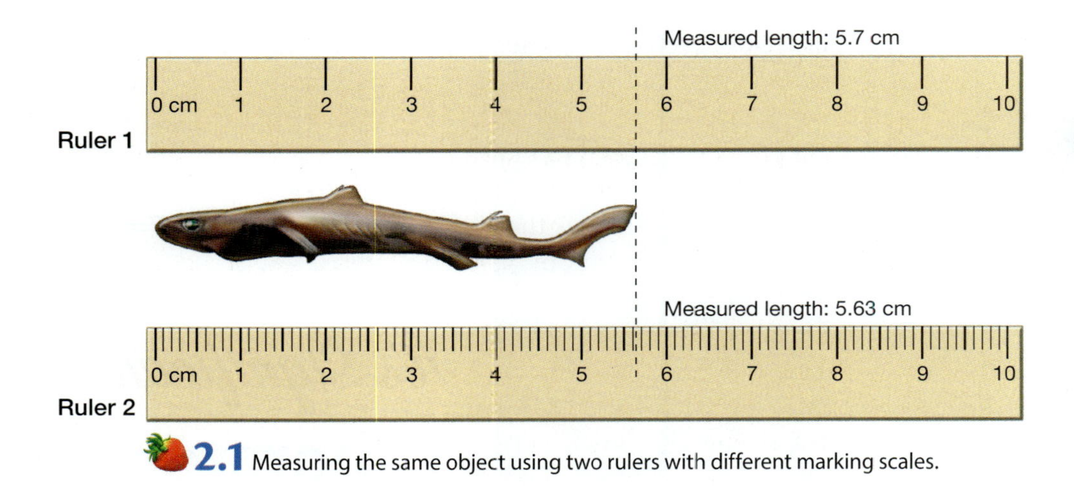

2.1 Measuring the same object using two rulers with different marking scales.

The **uncertainty** in a given measurement refers to the dispersion of values that could reasonably be recorded. The degree of uncertainty usually is reflected in the placement of the smallest digit. In general, we assume that the smallest digit in a measured value may vary by ± 1. Whenever you measure an object using an instrument with an analog scale, you should record the digits given to you directly by the measuring device *plus one estimated digit*. This final digit should represent your best guess about where the object falls relative to the markings on the instrument and is always the least certain digit.

When determining the final digit, try to estimate the value by mentally dividing the distance between the markings into 10 smaller increments. Following this rule, you might record the length of the dwarf lantern shark drawn in Figure 2.1 as 5.7 cm using Ruler 1, or 5.63 cm using Ruler 2.

In the first case, the 7 is an estimated digit that may vary with the person reading the ruler. In the second case, the 3 is the estimated digit, and it would be variable with the "eye" of the person taking the measurement. In both cases, the last digit gives you information about the uncertainty associated with the measurement. The second measurement indicates a higher resolution and degree of certainty because the uncertain digit resides in the hundredths place instead of the tenths place.

Many modern instruments, including certain balances and thermometers, provide measurements on a digital display. These devices are calibrated to estimate the final digit for the user. As such, you should always record *all* of the digits provided by such instruments and recognize that the final digit remains the least certain in the measured quantity.

The uncertainty of a measured quantity should not be confused with the **accuracy** of the value. Accuracy describes the degree of agreement between an experimental or measured quantity and the true value. It depends both on the quality of the measuring device and proper use of the tool. How can we know whether a measuring device is accurate?

Laboratory tools and equipment are tested for accuracy by using them to measure a **reference standard**, which is a sample of known or defined value. In analytical laboratories where quality control is vital, such as in a medical setting, instruments are tested for accuracy and calibrated frequently. For general use, you should not need to test basic measuring devices for accuracy.

However, it is important to recognize that not all laboratory tools are equally accurate because they may be intended for different purposes. Beakers, for example, are used primarily to hold and transfer materials. While they may have markings to indicate volume, these are meant only as approximations. Thus, a beaker would be a poor choice of tool to measure an accurate volume.

In addition to the appropriate number of digits in a measurement, each measured value must include an accompanying unit. After all, how can someone reading your measurement be expected to understand what quantity you measured or what scale you used if you do not include a unit?

In the United States, the U.S. Customary System of units is used for everyday measurements. These units include pounds, feet, and gallons. One major drawback of the Customary System is an inconsistency in the relationship between units used to measure large and small values of the same quantity.

For example, large masses may be measured in pounds while small masses may be measured in ounces. The relationship between these units is 1 pound = 16 ounces. Large or small distances may be measured using units of feet or inches, respectively. The relationship between these two units is 1 foot = 12 inches.

To avoid this lack of consistency, scientists most often use the *Système International d'Unités*, or SI system. The SI system is a modern version of the metric system, which employs a series of base units and prefixes. Table 2.1 lists commonly encountered units associated with the SI system and the corresponding quantities they are used to measure.

The SI system prefixes are used to indicate quantities that are larger or smaller than the base unit by a certain power of 10. For example, the prefix *kilo-* means "1,000 times larger." Thus, a kilometer is 1,000 times larger than a meter, which means that a kilometer is the same quantity as 1,000 meters. By contrast, the prefix *milli-* means "1,000 times smaller." Thus, dividing 1 meter by 1,000 provides a length that is equal to 1 millimeter, or one thousandth of a meter (0.001 m). Common SI system prefixes are listed in Table 2.2.

It often is necessary or convenient to convert a value measured or reported in one unit to an equivalent value reported in a different unit. For example, a procedure may call for 375 mg of salt, but it may be more convenient for you to recognize this quantity in grams because your balance displays grams.

To accomplish this mathematical conversion, we often use the "dimensional analysis," or "unit cancellation," method (Fig. 2.2) in which the initial quantity is multiplied by one or more **conversion factors** in order to transform it to an equivalent value in different units. A conversion factor is a fraction composed of two equal quantities expressed in different units, and dimensional analysis makes use of conversion factors that result in cancellation of all units except the desired unit after multiplication.

Equation 1 shows the conversion of 375 mg into gram units, a process that requires a single conversion factor based on the definitions of SI system prefixes presented in Table 2.2. Equation 2 illustrates a conversion of 375 mg into pound units, which requires a series of conversion factors based on the SI system prefixes and the standard relationship between kilograms and pounds (1 kg = 2.20 lb.).

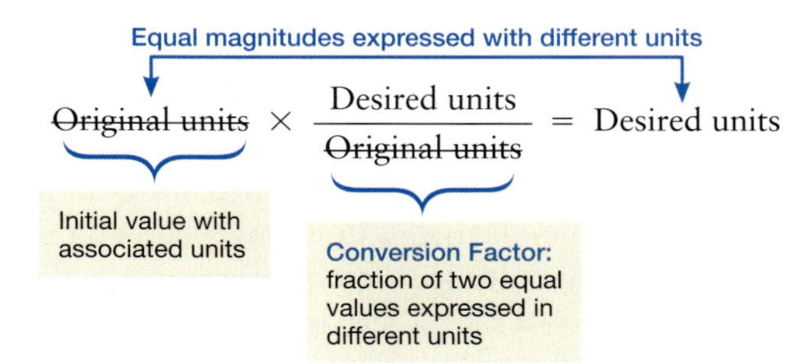

2.2 Dimensional analysis involves multiplying some initial quantity by one or more conversion factors. The conversion factors are fractions of two equal quantities expressed in different units and composed such that upon carrying out the mathematical operation, all units cancel except those desired in the final quantity.

$$375\ \cancel{\text{mg}} \times \frac{1\ \text{g}}{1000\ \cancel{\text{mg}}} = 0.375\ \text{g} \qquad \textbf{[Eq. 1]}$$

$$375\ \cancel{\text{mg}} \times \frac{1\ \cancel{\text{g}}}{1000\ \cancel{\text{mg}}} \times \frac{1\ \cancel{\text{kg}}}{1000\ \cancel{\text{g}}} \times \frac{2.20\ \text{lb.}}{1\ \cancel{\text{kg}}} = 8.25 \times 10^{-4}\ \text{lb.} \qquad \textbf{[Eq. 2]}$$

When carrying out calculations, it is important that the degree of uncertainty implied by the calculated value matches that of the original data. We use the concept of **significant figures** as a simple means of recognizing the uncertainty in a number. All digits in a numerical value are considered significant figures *except those used as placeholders*. Nonzero digits are *never* used as placeholders and therefore are *always significant*. Thus, the measurement from Ruler 1 in Figure 2.1 (5.7 cm) possesses two significant figures while the measurement from Ruler 2 (5.62 cm) possesses three.

There are two instances in which zeros might be placeholders:

1. Zeros that precede nonzero digits.
2. "Trailing zeros" that follow the nonzero digits of a large number and were not measured directly.

The first type of placeholder zero is illustrated in Figure 2.3. The zeros in the measured mass of the feather are not significant because they indicate only the magnitude of the feather's actual mass. The zeros themselves are not measured numbers; rather, the measured value is simply too small to occupy these numerical place values.

An example of "trailing zeros" is given in Figure 2.4. Without additional information, it is impossible to know whether the reported length of the border between the United States and Canada (5,061,000 m) was measured with a device that had a 1 m resolution or larger increments (e.g., 1,000 m resolution).

In the former case, the zero in the ones place would be the estimated digit in the measurement and therefore all three trailing zeros would be significant. In the latter case, the 1 in the thousands place would be the estimated digit. The trailing zeros would be recorded solely to indicate the scale of the measured digits. As placeholders, they would not count as significant figures.

It is incumbent upon the individual reporting this type of value to indicate the significance of the trailing zeros. In general, trailing zeros are assumed to be insignificant unless otherwise indicated explicitly. The most unambiguous way to indicate the significance of trailing zeros is to report the number in scientific notation. Only significant figures are included in the coefficient when scientific notation is employed. Figure 2.5 illustrates how scientific notation can be used to specify different numbers of significant figures in the border measurement.

It is worth noting that scientific notation is always the most explicit way to indicate the number of significant figures in a measurement. Consider again the feather example in Figure 2.3. Expressing the mass in scientific notation (4.2×10^{-3} g) unambiguously indicates that the value includes only two significant figures. In addition, a zero at the end of a decimal number (e.g., 21.30 m) is never an ambiguous trailing zero because this location never requires a placeholder. The only reason to record such a zero is to indicate a specific measurement of "0" in the hundredths position.

The "weakest link" principle is used to maintain the appropriate level of uncertainty in the value derived from a calculation. The final value is limited in terms of significant figures by the least certain number used in the calculation. Examples are provided in Figure 2.5.

2.3 Zeros that precede nonzero digits are not significant figures because they are placeholders. The zeros in the measured mass of the feather are not significant because they are not part of the measured value.

2.4 Trailing zeros of large numbers may or may not be placeholders, depending upon the resolution of the measuring device. Using scientific notation unambiguously identifies these zeros as significant or not. In the examples provided the placeholder zeros are shown in blue.

Multiplication and Division Rule

Final value is rounded so that it includes the same number of significant figures as the value in the calculation with the fewest significant figures.

If you measure the mass of one chocolate candy to be 3.72 g, how many candies would you calculate are in a jar that contains 8.3 oz. of candy?

$$8.3 \text{ oz.} \times \frac{28.3 \text{ g}}{1 \text{ oz.}} \times \frac{1 \text{ candy}}{3.72 \text{ g}} = 63.1424\ldots = 63 \text{ candies}$$

Weak Link (2 significant figures)

Stronger Link (3 significant figures)

Not significant Answer rounded to two significant figures

Note: Quantities in standard reference equalities (i.e., 28.3 g = 1 oz.) are considered exact and do not limit the number of significant figures in the conversion.

Addition and Subtraction Rule

Final value is rounded so that it includes the same number of decimal places as the value in the calculation with the fewest decimal places.

If the label of a candy jar says it contains 247.5 g of candy, and you measure the mass of the full jar of candy to be 624.82 g, what is the mass of the empty jar?

$$624.82 \text{ g} - 247.5 \text{ g} = 377.32 \text{ g} = 377.3 \text{ g}$$

Stronger Link (2 decimal places)

Weak Link (1 decimal place)

Not significant Answer rounded to one decimal place

2.5 The weakest link principle is used to determine the number of significant figures that should be included in a calculated value. Two different rules are applied to operations that involve multiplication and division and those that involve addition and subtraction.

TABLE **2.1** Commonly Measured Quantities and Corresponding Units Accepted Within the SI System

Measured Quantity	Unit
Length	Meter (m)
Mass	Gram (g)
Volume	Liter (L)
Temperature	Kelvin (K) or Degree Celsius (°C)
Time	Second (s)

TABLE **2.2** Most Common Prefixes in the SI System, Their Abbreviations, and Their Effect on the Value of the Base Unit

Metric Prefix	Abbreviations	Multiplier
mega	M	× 1,000,000
kilo	k	× 1,000
deci	d	÷ 10
centi	c	÷ 100
milli	m	÷ 1,000
micro	μ (or mc)	÷ 1,000,000

In this lab, you will practice using a variety of common laboratory tools to measure and report physical quantities with an appropriate indication of uncertainty. This means you must pay careful attention to the number of digits in the measured values you report! You also will explore resolution and accuracy as they relate to laboratory equipment and measurements. Finally, you will have an opportunity to explore and practice the concept of dimensional analysis.

Procedure

Materials

- ❑ Ruler
- ❑ Objects measured for length (3)
- ❑ Objects measured for mass (3)
- ❑ Small test tube
- ❑ 10 mL, 25 mL, and 50 mL graduated cylinders
- ❑ 50 mL and 400 mL beakers
- ❑ 50 mL Erlenmeyer flask
- ❑ 125 mL Erlenmeyer flasks (3)
- ❑ Celsius thermometer
- ❑ Hot plate
- ❑ Tongs

Part A: Examining the markings on a ruler and using it to make measurements in both SI and U.S. customary units.

Part B: Using a digital balance to measure mass and comparing the measurements obtained from different balances.

Part C: Examining the scales on various graduated cylinders and using them to correctly measure volume; measuring the volume of a test tube using different sizes of graduated cylinders; measuring 40 mL of distilled water, using different types of glassware.

Part D: Using a Celsius thermometer to measure the temperature of water samples that are warm, room temperature, and chilled; observing the effect of adding food coloring to each water sample.

Part A Measuring Length

1. Obtain a ruler and observe the markings on it.
2. What are the units associated with the numbered markings on the SI scale? Record your observation on the data sheet, page 25.
3. Determine the value, in the units recorded in Step 2, associated with the smallest lines marked on the SI scale. Record your observation on the data sheet.
4. What are the units associated with the numbered markings on the U.S. customary scale? Record your observation on the data sheet.
5. Determine the value, in the units recorded in Step 4, associated with the smallest lines marked on the U.S. customary scale. Record your observation on the data sheet.
6. Use the ruler to provide SI measurements of three objects provided by your instructor. Record the names of the objects and your measurements on the data sheet. Be sure to include units and the appropriate number of digits.
7. For each value you recorded on the data sheet, circle the least certain digit.
8. Use the ruler to measure the length of the box drawn below in both SI and U.S. customary units. Record your measurements on the data sheet.

9. Obtain measurements of the box taken by three of your lab mates and record them on the data sheet, page 25.

Part B Measuring Mass

1. Using a digital balance, measure the mass of three objects provided by your instructor. Record the names of the objects and your measurements on the data sheet, page 26.
2. Circle the least certain digit in each of your measurements.
3. Measure out *approximately* 250 mg of table sugar. Record the measured mass of your sample on the data sheet.
4. Transfer your sample to a different balance and record the mass on the data sheet.

Part C Measuring Volume

Measuring with Graduated Cylinders

1. A demonstration set of three graduated cylinders containing various amounts of colored liquid have been prepared for you.
2. Carefully observe the different values of the markings on each graduated cylinder.
3. On the data sheet, page 26, record the volume of liquid in each graduated cylinder. Be sure your measured values include units and the appropriate number of digits.

Resolution and Graduated Cylinders

1. Measure the volume of a small test tube by filling it with water and pouring the water into a 10 mL graduated cylinder. Record this volume on the data sheet, page 26.
2. Repeat Step 1 using a 25 mL and a 50 mL graduated cylinder. Record on the data sheet the volume you obtain using each. Be sure all of your measured values contain the appropriate number of digits.

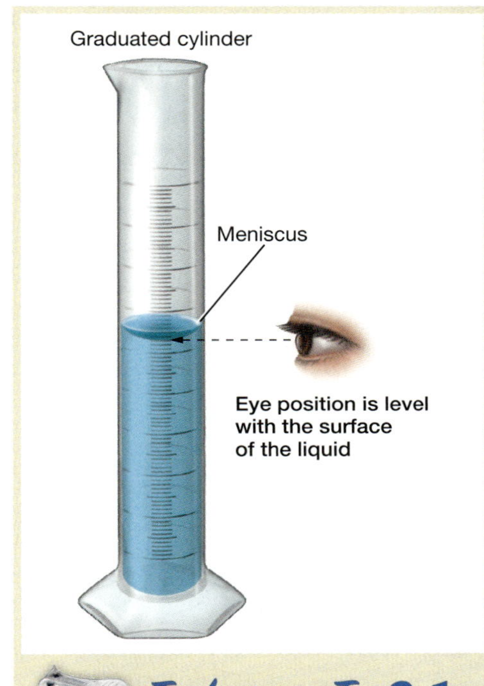

Technique Tip 2.1

Fill a graduated cylinder until it is within about 0.5 mL of the desired volume and "top-off" the volume with a pipette or dropper. Measure the volume by positioning your eye level with the liquid surface and reading the scale at the bottom of the meniscus (curve).

Accuracy: Graduated Cylinders vs. Beakers vs. Flasks

1. Record the mass of a 50 mL graduated cylinder on the data sheet, page 26.
2. Use the markings on the cylinder to measure 40 mL of distilled water. Record your volume measurement on the data sheet. Be sure to include the appropriate number of digits.
3. Reweigh the graduated cylinder, and record the mass of the cylinder with the water on the data sheet.
4. Calculate the mass of distilled water in the graduated cylinder and record this value on the data sheet.
5. Repeat Steps 1–4 using a 50 mL beaker in place of the graduated cylinder. Be sure your measured values contain the appropriate number of digits.
6. Repeat Steps 1–4 using a 50 mL Erlenmeyer flask in place of the graduated cylinder. Be sure your measured value contains the appropriate number of digits.

Part D Measuring Temperature

≡2

1. Label three 125 mL Erlenmeyer flasks A, B, and C.
2. Add water to the flasks until they are each approximately two-thirds full.
3. Leave Flask A on the lab bench.
4. Prepare an ice-water bath in a 400 mL beaker by half-filling it with an equal-parts mixture of ice and water. Place Flask B in the ice-water bath.
5. Heat Flask C on a hot plate until a Celsius thermometer indicates that the water temperature is approximately 70°C. When the water reaches temperature 70°C or higher, turn off the hot plate and use tongs or a hot mitt to move the flask to the lab bench.
6. Remove Flask B from the ice-water bath and place it on the lab bench.
7. Using a Celsius thermometer, measure the temperature of the water in each flask and record these values on the data sheet, page 27. Be sure your measurements contain the appropriate number of digits.
8. Add 2 drops of food coloring to each flask. Record your observations on the data sheet.

Name ______________________

Lab Partner ______________________

Lab Section ______________________ Date ______________________

Lab 2
Pre-Laboratory Exercise

1 Provide a term that matches each description below.

a SI base unit for mass. ______________________

b Describes the smallest difference between two objects that can be faithfully detected by a measuring device. ______________________

c Digits in a numerical value that are not placeholders. ______________________

d SI prefix meaning "one hundredth." ______________________

e A fraction composed of two equal quantities expressed in different units. ______________________

f Agreement between a measured value and the accepted or "true" value. ______________________

g Dispersion of values that could reasonably be reported for a measured quantity. ______________________

2 Determine the number of significant figures in each measurement below.

a 0.2530 g ______________________

b 150 g ______________________

c 1.30 g ______________________

d 0.00234 g ______________________

e 5843 g ______________________

3 Which of the measurements in Question 2 were taken on a device with the highest resolution? Explain.

____

____

4 When taking a measurement using a device with an analog scale, how do you determine the last digit in the number you report?

____

____

____

5 According to the ruler below, how long is this peanut? ______________________

0 cm 1 2 3 4 5 6 7 8 9 10

6 Explain why some people use the term *decimal system* as a synonym for the metric (or SI) system.

7 Describe some of the advantage(s) of the SI system over the U.S. Customary System of units.

8 One kilometer is the same distance as 0.621 miles.

a Write two conversion factors that may be derived from this information.

b If you run 3.2 miles, how many kilometers did you run?

9 Carry out each of the SI unit conversions below.

a 58 cg = ____________ g **b** 1.5 km = ____________ cm **c** 121 μL = ____________ dL

Name

Lab Partner

Lab Section   Date

Lab 2 DATA SHEET

Data and Observations

Part A Measuring Length

Ruler Observations

Procedure Reference	Description	Observation
Step 2	Units associated with numbered markings on SI scale	
Step 3	Value of smallest lines on the SI scale	
Step 4	Units associated with numbered markings on the U.S. customary scale	
Step 5	Values of smallest lines on the U.S. customary scale	

Ruler Measurements

Procedure Reference	Object	Length
Step 6	1.	
Step 6	2.	
Step 6	3.	
Step 8	Length of box in SI units	
Step 8	Length of box in U.S. customary units	

Lab Mates' Box Measurements

Lab Mate	SI Units	U.S. Customary Units
1		
2		
3		

Part B Measuring Mass

Procedure Reference	Object	Mass
Step 1	1.	
Step 1	2.	
Step 1	3.	
Step 3	Mass of sugar sample on Balance 1	
Step 4	Mass of sugar sample on Balance 2	

Part C Measuring Volume

Measuring with Graduated Cylinders

10 mL Cylinder	25 mL Cylinder	50 mL Cylinder

Resolution and Graduated Cylinders

10 mL Cylinder	25 mL Cylinder	50 mL Cylinder

Accuracy: Graduated Cylinders vs. Beakers vs. Flasks

Procedure Reference	Description	50 mL Graduated Cylinder	50 mL Beaker (Step 5)	50 mL Erlenmeyer Flask (Step 6)
Step 1	Mass: glassware (g)			
Step 2	Volume: water (mL)			
Step 3	Mass: glassware + water (g)			
Step 4	Calculated mass: water (g)			

Name ______________________________

Lab Partner ______________________________

Lab Section ____________________ Date ______________

Lab 2
DATA SHEET
(continued)

Part D Measuring Temperature

Flask	Temperature (°C) (Step 7)	Observation (Step 8)
A		
B		
C		

Reflective Exercises

1 When measuring the box in Part A (Step 8, p. 20), was it more convenient to use the ruler's SI or U.S. customary scale to determine the appropriate number of digits in your measurement? Explain your choice.

2 Explain why your measurement of the box in Part A (Step 8) might not be identical to the measurements taken by your lab mates. Assuming all the measurements were taken correctly, what is the maximum number of digits that should differ between your measurement and those from your lab mates?

3 In Part A (Step 8), you measured the box in both SI and U.S. customary units. Your two measurements describe the length of the same box, and the two values are therefore equal to each other.

a *Use your measurements* to write two *experimentally determined* conversion factors that represent the relationship between SI and U.S. customary units of length.

b Use one of your experimental conversion factors to determine a value in centimeters that is equivalent to 8.23 in. Show your calculation.

c Use one of your experimental conversion factors to determine a value in inches that is equivalent to 57.76 cm. Show your calculation.

d According to the universally accepted reference standard, 1 in. is equal to 2.54 cm. Use your experimentally determined conversion factor to convert 1 in. to its equivalent value in centimeters. Use your answer to evaluate the accuracy of your conversion factor. What does this tell you about the accuracy of your box measurements?

Name ______________________

Lab Partner ______________________

Lab Section ______________ Date ______________

Lab 2 DATA SHEET

(continued)

4 Paper Clip A is known to have a mass of 1.1865 g. Paper Clip B is known to have a mass of 1.1848 g. Suppose you are given one of these paper clips but not told which one. Could you use the balances in your lab to determine the identity of your paper clip? Use the data from Part B that you recorded on the data sheet, page 26, to justify your answer.

5 Which is the best choice of instrument to use when measuring a small volume of liquid, such as the volume of a small test tube: a 10 mL, 25 mL, or 50 mL graduated cylinder? Explain your choice.

6 The mass of exactly 40 mL of distilled water is exactly 40 g. Based on this information, which laboratory measuring device should you use to obtain the most accurate volume measurements: a graduated cylinder, an Erlenmeyer flask, or a beaker? Use the data from Part C that you recorded on the data sheet, page 26, to justify your answer.

7 Record the Celsius temperatures according to each thermometer shown below.

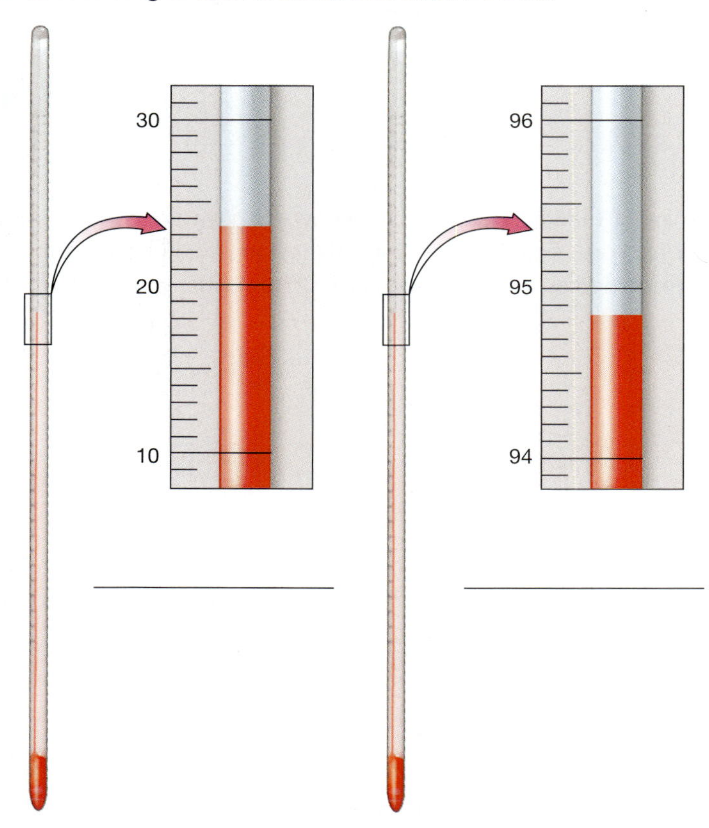

8 Portions of two graduated cylinders are shown below. Provide a measurement of the volume of liquid in each cylinder.

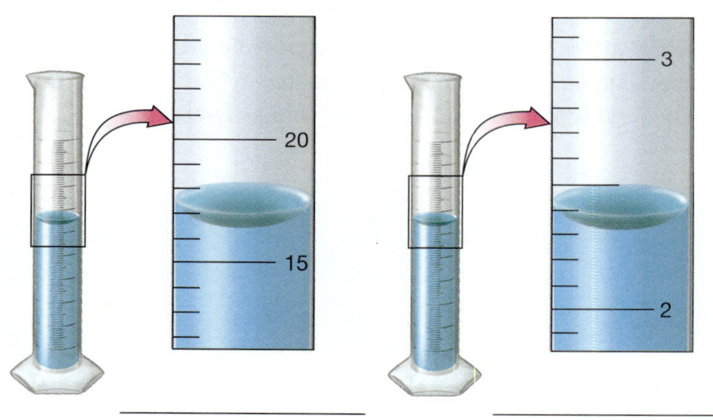

9 We typically think about temperature as a gauge of how hot or cold a substance feels relative to some standard. At the particle level, we can think of temperature as a measure of the average speed at which the particles that make up a substance are moving. Warmer substances contain faster moving particles. Were the observations from Part D that you recorded on the data sheet, page 27, consistent with this idea? Explain.

Name ______________________________

Lab Partner ______________________________

Lab Section ____________________ Date ____________

Lab 2
DATA SHEET
(continued)

10 Blue whales are the largest animals known to have existed. The largest recorded specimen measured 32.9 m in length.

a How many centimeters long was this whale? Show your calculation.

b How many kilometers long was this whale? Show your calculation.

c In the United States, school buses are restricted to no more than 45 ft. in length. Was the whale longer or shorter than the maximum school bus length? [1 m = 3.28 ft.] Show your work.

11 A baker is following a European recipe for making a certain type of bread. The recipe calls for about 710 mL of honey. The baker only has a measuring cup that uses U.S. customary units. How much honey should the baker measure out for the bread? Circle the letter for the correct answer and show your work to justify your answer.

Useful Relationships
29.6 mL = 1 fluid ounce
8 fluid ounces = 1 cup

▲ *Liquids that do not mix (such as oil and water) layer based on density.*

Lab 3

Be Precise

Density and Measurement Precision

Density, the mass of a substance in a given volume, is a characteristic physical property of solids and liquids. Mathematically, density may be calculated using the formula in Equation 1. Thus, to experimentally determine the density of a substance, you must measure both its mass and volume.

$$\text{Density} = \frac{\text{Mass}}{\text{Volume}} \qquad \textbf{[Eq. 1]}$$

To have confidence in an experimentally determined density value, you need to make use of the most accurate measuring devices available in your lab. Accurate mass measurements are readily made with an electronic balance, and very accurate liquid volumes may be measured by using either measuring or volumetric pipettes. There are a variety of pipette styles, and your instructor may demonstrate the proper use of those available in your lab.

The volume of a solid object with a regular shape, such as a cube, may be calculated by multiplying the measured length, width, and height. However, if the solid has an irregular shape, taking these measurements may be impractical. Instead, the volume of an irregularly shaped solid can be measured indirectly by **displacement**. This process involves immersing the object in a liquid, typically water. The volume of water displaced (i.e., the change in the measured volume of liquid) is equal to the volume of the immersed solid.

Some measurements, including volume by pipette or displacement, require careful manipulation of materials. To avoid drawing conclusions from data in which there were unnoticed manipulation errors, such measurements are often repeated multiple times in order to instill confidence in the results. The degree of agreement between multiple measurements of the same quantity is called the **precision** of the data.

As a qualitative indicator of repeatability, high precision provides confidence in the quality of the measurement process *without any consideration of accuracy* (*correctness*). To illustrate this concept, consider an archer who repetitively lands arrows in a cluster far from the bull's-eye (Fig. 3.1). Although the precision is very

Objectives

After completing this lab you should be able to:

- Calculate densities of solids and liquids from measurements of mass and volume.
- Calculate mass and volume quantities by applying the definition of density.
- Describe the relationship between density and layering of objects that do not mix.
- Discuss the concepts of measurement accuracy and precision and their relationship to confidence in a measured quantity.

high, the archer has poor accuracy because the arrow cluster is not close to the center of the target (i.e., the "correct" value). Similarly, data obtained by measurement can be precise without being accurate.

The mathematical **mean** (or average) is used to represent the combined information obtained by the series of measurements. The value of the mean can be visualized as the point around which the arrows are clustered in Figure 3.1.

An **outlier** is a single measurement that is significantly different from the others in a series. Outliers are most likely the result of mistakes made while taking an individual measurement. In general, it is acceptable to ignore obvious outliers and repeat the measurement when the procedure calls for you to obtain an average of multiple measurements.

In this lab, you will take a series of mass and volume measurements and use them to calculate the average densities of a mystery solid sample and a mystery liquid sample. Since density is characteristic of a substance, the density of a particular sample provides evidence of its identity. In addition, an object's density is responsible for some important physical behaviors.

For example, the ability of one substance to float on another is due, in part, to differences in their densities. If two liquids are **immiscible** (i.e., do not mix with each other), the higher-density liquid will sink to the bottom while the lower-density liquid floats on top of it.

Similarly, a solid that is more dense than a certain liquid will sink in that liquid while a solid that is less dense than the liquid will float. You will explore this concept by measuring the densities of four immiscible liquids and examining the ability of one to float on top of another.

3.1 The archer has high precision because the arrows are clustered around the same position. However, the archer's accuracy is low because the arrow cluster is far from the bull's-eye. The mean value (or average) can be thought of as the point around which the arrows are clustered.

Procedure

What Am I Doing?

Part A: Determining the density of a mystery liquid by measuring the mass of an accurately measured 10 mL sample of the liquid; repeating the measurement two times allows the calculation of a mean value from the three trials.

Part B: Determining the density of a mystery solid by measuring its mass and its volume by displacement; repeating the measurement two times allows the calculation of a mean value from the three trials.

Part C: Measuring the mass and volume of four different liquids in order to calculate their densities; combining the immiscible liquids into a single test tube in order to examine the relationship between density and position in the mixture.

Materials

- ❑ 50 mL beaker
- ❑ 50 mL Erlenmeyer flask
- ❑ 10 mL volumetric or measuring pipette
- ❑ 10 mL and 25 mL graduated cylinders
- ❑ 50 mL or 100 mL graduated cylinder
- ❑ Tongs or string for lowering metal solid
- ❑ Large test tube (25 x 200 mm)
- ❑ Long stem funnel

Part A Identifying a Mystery Liquid by Density Determination

3

1. Pour about 30 mL of a mystery liquid into a small beaker, and record its identification code on the data sheet, page 41.
2. Record the mass of an empty 50 mL Erlenmeyer flask on the data sheet.
3. Using a pipette, transfer 10 mL of the mystery liquid to the flask. Record the volume of the liquid sample on the data sheet. Be sure to use proper measuring technique for the pipette available to you.
4. Record the mass of the flask containing the liquid on the data sheet.
5. Determine the mass of the liquid by difference and calculate the density of the sample.
6. Dispose of the liquid as directed by your instructor.
7. Repeat Steps 2–6 two times, and then calculate the average density of the sample based on the data obtained in your three trials. Note that you do not need to dry the Erlenmeyer flask between each trial.
8. Your mystery liquid is one of the substances listed in Table 3.1. Based on your data, report the most probable identity of your sample.

Part B Identifying a Mystery Solid by Density Determination

Two procedures for measuring the density of a solid sample are provided. You should choose the procedure that matches the type of solid samples available in your laboratory: large, single-piece; or small pieces.

Procedure for Solids Available as a Large, Single Piece

1. Obtain a mystery solid, and record its identification code on the data sheet, page 42.
2. Measure and record the mass of the solid on the data sheet.
3. Obtain a graduated cylinder into which you will be able to easily lower your solid. Half-fill the cylinder with distilled water and record the initial volume on the data sheet.
4. Carefully lower the solid into the half-filled cylinder using tongs or a piece of string.
5. Record the new volume on the data sheet, and then calculate the volume of the solid by difference.
6. Calculate the density of the solid sample.

SAFETY NOTE

Use caution: Dropping or sliding the metal into the cylinder may break the glass!

7. Repeat Steps 2–6 two times, and then calculate the average density of the sample based on the data obtained in your three trials. Note that you should dry the solid with paper towels between trials in order to obtain accurate mass measurements.
8. Your sample is one of the substances listed in Table 3.2. Based on your data, report the most probable identity of your sample.

≡3

Procedure for Solids Available in Small Pieces

1. Record the identification code for your mystery solid on the data sheet, page 42.
2. Measure out approximately 40 g of the solid pieces in a weigh boat or small beaker. Record the mass measurement on the data sheet.

TABLE 3.1 ■ Densities of Some Common Liquids

Substance	Density (g/mL)
Isopropanol	0.781
Mineral oil	0.830
Olive oil	0.911
Seawater	1.02
Ethylene glycol	1.11

TABLE 3.2 ■ Densities of Some Common Solids

Substance	Density (g/mL)
Aluminum	2.70
Zinc	7.14
Tin	7.27
Iron	7.87
Brass	8.59
Copper	8.96
Silver	10.5
Lead	11.3

3. Add about 10 mL of distilled water to a 25 mL graduated cylinder and record the initial volume on the data sheet.
4. Carefully add the solid pieces to the cylinder.
5. Record the new volume on the data sheet, and then calculate the volume of the solid by difference.
6. Calculate the density of the solid sample.
7. Repeat Steps 2–6 two times, and then calculate the average density of the sample based on the data obtained in your three trials.
8. Your sample is one of the substances listed in Table 3.2. Based on your data, report the most probable identity of your sample.

Part C Density and Layering

1. Record the mass of a clean, dry 10 mL graduated cylinder on the data sheet, page 42.
2. Add about 5 mL of maple syrup to the graduated cylinder. On the data sheet, record the measured volume of the liquid with the appropriate number of digits.
3. Record the mass of the graduated cylinder with the syrup on the data sheet.
4. Determine the mass of the syrup by difference and calculate the density of the liquid.
5. Pour the syrup into a clean, dry 25 × 200 mm test tube.
6. Record the mass of another clean, dry 10 mL graduated cylinder on the data sheet.
7. Add about 5 mL of liquid laundry detergent to the graduated cylinder, and on the data sheet record the measured volume of the liquid with the appropriate number of digits.
8. Record the mass of the graduated cylinder with the laundry detergent on the data sheet.
9. Determine the mass of the sample by difference and calculate the density of the liquid.
10. Add the detergent to the test tube containing the syrup using a long-stem funnel tilted so that the stem touches the side of tube (see Technique Tip 3.1). Add the detergent slowly such that it runs gently down the side of the cylinder.
11. Repeat Steps 6–9 using approximately 5 mL of vegetable oil. Slowly add the vegetable oil to the test tube that contains the syrup and detergent using the same procedure described in Step 10.
12. Repeat Steps 6–9 using approximately 5 mL of rubbing alcohol. Slowly add the alcohol to the test tube that contains the syrup, detergent, and vegetable oil using the same procedure described in Step 10.
13. On the data sheet, page 42, record the location of each liquid layer on the test tube drawing.
14. The contents of the test tube may be poured down the drain unless otherwise directed by your instructor.

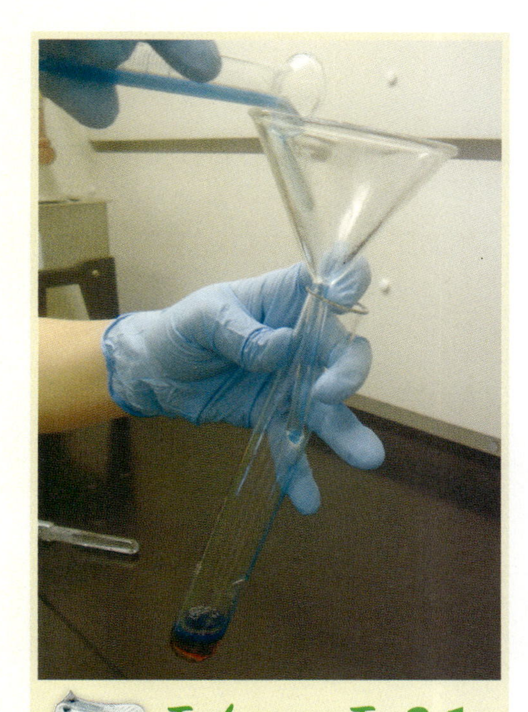

Technique Tip 3.1

To ensure the liquid runs slowly down the side of the test tube, add it through a long stem funnel that is tilted so the stem touches the side of the test tube.

Name ______________________

Lab Partner ______________________

Lab Section ______________ Date ______________

Lab 3 Pre-Laboratory Exercise

1 Provide a term that matches each description below.

a The agreement between several measurements of the same quantity. ______________

b Ratio between the mass and volume of a substance. ______________

c In a series of measurements of the same quantity, one that is significantly different from the others. ______________

d Mathematical value reported when a quantity is measured multiple times. ______________

e Term describing two liquids that do not mix together. ______________

2 In the procedure for Part A, why is it not necessary to dry the Erlenmeyer flask between each trial?

3 Name and describe the procedure used to measure the volume of the solid metal object in this experiment.

4 A fishing sinker has a mass of 31.560 g and displaces 11.7 mL of water. What is the density and identity of the sinker? Show your work to justify your answer.

5 A student attempting to identify a mystery sample of liquid collects the data in the table provided below. He knows the liquid must be one of those in Table 3.1. Can he successfully identify the liquid from his data? If so, provide the identity of the mystery liquid. If not, describe how he might be able to correct his problem and successfully identify the liquid.

Procedure Reference	Value	Trial 1	Trial 2	Trial 3
Step 2	Mass: flask	121.976 g	125.475 g	123.231 g
Step 3	Volume: mystery liquid	10.0 mL	10.0 mL	10.0 mL
Step 4	Mass: flask + mystery liquid	130.966 g	133.455 g	132.431 g
Step 5	Mass: mystery liquid			
Step 5	Density: mystery liquid			
Step 7	Average density: mystery liquid			
Step 8	Proposed identity: mystery liquid			

Name ______________________________

Lab Partner ______________________________

Lab Section ____________________ Date ____________

Lab 3 DATA SHEET

Data and Observations

Part A Identifying a Mystery Liquid by Density Determination

Mystery Liquid Identification Code: ____________

Procedure Reference	Value	Trial 1	Trial 2	Trial 3
Step 2	Mass: flask			
Step 3	Volume: mystery liquid			
Step 4	Mass: flask + mystery liquid			
Step 5	Mass: mystery liquid			
Step 5	Density: mystery liquid			
Step 7	Average density: mystery liquid			
Step 8	Proposed identity: mystery liquid			

Part B Identifying a Mystery Solid by Density Determination

Mystery Solid Identification Code: ______________

Procedure Reference	Value	Trial 1	Trial 2	Trial 3
Step 2	Mass: solid			
Step 3	Initial volume: water			
Step 5	Final volume: water			
Step 5	Volume: mystery solid			
Step 6	Density: mystery solid			
Step 7	Average density: mystery solid			
Step 8	Proposed identity: mystery solid			

Part C Density and Layering

Procedure Reference	Value	Maple Syrup	Detergent	Vegetable Oil	Rubbing Alcohol
Step 1 or 6	Mass: cylinder				
Step 2 or 7	Volume: liquid				
Step 3 or 8	Mass: cylinder + liquid				
Step 4 or 9	Mass: liquid				
Step 4 or 9	Density: liquid				

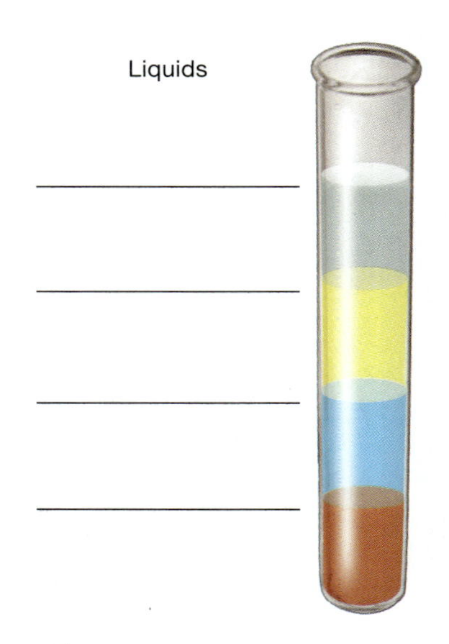

Name ______________________________

Lab Partner ______________________________

Lab Section ____________________ Date ____________

Lab 3
DATA SHEET
(continued)

Reflective Exercises

1 In the table below, make a claim regarding the identity of your mystery liquid (Part A). Describe your evidence and provide a rationale for your claim.

Claim	
Evidence	**Rationale**

2 In the table below, make a claim regarding the identity of your mystery solid (Part B). Describe your evidence and provide a rationale for your claim.

Claim	
Evidence	**Rationale**

3 Describe the accuracy of your density measurements assuming that you correctly identified your mystery solid and liquid.

4 How would you describe the precision of the data you collected in Parts A and B? How does this influence your confidence in the identities of the mystery liquid and solid you reported?

5 Think carefully about the ways in which the procedure required you to manipulate the mystery liquid and solid in order to obtain the data in Parts A and B. What sources of error might be present in each method? Do you think your data is more *accurate* for the density of the liquid or for the density of the solid? Justify your answer.

6 When performing Part A, a student did not fully drain the volumetric pipette before taking the mass measurement. Would his calculated density be higher, lower, or the same as that of a student who did not make this mistake? Explain your choice.

7 When performing Part B, a student's metal object was not fully submerged when she measured the displacement of water. Would her calculated density be higher, lower, or the same as that of a student who did not make this mistake? Explain your choice.

Name ______________________

Lab Partner ______________________

Lab Section ______________ Date ______________

Lab 3
DATA SHEET
(continued)

8 A drinking glass has a mass of 325.2 g. A 109 mL sample of milk is added to the glass. The mass of the glass with the milk is found to be 437.5 g. What is the density of the milk? Show your work to justify your answer.

9 A jeweler offers to sell you a ring he claims is pure gold, which has a density of 19.3 g/mL. Before buying, you ask to test the jeweler's claim. You find that the ring has a mass of 12.7 g, and when you add it to a graduated cylinder containing 10.2 mL of water, the volume rises to 12.1 mL.

a Is his claim that the ring is made of pure gold true? Show your work to justify your answer.

b What volume of water (in mL) should be displaced by a 12.7 g ring of pure gold? Show your work to justify your answer.

10 Given the densities of the following liquids and solids, label the picture of the test tube with the appropriate position of each substance. Assume the liquids do not mix.

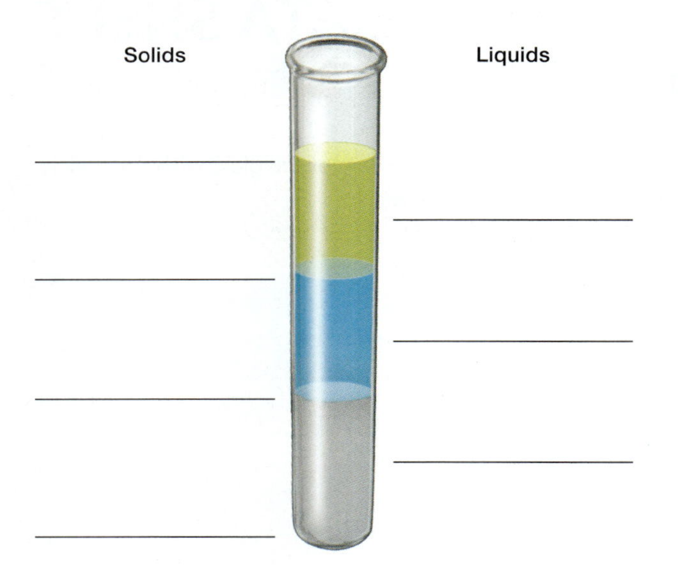

Substance	Rubbing Alcohol
Seawater (liquid)	1.03
Mercury (liquid)	13.6
Gasoline (liquid)	0.69
Ice	0.92
Cork	0.24
Silver ring	10.5
Gold ring	19.3

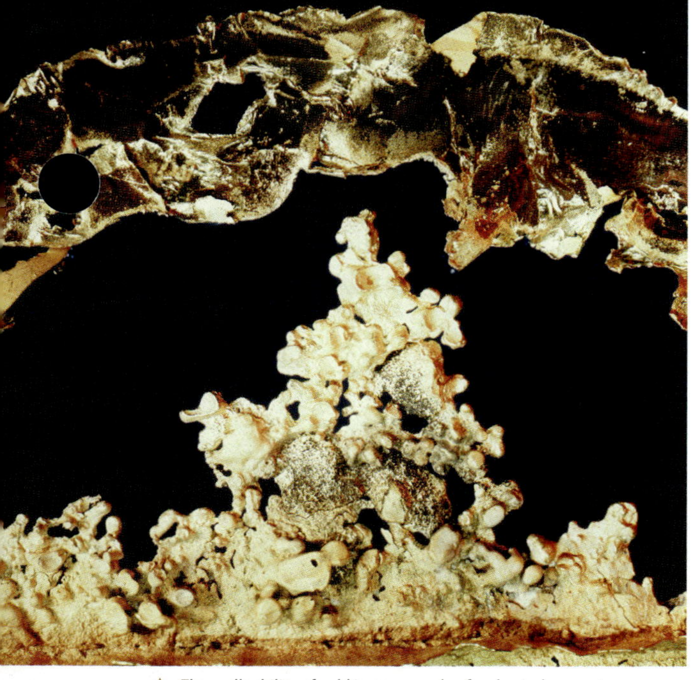

▲ *The malleability of gold is an example of a physical property.*

Lab 4

The More Things Change
Chemical and Physical Properties

We often use the *properties* of a material to distinguish it from other substances. For example, if you are offered a cup of water and a cup of vinegar, you might use the smell of each substance to decide which you would prefer to drink. Chemists divide properties into two major categories: physical properties and chemical properties.

Physical properties are those that are displayed without changing the identity of the substance. Examples include smell, color, physical state (solid, liquid, or gas), melting point, boiling point, density, and solubility. Physical properties often are used to help determine a substance's identity.

For example, both acetone and ethyl acetate are clear, colorless liquids, but the two can be distinguished by their solubility properties. Acetone is **miscible** with water, which means that the two liquids form a completely uniform mixture. Ethyl acetate is **immiscible** with water and forms a separate layer when the two are combined. You may be more familiar with the corresponding terms **soluble** and **insoluble**, which refer to the ability or inability of a *solid* to fully dissolve in a liquid.

Boiling point is another characteristic physical property often used to help establish the identity of a liquid. Although technically more complex, we can think of the boiling point as the temperature at which we observe a liquid boiling (rapidly turning into a gas throughout the sample). As long as two substances have a measurable difference in boiling point, this property may be used to differentiate between them.

Figure 4.1 illustrates a simple apparatus for measuring the boiling point of a small volume of liquid. As the sample is heated and begins to boil, its vapors travel up the tube, cool, condense back into the liquid phase, and run back into the bottom of the tube. This process is called **refluxing**. The recondensed vapors can be observed on the walls of the test tube as a shimmering ring of liquid that slowly rises up the tube as the glass warms.

To determine the boiling point, the liquid sample should be heated until the ring of condensate is 1–2 cm above the bulb of the thermometer. As the vapors cover the bulb, the temperature should increase until a maximum, stable point is reached. This is considered the boiling point.

Objectives

After completing this lab you should be able to:

- Distinguish between physical and chemical changes.
- Describe how a chemical change may influence a physical property such as solubility.
- Measure the boiling point of a liquid.
- Describe how a physical property such as boiling point may be used to help identify a substance.

It is important to ensure that the thermometer bulb is not touching the walls of the test tube, which would produce a false reading. In addition, the heating process should be carried out slowly. Rapid, excessive heat may cause superheating of the vapors or radiate sufficient heat through the tube walls that a false reading or continuous temperature rise are observed.

Chemical properties are those that are displayed by a substance when it changes composition. For example, combining baking soda and vinegar produces new substances, one of which is carbon dioxide gas that causes the mixture to bubble vigorously. On the other hand, adding table salt to vinegar does not produce a chemical change, and there is no evidence of a transformation in identity when the two chemicals interact. The chemical components of the salt simply dissolve and become dispersed throughout the vinegar, which is a physical alteration in the states of the two substances.

It sometimes can be difficult to distinguish between a **physical change** (a change involving physical properties) and a **chemical change** (a change involving chemical properties). Both types of changes may produce evidence of the transformation, including changes in color, temperature, or physical form. Fortunately, there is a good rule of thumb to keep in mind: If a substance has undergone a physical change, restoring the original conditions will cause the substance to revert to its original state. This is not true for a chemical change.

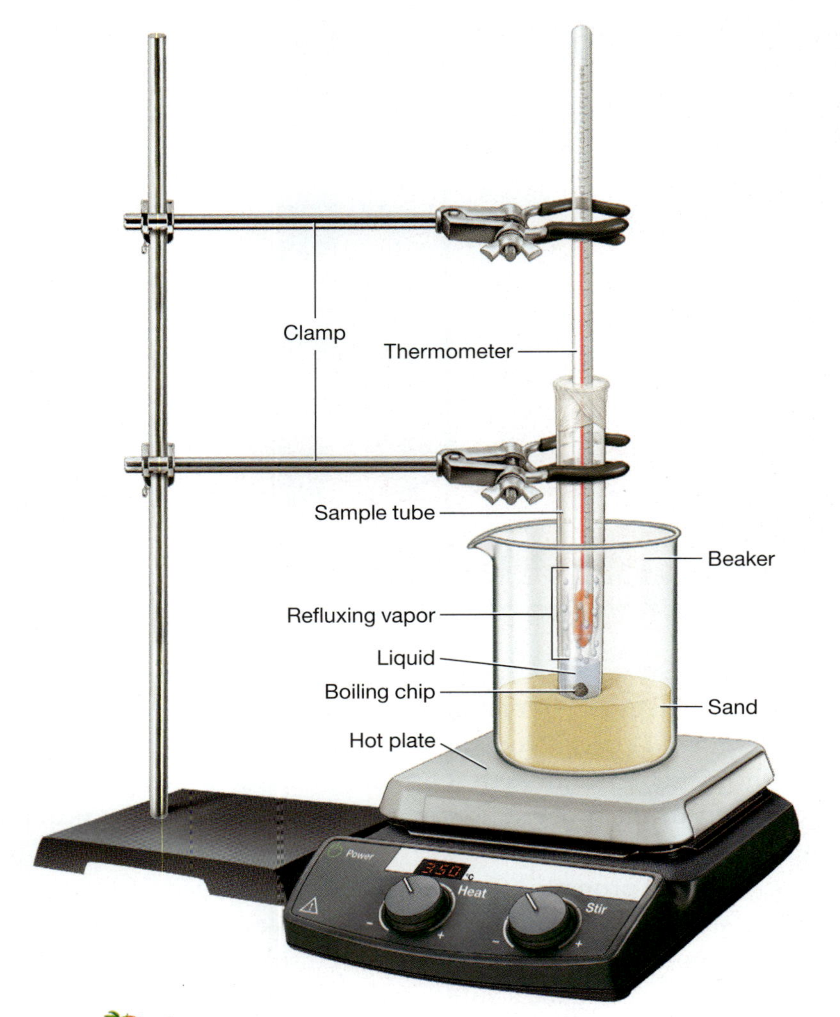

4.1 Apparatus for small scale determination of boiling point.

For example, removing ice cubes from a cold freezer will cause them to melt into liquid water, which is the same substance but in a different physical form. Restoring the original conditions by returning the water to the freezer will cause it to revert to solid ice. On the other hand, when a strip of bacon is heated on the stove and then returned to room temperature, the crispy, cooked bacon is not identical to the original uncooked meat (Fig. 4.2).

The rule of thumb is based on the fact that a chemical change involves transforming a material into a new substance while a physical change does not. Thus, re-establishing the original conditions should allow a substance that has undergone a physical change to revert to its original state. By contrast, a substance that has undergone a chemical change would not revert to the original form because it is no longer the same substance!

Of course, care must be exercised when applying this rule because it is not always possible to restore the exact original conditions. For example, the carbon dioxide that bubbles away as a gas upon mixing baking soda and vinegar cannot be easily added back to the mixture. In cases where the rule of thumb cannot be reliably applied, chemical changes can be identified by observations that indicate the production of a new substance with physical properties that differ from those of the starting material.

In this laboratory exercise, you will explore the physical and chemical properties of various substances. With practice, you will learn to use careful observation and experimentation to determine whether an action results in a chemical or physical change. You also will find that differences in carefully measured physical properties may be used to help determine the identity of a substance.

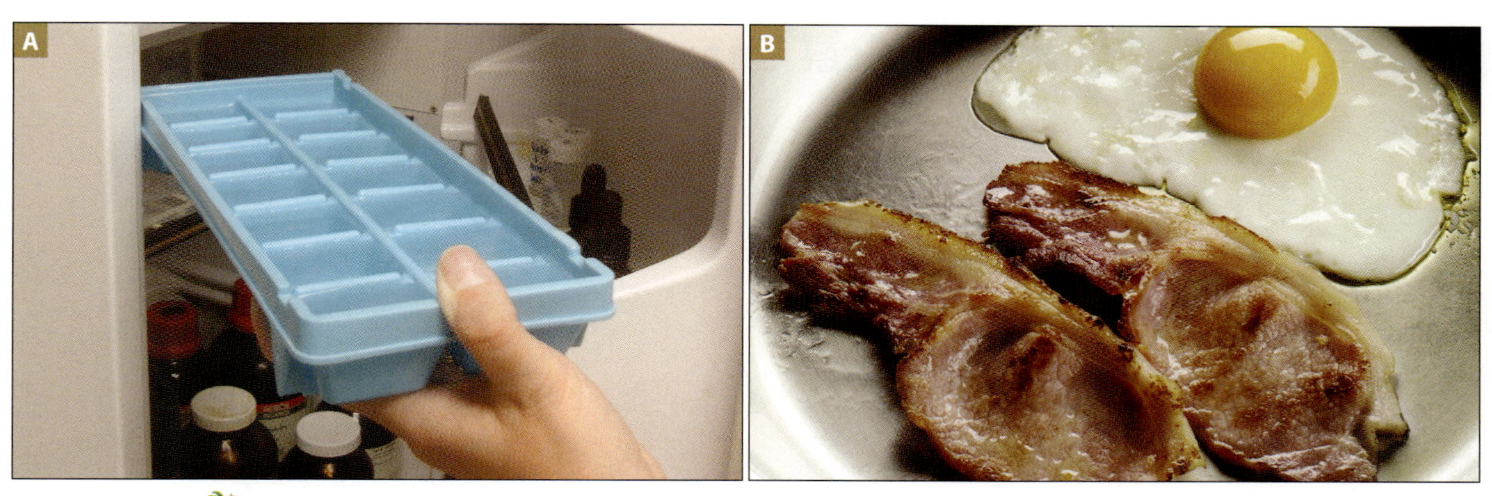

4.2 Applying the rule of thumb: (A) melting ice is a physical change; (B) cooking bacon is a chemical change.

4

Materials

- ❑ Test-tube holder
- ❑ Test tubes (6)
- ❑ Bunsen burner
- ❑ Microspatula
- ❑ Hot plate
- ❑ Ring stand
- ❑ Clamps (2)
- ❑ 250 mL beaker
- ❑ Large test tube (25 × 200 mm)
- ❑ Thermometer
- ❑ Parafilm
- ❑ Striker

Procedure

Part A: Observing an instructor demonstration and interpreting the observed changes as *physical* or *chemical* in nature.

Part B: Determining whether the addition of heat to two solids causes *physical* or *chemical* changes; examining the physical property of solubility and exploring whether changes in solubility relate to change in the *physical* or *chemical* nature of substance; exploring whether the mixing of chemicals results in *physical* or *chemical* changes.

Part C: Determining the boiling point of methanol and a mystery sample; using the boiling point of the mystery sample to determine its identity.

Part A Demonstrations

1. Observe a wooden applicator stick. Your instructor will heat it in a flame until it ignites. After the stick has been extinguished and cooled, carefully observe it again. Record your observations on the data sheet, page 55, and classify any change as *physical* or *chemical*.

2. Watch as your instructor heats a few iodine crystals in a flask that contains a test tube full of ice. Record your observations on the data sheet and classify any change as *physical* or *chemical*.

Part B Physical and Chemical Changes

Adding Heat

1. Add about a rice grain-sized amount of table sugar (sucrose) to a clean, dry test tube.
2. Using a test-tube holder, gently heat the solid by passing the test tube through the flame of a Bunsen burner several times. Continue heating the solid until a change in appearance is noted.
3. Allow the test tube to cool to room temperature.
4. Record your observations on the data sheet, page 55, and determine whether the change is *physical* or *chemical* in nature.
5. Using a clean test tube, repeat Steps 1–4 using a rice grain-sized quantity of salicylic acid.
6. Dispose of the test tube contents as directed by your instructor.

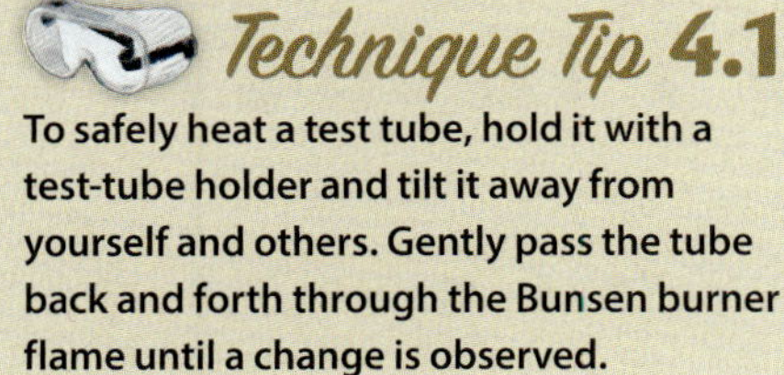

Technique Tip 4.1

To safely heat a test tube, hold it with a test-tube holder and tilt it away from yourself and others. Gently pass the tube back and forth through the Bunsen burner flame until a change is observed.

Solubility

1. Label five test tubes 1–5. Be sure all of the test tubes are clean and dry.
2. Add a rice grain-sized amount of table sugar and about 2 mL of distilled water to Test Tube 1. Gently swirl the test tube, and record your observations on the data sheet, page 55. Indicate whether the sugar is soluble or insoluble in water.
3. Add a rice grain-sized amount of table sugar and about 2 mL of acetone to Test Tube 2. Gently swirl the test tube, and record your observations on the data sheet. Indicate whether the sugar is soluble or insoluble in acetone.
4. Add about 2 mL of olive oil and about 2 mL of distilled water to Test Tube 3. Gently swirl the test tube, and record your observations on the data sheet. Indicate whether the olive oil is miscible or immiscible with water.
5. Add about 2 mL of olive oil and about 2 mL of acetone to Test Tube 4. Gently swirl the test tube, and record your observations on the data sheet. Indicate whether the olive oil is miscible or immiscible with acetone.
6. Add a pea-sized amount of benzoic acid and about 3 mL of distilled water to Test Tube 5. Gently swirl the test tube, and record your observations on the data sheet. Indicate whether the benzoic acid is soluble or insoluble in water.
7. Add 10 drops of 6 M ammonium hydroxide (NH_4OH) to Test Tube 5. Gently swirl the test tube. Record your observations on the data sheet, and indicate whether the addition of NH_4OH generated a soluble or insoluble substance.
8. Add 30 drops of 6 M hydrochloric acid (HCl) to Test Tube 5. Gently swirl the test tube. Record your observations on the data sheet, and indicate whether addition of the HCl generated a soluble or insoluble substance.
9. Dispose of the test tube contents as directed by your instructor.

Mixing Chemicals

1. Label four test tubes 1–4.
2. Add about 2 mL of sodium bicarbonate ($NaHCO_3$) solution to Test Tube 1.
3. Add about 2 mL of sodium sulfate (Na_2SO_4) solution to Test Tube 2.
4. Slowly add about 2 mL of vinegar each to Test Tubes 1 and 2.
5. Record your observations on the data sheet, page 56, and indicate whether or not there is evidence to indicate that a *chemical change* took place in each test tube.
6. Add about 2 mL of sodium hydroxide (NaOH) solution to Test Tube 3.
7. Add about 2 mL of sodium sulfate (Na_2SO_4) solution to Test Tube 4.
8. Add about 2 mL of copper(II) chloride ($CuCl_2$) solution each to Test Tubes 3 and 4.
9. Record your observations on the data sheet, and indicate whether or not there is evidence to indicate that a *chemical change* took place in each tube.
10. Dispose of the test tube contents as directed by your instructor.

Part C Boiling Point and Identity

1. Begin assembling the apparatus for boiling point determination (see Fig. 4.1) by heating about 1 in. of sand in a 250 mL beaker to approximately 100°C on a hot plate. (A boiling-water bath also may be used, but the steam often obscures a clear view of the refluxing liquid.)
2. Determine the boiling point of methanol:
   - **a** Add a boiling chip and about 2 mL of methanol to a large test tube (25 × 200 mm).
   - **b** Cover the tube with parafilm and use a sharp implement, such as the narrow end of a microspatula, to cut a slit in the center of the covering.
   - **c** Clamp the test tube to a ring stand above the sand bath.
   - **d** Separately clamp a thermometer above the tube.
   - **e** When the sand bath is heated, lower the test tube into the bath so sand is even with the surface of the liquid.
   - **f** Lower the thermometer through the slit in the parafilm until the bottom of the bulb is about 1 cm above the liquid surface.
   - **g** The liquid should soon begin to boil and you should see evidence of the refluxing vapors. When the thermometer reading stops rising and is stable for at least 1 minute, record the boiling point on the data sheet, page 56.
   - **h** Raise the thermometer and remove the test tube from the sand bath. Allow the liquid to cool, and then dispose of it as directed by your instructor.

3. Obtain a mystery liquid and record its identification code on the data sheet.

4. Determine the boiling point of your mystery liquid by repeating Step 2 with a *dry* tube, substituting the mystery liquid for methanol. Record the boiling point of your mystery liquid on the data sheet, and use Table 4.1 to propose an identity for the substance.

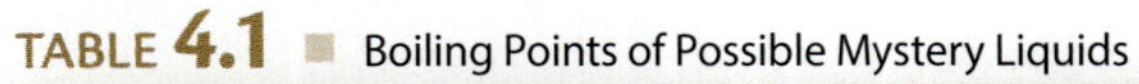

TABLE 4.1 ■ Boiling Points of Possible Mystery Liquids

Substance	Boiling Point
Acetone	56.1°C
Methanol	64.5°C
Hexane	68.7°C
Ethanol	78.2°C
Isopropanol	82.2°C

Name ______________________________

Lab Partner ______________________________

Lab Section ____________________ Date ____________

Lab 4 Pre-Laboratory Exercise

1 Provide a term that matches each description below.

a Characteristic of a substance that can be observed without changing the composition of the substance. ____________

b Process of vapors from a boiling liquid condensing and returning to the boiling sample. ____________

c Characteristic of a substance that cannot be observed without altering the identity of the substance. ____________

d Describes a solid that does not dissolve in a liquid. ____________

e Transformation in which one or more new substances is produced. ____________

f Describes a liquid that mixes completely with another liquid. ____________

2 Classify each of the following as either a *physical* or *chemical* change.

a Steam condensing above a pot of boiling water. ____________

b Propane burning in a grill. ____________

c A banana peel turning black. ____________

d Baking a cake. ____________

e Snow melting. ____________

f Cocoa dissolving in milk. ____________

g A copper statue turning green. ____________

3 Chemical and physical changes sometimes result in the production of a gas. Imagine such a scenario and describe two observations that might lead you to conclude that a gas was produced.

a ____________

b ____________

4 Explain how a physical property, such as boiling point, can help you distinguish one substance from another.

____

____

____

5 When a substance undergoes a chemical change, the product exhibits different physical properties than the starting material. Describe three changes in physical properties you might observe when two liquid solutions are mixed and a chemical change takes place.

a ______________________________

b ______________________________

c ______________________________

6 Describe the proper position of the thermometer used in a small scale boiling point determination. Why is this position critical to the experiment?

Name ______________________________

Lab Partner ______________________________

Lab Section ____________________ Date ______________

Lab 4 DATA SHEET

Data and Observations

Part A Demonstrations

Procedure Reference	Substance	Observations	Type of Change
Step 1	Wooden applicator stick		
Step 2	Iodine crystals		

Part B Physical and Chemical Changes

Adding Heat

Procedure Reference	Experiment	Observations	Type of Change
Step 4	Sugar + heat		
Step 5	Salicylic acid + heat		

Solubility

Test Tube	Mixture	Observations	Miscible/Immiscible or Soluble/Insoluble
1	Sugar + water		
2	Sugar + acetone		
3	Olive oil + water		
4	Olive oil + acetone		
5	Benzoic acid + water		
5	Benzoic acid + water + NH_4OH		
5	Benzoic acid + water + NH_4OH + HCl		

Mixing Chemicals

Test Tube	Mixture	Observations	Evidence of Chemical Change (Yes/No)
1	Sodium bicarbonate ($NaHCO_3$) + vinegar		
2	Sodium sulfate (Na_2SO_4) + vinegar		
3	Sodium hydroxide (NaOH) + copper(II) chloride ($CuCl_2$)		
4	Sodium sulfate (Na_2SO_4) + copper(II) chloride ($CuCl_2$)		

Part C Boiling Point and Identity

Procedure Reference	Substance	ID Code	Measured Boiling Point	Identity
Step 2	Methanol			
Steps 3–4	Mystery sample			

Name ______________________________

Lab Partner ______________________________

Lab Section ____________________ Date ____________

Lab 4
DATA SHEET
(continued)

Reflective Exercises

1 In the table below, present claims regarding whether the changes that occurred upon the addition of heat to both sugar and salicylic acid were physical or chemical. Describe your evidence and provide rationales for your claims.

Claim: Sugar	
Evidence	**Rationale**
Claim: Salicylic Acid	
Evidence	**Rationale**

2 The dissolution of table salt in water is a physical change. Describe a simple experiment based on the "rule of thumb" described in this lab that would demonstrate this fact.

3 In the Part B solubility experiments, the contents of Test Tubes 1–4 form uniform mixtures if the two chemicals share certain structural similarities that cause them to be strongly attracted to each other.

a Based on this information, do these solubility results illustrate chemical or physical properties? Explain your answer.

b Lactose is a solid with a structure similar to table sugar (sucrose). In which liquid is lactose more likely to dissolve, water or acetone? ______________________

c Arachidonic acid is a liquid with structural similarities to the components of olive oil. With which liquid is arachidonic acid more likely to be miscible, water or acetone? ______________________

4 In Test Tube 5 of the Part B solubility experiments, you first examined the solubility of benzoic acid in water and then studied the effects of added chemicals.

a Is benzoic acid soluble or insoluble in water? ______________________

b After you added NH_4OH, was the substance in Test Tube 5 soluble or insoluble in water? Which type of change—chemical or physical—does this suggest was caused by the addition of NH_4OH? Explain your answer.

c Did the addition of HCl to Test Tube 5 cause a chemical or physical change? Explain your answer.

Name ______________________________

Lab Partner ______________________________

Lab Section ____________________ Date ____________________

Lab 4
DATA SHEET
(continued)

5 Based on the observations you recorded on the data sheet from Part B (pp. 55–56), is the following statement true or false? "A chemical change occurs *only* when two different chemicals are mixed together." Use your data to justify your answer.

6 Based on your observations from Part B, is the following statement true or false? "A chemical change *always* takes place when two different chemicals are mixed together." Use your data to justify your answer.

7 The standard reference value for the boiling point of methanol is 64.5°C. Describe the accuracy of the boiling point you measured for methanol and identify sources of error that may account for any difference between your measured value and the reference value.

8 In Part C, you used a physical property, the boiling point, to identify a liquid from the possibilities given in Table 4.1. Do you think that using a single physical property is a reliable way to identify a substance with complete certainty? Explain your answer.

▲ A calorimeter allows scientists to measure the energy content of a sample.

Lab 5

Counting Calories
Calorimetry and Specific Heat

Objectives

After completing this lab you should be able to:

- Describe the components and operating principles of a calorimeter.
- Measure the energy content of a food sample.
- Measure the specific heat of a metal object.
- Apply calorimetry data to calculations involving energy transfer in a calorimeter.

Whenever a substance undergoes a chemical or physical change, there is a corresponding exchange of energy between the substance and its surroundings. A large part of this energy exchange can be detected as a change in temperature, which results when energy flows from a warmer substance to a cooler one. This type of energy exchange between objects is what we call **heat**. When heat flows into an object, the kinetic energy of its particles increases and its temperature will increase proportionally. Conversely, heat flowing out of an object causes a proportional decrease in its temperature.

The amount of heat exchange required to generate a given temperature change in a sample depends upon two factors: (1) the quantity of substance present, and (2) the intrinsic ability of the substance to absorb thermal energy. The first factor is readily measured by mass, and the second is described by a physical property called the **specific heat capacity** (often simply referred to as the **specific heat**). The specific heat of a substance indicates the quantity of energy that must be absorbed for one gram of the substance to exhibit a temperature increase of one degree Celsius. The corresponding units of specific heat (J/g°C or cal/g°C) reflect this definition.

Water, for example, has a specific heat of 4.184 J/g°C (or 1.00 cal/g°C). This is a remarkably large value. Vegetable oil, by contrast, has a far lower specific heat of 1.97 J/g°C. You may have noticed while cooking that oil heats up much faster than water, and the difference in specific heat values explains this observation! Specific heats for some common metals are given in Table 5.1.

Equation 1 describes the mathematical relationship between the amount of heat absorbed or released (q) and the temperature change of a substance (ΔT). As described above, the energy value is directly proportional to both the mass and specific heat of the sample. A larger mass and/or specific heat means that more heat is needed to increase a substance's temperature.

$$q = m \times c \times \Delta T \qquad \text{[Eq. 1]}$$

q = Amount of heat exchanged
m = Mass of the substance
c = Specific heat of the substance
ΔT = Temperature change of the substance

As an example, let us assume that you observed 12 g of water increase in temperature from 61°C to 85°C. The temperature *increase* indicates that the water *absorbed* heat, and the exact amount of heat may be calculated by inputting the appropriate quantities into Equation 1:

≡5

$$q = 12\,\cancel{g} \times 4.184\,\frac{J}{\cancel{g\,°C}} \times 24\,\cancel{°C} = 1{,}200\text{ J}$$

In the above example, the *source* of the heat absorbed by the water is not specified. If the experimental conditions were controlled such that there could only be one heat source, then the quantity of heat absorbed by the water (q) must be equal to the quantity lost by the source. **Calorimetry** is an experimental technique that takes advantage of this fact to measure such heat transfers and thereby determine the energy content of a sample.

Figure 5.1 presents a schematic of a typical **calorimeter**, the device commonly used to perform calorimetry experiments. The sample of interest is placed in a chamber filled with oxygen, which is then immersed in a water bath. The large specific heat value of water makes it an ideal medium because it can absorb a significant quantity of energy before the temperature rises so much that it begins to change phase (i.e., boil). Upon ignition, the sample undergoes combustion and releases energy in the form of heat, which flows into the water and causes an increase in the water temperature.

If the mass of water and its temperature change are measured, Equation 1 may be used to calculate the quantity of heat absorbed by the water. Since the heat flowing into the water comes from the sample, this calculation also provides the quantity of heat released by the sample. Experiments of this kind are used to measure the energy content of foods (i.e., Calories) listed on nutrition labels.

In this experiment, you will construct two rudimentary calorimeters. In Part A, you will construct a calorimeter to measure the energy content of a food sample. In Part B, another calorimeter will be used to determine the specific heat of a metal object. You will use your measurement of this physical property to determine the identity of the metal sample.

TABLE 5.1 ■ Specific Heat Values for Some Common Metals

Metal	Specific Heat (J/g°C)
Aluminum	0.90
Stainless steel	0.48
Iron	0.45
Zinc	0.39
Silver	0.24
Lead	0.13

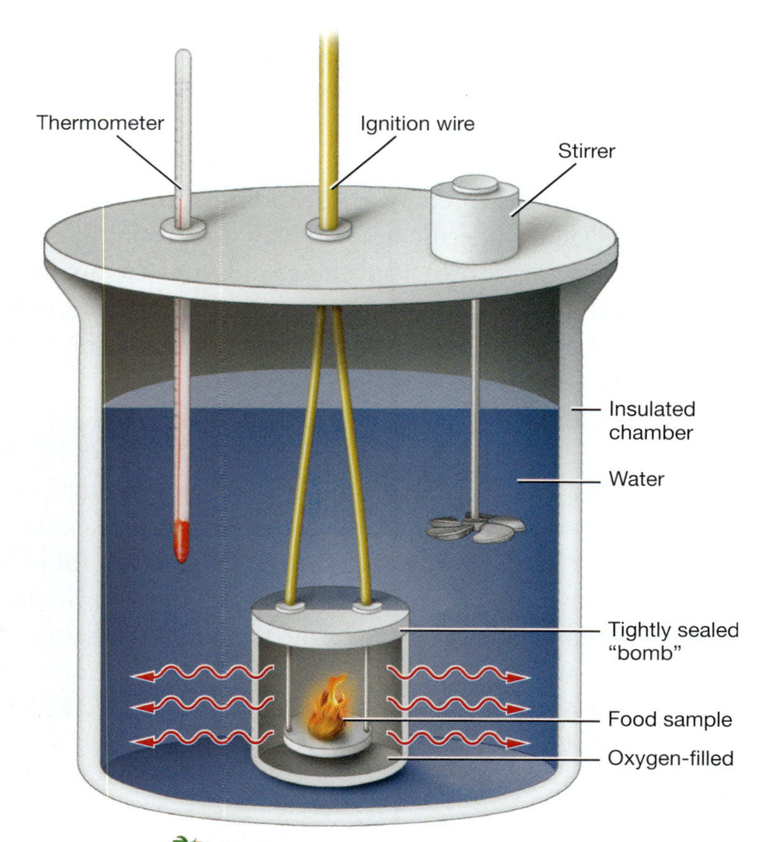

5.1 Schematic of a typical calorimeter.

Food Calorimetry

You will construct the calorimeter shown in Figure 5.2 to measure the energy content of a food sample. The major difference between your calorimeter and more accurate commercial calorimeters is that your food sample will undergo combustion in open air beneath the water bath rather than in a sealed chamber immersed in a water bath. Nevertheless, the same principle applies. The energy released by the food sample as heat will be absorbed by the water, causing a temperature increase in the water bath.

The heat released by the food sample (q_{food}) is equal to the heat gained by the calorimeter water (q_w) and may be calculated by inputting the data gathered for the water into Equation 1 as shown in Equation 2.

$$q_{food} = q_w = m_w \times c_w \times \Delta T_w \qquad \textbf{[Eq. 2]}$$

m_w = Mass of calorimeter water

c_w = Specific heat of water

ΔT_w = Temperature change of calorimeter water

It is likely that your food sample will not burn completely and that ashes will remain after the flame extinguishes itself. The mass of food that underwent combustion (m_{food}) is the difference between the mass of the original food sample and the mass of ashes remaining. To account for the mass of food that actually burned when you report the energy content of the sample, simply divide q_{food} by m_{food} (Eq. 3) and report the energy content per gram of food.

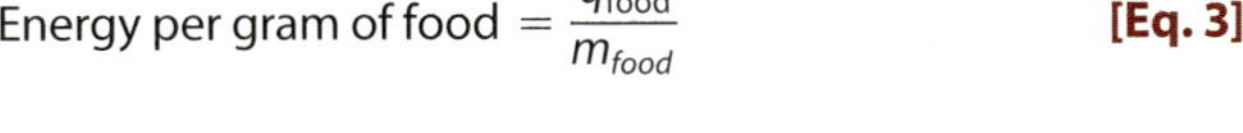

$$\text{Energy per gram of food} = \frac{q_{food}}{m_{food}} \qquad \textbf{[Eq. 3]}$$

5.2 Rudimentary food calorimeter.

Determination of Specific Heat

In order to solve for the value of specific heat, Equation 1 may be rearranged to Equation 4.

$$c = \frac{q}{m \times \Delta T} \qquad \textbf{[Eq. 4]}$$

Thus, it is necessary to measure three quantities in order to calculate the specific heat of an object. Mass (m) and temperature change (ΔT) may be measured with a scale and thermometer, respectively. The heat exchanged (q) by the sample must be determined by calorimetry. In order to calculate the specific heat of a mystery metal, you will construct a calorimeter using an insulated Styrofoam cup, a thermometer, and water as shown in Figure 5.3.

In the experiment, you will first heat the mystery metal and then quickly transfer it into the water-filled calorimeter. As the metal transfers heat to the calorimeter water, the water temperature will rise until there is no longer a temperature difference between it and the metal. The quantity of energy lost by the metal (q_m) is equal to that gained by the calorimeter water (q_w) and can be calculated by applying the data to Equation 1 as shown in Equation 5:

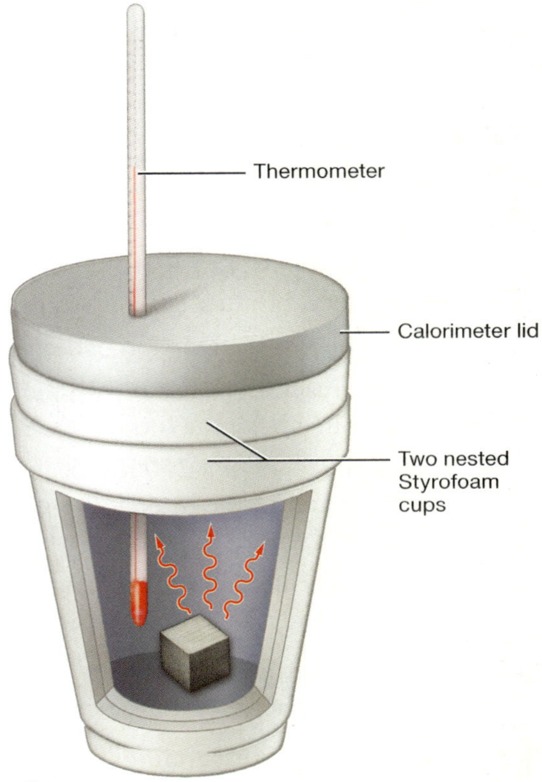

5.3 Schematic of a Styrofoam cup calorimeter.

$$q_m = q_w = m_w \times c_w \times \Delta T_w \quad \text{[Eq. 5]}$$

m_w = Mass of calorimeter water

c_w = Specific heat of water

ΔT_w = Temperature change of calorimeter water

The calculated value of q_m, the measured mass of the metal (m_m), and the temperature change of the metal (ΔT_m) may then be input to Equation 4 to calculate the specific heat of the metal (c_m), as shown in Equation 6:

$$c_m = \frac{q_m}{m_m \times \Delta T_m} \quad \text{[Eq. 6]}$$

Note that the temperature change for the metal (ΔT_m) is *not* identical to that of the water (ΔT_w). The metal will be heated in a boiling-water bath, and its initial temperature will be equal to the temperature of the boiling water. The final temperature of the metal will be the same as the final temperature of water in the calorimeter.

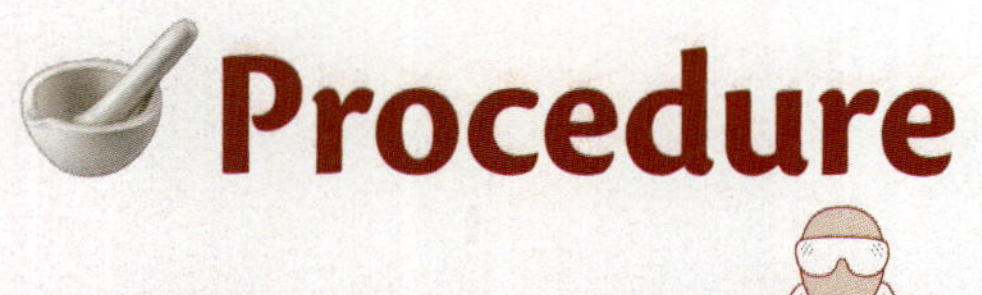

Procedure

What Am I Doing?

Part A: Measuring the temperature change in a sample of water upon absorption of heat from a food sample undergoing combustion in order to calculate the energy content of the food.

Part B: Measuring the temperature change in a sample of water upon absorption of heat from a hot metal object in order to calculate the specific heat of the metal.

Materials

- ❑ Ring stand
- ❑ Iron ring
- ❑ 250 mL and 400 mL beakers
- ❑ Wire gauze
- ❑ Reaction pedestal
- ❑ Thermometer
- ❑ Small clamp
- ❑ Lighter or other flame source
- ❑ Aluminum foil
- ❑ Tongs
- ❑ Styrofoam cups with tops
- ❑ Hot plate

Part A Food Calorimetry

1. Set up a calorimetry apparatus as illustrated in Figure 5.2 by following these steps:
   - **a** Clamp an iron ring onto a ring stand, and place wire gauze on top of it.
   - **b** Select a food product and record its identity and mass on the data sheet, page 69.
   - **c** Pin the food to the reaction pedestal provided.
   - **d** Place the reaction pedestal with the attached food beneath the iron ring. Lower the ring such that it sits no more than 2–3 in. above the food.
   - **e** Record the mass of a 250 mL beaker on the data sheet. Add approximately 100 mL of room temperature distilled water to the beaker and record the mass of the beaker with the water on the data sheet. Place the beaker on the wire gauze.
   - **f** Clamp a thermometer to the ring stand above the beaker. Lower the thermometer into the beaker so that the bulb is as close as possible to the center of the water column.
2. Monitor the thermometer for any temperature changes for several minutes. Proceed to Step 3 only when the temperature of the water in the calorimeter is stable. Record the initial temperature of the water on the data sheet.
3. Build a "tent" of aluminum foil around the calorimeter in order to prevent too much loss of heat to the air. You need to keep a small opening at the seam of your tent in order to ignite the sample and then allow oxygen to reach the flame. An example is shown in Figure 5.4.

4. Hold a flame to the food for a few seconds in order to ignite it. As soon as the food burns on its own, remove the flame.
5. Allow the food to combust until the flame extinguishes itself. On the data sheet, record the maximum temperature attained by the water in the calorimeter.
6. On the data sheet, record the mass of the remaining ash.
7. Fill in the calculations table on the data sheet. Use your data to calculate the energy released by combustion of the food both in units of joules and calories, and then calculate the corresponding energy content per gram of food.
8. Repeat Steps 1–7 with a second food sample as required by your instructor.

5.4 Building an aluminum foil "tent" around your calorimeter.

5

Part B Determination of Specific Heat

1. Prepare a boiling-water bath on a hot plate using a half-filled 400 mL beaker.
2. Obtain a mystery sample of metal and record its identification code on the data sheet, page 70.
3. Measure the mass of the metal and record the value on the data sheet.
4. Using tongs, gently lower the metal into the boiling-water bath and allow it to heat up in the boiling water for at least 10 minutes.
5. While the metal is warming, prepare a coffee cup calorimeter (see Fig. 5.3) by following these steps:
   - **a** Nest two Styrofoam cups together to create the calorimeter base. Record the mass of the calorimeter base on the data sheet, page 70.
   - **b** Add enough room temperature distilled water to the calorimeter such that the metal will be completely covered upon addition. On the data sheet, record the mass of the calorimeter with the water.
   - **c** Place the top with the thermometer on the base and monitor it for any temperature changes for several minutes. Proceed to Step 6 only when the temperature of the water in the calorimeter is stable.
6. On the data sheet, record the initial temperature of the water in the calorimeter.
7. Once the metal has been heated in the boiling water for at least 10 minutes, measure the temperature of the boiling-water bath and record the value on the data sheet.
8. Use tongs to quickly transfer the metal into the calorimeter. Try to shake off any excess water on the metal during this transfer process, and ensure that it is fully submerged after you add it to the calorimeter.
9. Cover the calorimeter tightly and observe the change in water temperature. On the data sheet, record the final temperature of the water in the calorimeter once it reaches a maximum value.
10. Repeat Steps 3–9 to obtain a second set of data on your metal sample.
11. Fill in the calculations table on the data sheet, page 71. Use each data set to calculate a value for the specific heat of the metal.
12. Use Table 5.1 (p. 62) to determine the identity of the mystery metal.

Name ______________________

Lab Partner ______________________

Lab Section ______________________ Date ______________

Lab 5
Pre-Laboratory Exercise

1 Provide a term or unit that matches each description below.

a Technique that measures the heat transfer between objects. ______________

b Unit associated with energy (or heat exchange). ______________

c Intrinsic property of a material that describes the quantity of heat required to cause a given temperature change for a certain mass of the substance. ______________

d Flow of energy from a warmer object to a cooler one. ______________

e Device used to measure the heat transfer between objects. ______________

f Unit associated with specific heat. ______________

2 Why is it necessary to weigh the food sample before and after combustion in Part A?

3 In Part B, the temperature of the water in the calorimeter increases when the metal object is added.

a What is the source of heat causing the increase in water temperature?

b Why is it critical that the metal object be fully immersed in the calorimeter water for the experiment to be valid?

c Why must the thermometer be immersed in the water but not touching the metal object for the experiment to be valid?

4 Why is it necessary to measure the mass of the water used in the two calorimeters?

5 If you were going to build a picnic table to use in a very sunny back yard, would it be better to construct the table out of wood, aluminum, or iron. Note that the specific heat of wood is approximately 1.4 J/g°C. Justify your answer.

Name ______________________________

Lab Partner ______________________________

Lab Section ____________________ Date ____________

Lab 5 DATA SHEET

Data and Observations

Part A Food Calorimetry

Food Calorimetry Data

Procedure Reference	Quantity	Trial 1	Trial 2
Step 1	Food identity		
Step 1	Mass: food sample (g)		
Step 1	Mass: empty beaker (g)		
Step 1	Mass: beaker + water (g)		
Step 2	Initial water temperature (°C)		
Step 5	Final water temperature (°C)		
Step 6	Mass: food ash (g)		

Food Calorimetry Calculations—Trial 1

Variable	Quantity	Calculation/Source	Value
m_w	Mass: calorimeter water (g)	=	
m_{food}	Mass: food combusted (g)	=	
ΔT_w	Temperature change: calorimeter water (°C)	=	
$q_w = q_{food}$	Energy gained by water/lost by food (cal)	=	
N/A	Energy per gram of food (cal/g)	=	
$q_w = q_{food}$	Energy gained by water/lost by food (J)	=	
N/A	Energy per gram of food (J/g)	=	

Food Calorimetry Calculations—Trial 2

Variable	Quantity	Calculation/Source	Value
m_w	Mass: calorimeter water (g)	=	
m_{food}	Mass: food combusted (g)	=	
ΔT_w	Temperature change: calorimeter water (°C)	=	
$q_w = q_{food}$	Energy gained by water/lost by food (cal)	=	
N/A	Energy per gram of food (cal/g)	=	
$q_w = q_{food}$	Energy gained by water/lost by food (J)	=	
N/A	Energy per gram of food (J/g)	=	

Part B Determination of Specific Heat

Specific Heat Data

Procedure Reference	Quantity	Trial 1	Trial 2
Step 2	Mystery metal identification code		
Step 3	Mass (m_m): metal sample (g)		
Step 5	Mass: empty calorimeter (g)		
Step 5	Mass: calorimeter + water (g)		
Step 6	Initial temperature: calorimeter water (°C)		
Step 7	Temperature: boiling-water bath (°C)		
Step 9	Final temperature: calorimeter water (°C)		

Name ______________________________

Lab Partner ______________________________

Lab Section ____________________ Date ____________

Lab 5
DATA SHEET
(continued)

Specific Heat Calculations—Trial 1

Variable	Quantity	Calculation/Source	Value
m_w	Mass: calorimeter water (g)	=	
ΔT_w	Temperature change: calorimeter water (°C)	=	
ΔT_m	Temperature change: mystery metal (°C)	=	
$q_w = q_m$	Heat gained by water/lost by metal (J)	=	
c_m	Specific heat of metal (J/g°C)	=	

Specific Heat Calculations—Trial 2

Variable	Quantity	Calculation/Source	Value
m_w	Mass: calorimeter water (g)	=	
ΔT_w	Temperature change: calorimeter water (°C)	=	
ΔT_m	Temperature change: mystery metal (°C)	=	
$q_w = q_m$	Heat gained by water/lost by metal (J)	=	
c_m	Specific heat of metal (J/g°C)	=	

Proposed identity of mystery metal: ____________________

Reflective Exercises

1 Describe your level of confidence in the results you obtained for Part A. Discuss any sources of error you can identify from the procedure you used to determine the energy content of food.

2 A student performed a food calorimetry experiment on a cracker with a mass of 16.23 g. At the end of the experiment, the remaining ash had a mass of 1.39 g.

a If the temperature of 99.68 g water increased from 23.5°C to 72.9°C during the experiment, what is the caloric content of the cracker in calories per gram? Show your work.

b The "Calories" reported on nutrition labels are actually kilocalories (1 kilocalorie = 1,000 calories). What is the energy content of this student's cracker in Calories per gram? Show your work.

Name ______________________________

Lab Partner ______________________________

Lab Section ____________________ Date ______________

Lab 5 DATA SHEET

(continued)

3 In the table below, present a claim regarding the identity of your Part B mystery metal. Describe your evidence and provide a rationale for your claim.

Claim	
Evidence	**Rationale**

4 In Part B, you measured the initial temperature of the metal indirectly. What assumption was implicit in taking this measurement? Based on your observations, do you believe this was a valid assumption? Fully explain your answers.

____

____

____

____

____

____

5 The procedure used in Part B is a rather unsophisticated means of determining the specific heat of a metal. Identify three sources of experimental error that might be expected when using this procedure.

1. ______________________________

2. ______________________________

3. ______________________________

6 Why was it a good idea to repeat the experiment in Part B? Describe the accuracy and precision of the data you obtained. Do you think that additional repetitions of the experiment would increase your confidence in the identity of the metal? Why or why not?

7 A metal block has a mass of 62.48 g. The metal block was placed in boiling water with a measured temperature of 98.9°C for 10 minutes. It was then transferred to a Styrofoam calorimeter containing 86.23 g of water at 22.5°C. After several minutes, the water in the calorimeter reached a maximum temperature of 32.9°C. Calculate the specific heat of the metal and use Table 5.1 (p. 62) to identify it.

8 A student performing Part B of this lab did not add enough water to her calorimeter, and part of the metal sample was not submerged during her experiment. She recognized her error at the end of the experiment and did not have time to repeat it. She calculated the specific heat of her metal to be 0.48 J/g°C. Is it more likely that her sample was iron or stainless steel? Explain your answer.

Name ____________________

Lab Partner ____________________

Lab Section ____________________ Date ____________________

Lab 5 DATA SHEET

(continued)

9 Explain why it is reasonable to use a poorly insulated glass beaker to hold the calorimeter water in Part A of this experiment while Part B requires the use of well-insulated Styrofoam cups.

10 The specific heat of liquid ethanol is 2.44 J/g°C, and its boiling point is 78°C. Ethanol was used in place of water in a food calorimeter. The mass of the ethanol was 168 g and the initial temperature was 25°C. The food transferred 24,185 J to the ethanol. Use this information to answer the questions that follow.

a Did the temperature of the ethanol increase or decrease? ____________________

b What was the temperature change (ΔT) of the ethanol?

c Considering your answer to b, explain why water is used as the heat-absorbing medium in calorimeters rather than some other liquid, such as ethanol.

▲ *Copper, mercury, and magnesium are all elements on the periodic table.*

Lab 6

All in the Family
Elements and the Periodic Table

Objectives

After completing this lab you should be able to:

- Describe the periodic law.
- Apply the periodic law to identify elements that share similar properties.
- Properly perform flame tests.
- Use chemical changes to identify elements that share similar properties.
- Use chemical changes to distinguish between different elements.
- Apply data obtained in this experiment to identify metal and halogen components of a mystery sample.

The **periodic law** states that the physical and chemical properties of the elements recur in a repeating pattern when they are ordered by increasing atomic number. On the **periodic table**, elements are arranged by increasing atomic number into horizontal rows and vertical columns. The rows are called **periods**, and the columns are called **groups**.

This arrangement yields an important pattern: elements with similar properties are located within the same group. The pattern is a direct result of the fact that elements within a group possess the same number of **valence electrons**, those found in the outermost energy level of the atom. It is the valence electrons that are primarily responsible for an element's reactivity.

For that reason, a group on the periodic table can be thought of as a "family" of elements that share many chemical properties. Some of the groups on the periodic table provided on the inside of the front cover are labeled with the family name for elements found within that group.

Because they share similar chemical properties, we often can identify which elements belong to the same family (i.e., group) by examining some of their chemical reactions. In this laboratory exercise, you will examine the reactivity of elements in three families: alkali metals, alkaline earth metals, and halogens. Like brothers and sisters, elements within a family still have individual characteristics that make them unique. In addition to identifying "family traits," you will use your experiments to identify distinguishing characteristics of the individual elements within a group.

It is important to note that you will *not* use pure samples of elements during this lab. Pure alkali metals, for example, can explode when they come into contact with water in the air! During the lab, you will use aqueous (water-based) solutions that contain **ionic**, or charged, forms of the elements. For example, you will use a "sodium ion solution" rather than elemental sodium. The periodic law applies to these ions in the same way that it applies to the parent atoms because elements within a group form ions that bear identical charges.

In Part A, you will perform a flame test on six solutions, each containing a single alkali or alkaline earth metal ion. A flame test involves placing a small amount of solution on a metal loop and holding the loop in the tip of a flame. As the substance is heated, the color of the flame is observed (Fig. 6.1). Each metal ion produces a different color, providing information that should help you distinguish between the six metal ions you examine.

Flame tests can be difficult to read, so you may wish to repeat your flame tests several times in order to verify your observations. In addition, sodium often is present as an impurity in solutions of various alkali and alkaline earth metal ions. Sodium produces an orange-yellow flame test. When present as an impurity, the sodium should produce only a weak orange-yellow color, and the color of the primary component of the solution should dominate. A true sample of sodium should produce an intense orange-yellow color during a flame test.

6.1 Flame tests involve holding a sample in a flame and observing the color produced.

In Part B, you will examine the behavior of six metal ions in three separate chemical reactions. For example, you will observe the separate reactions of a barium ion solution with solutions of ammonium carbonate, ammonium hydrogen phosphate, and ammonium sulfate. You may observe no reaction for the barium ions with these substances, or you may observe the formation of a white solid.

A solid formed when two solutions undergo a chemical reaction is called a **precipitate**. The formation of a precipitate generally causes the solution to turn cloudy because the grains of solid are very tiny and remain suspended in the liquid (Fig. 6.2). Large crystals are rarely observed during a precipitation reaction.

6.2 When a precipitate forms, the solution changes appearance from clear to cloudy.

You will be able to use your results from Part B to classify the six metal ions into two families based on similar patterns of reactivity. For example, if barium and potassium ions both form white solids when mixed with the same substances, then they belong in the same chemical family. If they display different reactivities, then they belong in different families.

In Part C, you will examine the distinct behavior of the three halogen ions under identical reaction conditions. When mixed with bleach (sodium hypochlorite), the halogen ions undergo a similar reaction but each exhibits a different color. Thus, observing the color produced by this reaction should allow you to distinguish between the three halogen ions.

In Part D, you will apply the testing procedures from Parts A through C to a mystery sample that contains a metal ion and a halogen ion. You should be able to identify the components of the mystery sample by comparing your results from Part D with your data from Parts A through C.

Procedure

What Am I Doing?

Part A: Observing and recording characteristic colors produced in a flame by six different metal ions.

Part B: Observing and recording whether or not a precipitate forms when six different metal ions are each mixed with three different chemicals.

Part C: Observing and recording characteristic colors produced by three different halogen ions when mixed with a certain combination of chemicals.

Part D: Repeating the experiments from Parts A through C on a mystery sample containing one of the six metal ions and one of the six halogen ions in order to identify the components of the sample.

Materials

- ❑ Test tubes (6)
- ❑ Flame test wire
- ❑ Bunsen burner
- ❑ Hot plate
- ❑ 50 mL and 250 mL beakers

6

Part A Flame Tests for Metal Ions

1. Label six test tubes 1–6.
2. To each test tube, add 10 drops of the appropriate metal ion sample as indicated below:
   - Test Tube 1: Barium ion solution
   - Test Tube 2: Calcium ion solution
   - Test Tube 3: Lithium ion solution
   - Test Tube 4: Sodium ion solution
   - Test Tube 5: Potassium ion solution
   - Test Tube 6: Strontium ion solution

SAFETY NOTE

Hydrochloric acid (HCl) is a strong acid. ***Handle with caution!*** Certain metal ions are toxic. Use caution and follow all instructor safety precautions when handling these chemicals!

3. Clean a flame test wire by dipping it into a small amount of 6 M HCl and holding it in the flame of a Bunsen burner. Repeat this step two times.
4. Dip the test wire into Test Tube 1. Hold the wire loop in the tip of the burner flame. On the data sheet, page 85, record the color of the flame you observe.
5. Repeat Steps 3 and 4 on the samples in Test Tubes 2–6. On the data sheet, record the color of the flame produced as each sample is heated. **Do not discard the solutions in the test tube. You will use them again in Part B.**

Part B Reactions of Metal Ions

1. Prepare a boiling-water bath on a hot plate using a half-filled 250 mL beaker. Retain the water bath for Parts B and D.
2. To each of the test tubes used for the flame tests in Part A, add 5 drops of ammonium carbonate [$(NH_4)_2CO_3$] solution. If a solid forms (the solution changes appearance from clear to cloudy as in Fig. 6.2), record *ppt* (precipitate) in the table on the data sheet, page 85. If no solid forms, record *NR* (no reaction) in the table.
3. Dispose of the test tube contents as directed by your instructor and rinse the tubes with distilled water.
4. Refill each test tube with 10 drops of the appropriate metal ion solution (listed in Part A, Step 2).
5. To each test tube, add 5 drops of ammonium hydrogen phosphate [$(NH_4)_2HPO_4$] solution. Record your observations as *ppt* or *NR* on the data sheet.
6. Dispose of the test tube contents as directed by your instructor, and rinse the tubes with distilled water.
7. Refill each test tube with 10 drops of the appropriate metal ion solution (listed in Part A, Step 2).
8. To each tube, add 5 drops of ammonium sulfate [$(NH_4)_2SO_4$] solution. For any test tube in which you observe no precipitate formation, warm the solution by placing the test tube in the boiling-water bath for up to 5 minutes. Record your observations as *ppt* or *NR* on the data sheet, including whether additional heat was necessary (+ *heat*).
9. Dispose of the test tube contents as directed by your instructor.

≡6

Part C Halogen Ion Reactions

1. Label three test tubes 1–3.
2. To each test tube, add 10 drops of the appropriate halogen ion sample as indicated below:
   - Test Tube 1: Chlorine ion
   - Test Tube 2: Bromine ion
   - Test Tube 3: Iodine ion
3. To each test tube, add 20 drops of hexane, 1 drop of nitric acid (HNO_3), and 5 drops of bleach (sodium hypochlorite).
4. Gently agitate the contents of each test tube, and read the halogen test result by observing the color of the upper hexane layer (Fig. 6.3). Record the color of the hexane layer on the data sheet, page 86.
5. Dispose of the solution as directed by your instructor.

Nitric acid is a strong acid. ***Handle with caution!*** Hexanes and the elemental halogens produced during this test are flammable and inhalation hazards. Perform this test in a fume hood or appropriately ventilated area! Do not perform this test in the presence of open flames!

6.3 To read the halogen test result, observe the color of the upper hexane layer.

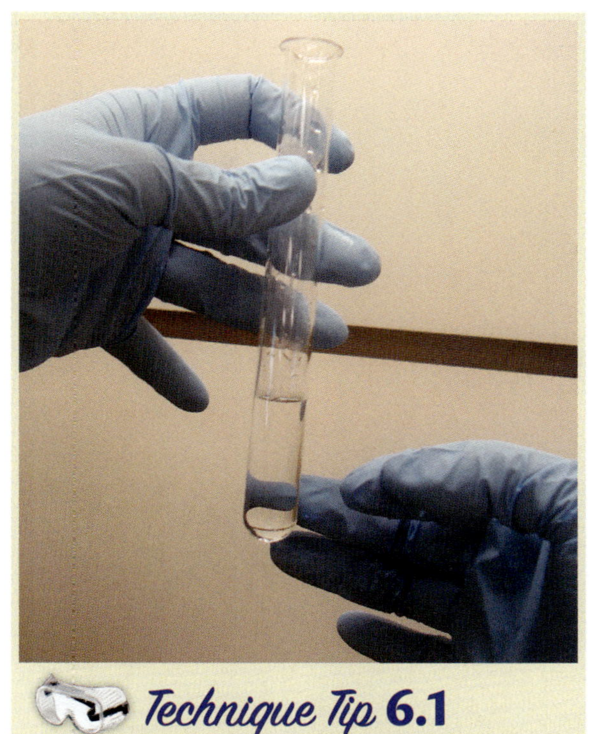

***Technique Tip* 6.1**

To safely agitate a test tube, hold it between your thumb and split first two fingers. Flick the test tube using a finger from the opposite hand.

Part D Determining the Ion Components in a Mystery Sample

1. Record the identification code of your mystery solution on the data sheet, page 86.
2. Label four test tubes 1–4, and add 10 drops of the mystery sample into each.
3. Use Test Tube 1 to perform a flame test, and record your observation on the data sheet.
4. Add 5 drops of ammonium carbonate [$(NH_4)_2CO_3$] solution to Test Tube 1, and record your observation on the data sheet.
5. Add 5 drops of ammonium hydrogen phosphate [$(NH_4)_2HPO_4$] solution to Test Tube 2, and record your observation on the data sheet.
6. Add 5 drops of ammonium sulfate [$(NH_4)_2SO_4$] solution to Test Tube 3. If no precipitate forms immediately, warm the solution by placing the test tube in the boiling-water bath for up to 5 minutes. Record your observation on the data sheet.
7. Add 20 drops of hexane, 1 drop of nitric acid (HNO_3), and 5 drops of bleach (sodium hypochlorite solution) to Test Tube 4. Gently agitate the contents of the test tube, and record the color of the hexane layer on the data sheet.
8. Use your data to determine which metal ion and halogen ion are likely to be present in your mystery solution.

Name ______________________________

Lab Partner ______________________________

Lab Section ____________________ Date ____________

Lab 6 Pre-Laboratory Exercise

1 Provide a term that matches each description below.

a Horizontal row on the periodic table. ____________________

b Test that involves holding a sample in a flame. ____________________

c Insoluble solid formed during a reaction in solution. ____________________

d Charged form of an atom. ____________________

e Column on the periodic table. ____________________

f States that the properties of elements repeat in a recurring pattern when ordered by increasing atomic number. ____________________

2 The ions of six metals are investigated in this experiment. Name the six metals.

____________________ ____________________

____________________ ____________________

____________________ ____________________

3 When performing a flame test, which element is often present as an impurity? How can the presence of this element as an impurity be distinguished from its presence as a primary component of a solution?

4 The ions of three halogens are investigated in this experiment. Name the three halogens.

5 The halogen test is observed in a mixture of hexane and water, which separate into layers. Are the test results observed in the hexane or the water layer? Is this the upper layer or the lower layer?

Name ______________________________

Lab Partner ______________________________

Lab Section ____________________ Date ______________

Lab 6 DATA SHEET

Data and Observations

Part A Flame Tests for Metal Ions

Metal Ions Flame Test

Test Tube	Solution Tested	Observations
1	Barium ion	
2	Calcium ion	
3	Lithium ion	
4	Sodium ion	
5	Potassium ion	
6	Strontium ion	

Part B Reactions of Metal Ions

Observation After Substance Added

Test Tube	Solution Tested	$(NH_4)_2CO_3$	$(NH_4)_2HPO_4$	$(NH_4)_2SO_4$
1	Barium ion			
2	Calcium ion			
3	Lithium ion			
4	Sodium ion			
5	Potassium ion			
6	Strontium ion			

Part C Halogen Ion Reactions

Halogen Ion Reaction Results

Test Tube	Solution Tested	Observations
1	Chlorine ion	
2	Bromine ion	
3	Iodine ion	

Part D Determining the Ion Components in a Mystery Sample

Observations

Sample Identification Code	Flame Test	$(NH_4)_2CO_3$ added	$(NH_4)_2HPO_4$ added	$(NH_4)_2SO_4$ added	Halogen Ion Test

Metal ion present ______________________

Halogen ion present ______________________

Reflective Exercises

1 Fireworks contain chemicals and gunpowder packed together in a rocket shell. Based on your data from Part A, which element ions would you expect to find in fireworks that produce the following colors?

a violet ______________________

b bright red ______________________

c green ______________________

d bright orange ______________________

2 Group the six metal ions tested in Part B into two "families" based on your observations of their reactivities with $(NH_4)_2CO_3$, $(NH_4)_2SO_4$, and $(NH_4)_2HPO_4$.

Family 1	Family 2
______________________	______________________
______________________	______________________
______________________	______________________

3 What is the group number on the periodic table that corresponds to each family in Question 2? What is the group name associated with each family?

	Family 1	Family 2
Group Number	______________________	______________________
Group Name	______________________	______________________

Name ____________________

Lab Partner ____________________

Lab Section ____________________ Date ____________________

Lab 6 DATA SHEET

(continued)

4 Using the periodic table, provide the names and symbols of elements that match each description below.

a Element with properties similar to sulfur. ____________________

b Two halogens. ____________________

c Noble gas in period 2. ____________________

d Alkali metal in period 4. ____________________

e Element with properties similar to silver. ____________________

f Alkaline earth metal in period 5. ____________________

5 In the table below, make a claim regarding the identities of the metal and halogen ions in your mystery sample (Part D). Describe your evidence and provide a rationale for your claim.

Claim: Metal Ion	
Evidence	**Rationale**
Claim: Halogen Ion	
Evidence	**Rationale**

6 A mystery solution comprised of one metal and one halogen ion gave the flame test shown in the image to the right. The solution produced a white precipitate upon addition of ammonium carbonate and ammonium hydrogen phosphate, but required heat for a precipitate to form upon addition of ammonium sulfate. A halogen ion test gave a purple color in the hexane layer. Identify the metal and the halogen ions present in the solution.

a Metal ion present ______________________

b Halogen ion present ______________________

7 A mystery solution comprised of one metal and one halogen ion gave the flame test shown in the image to the right. The solution produced no precipitates upon addition of ammonium carbonate, ammonium hydrogen phosphate, or ammonium sulfate. A halogen ion test gave a yellow-orange color in the hexane layer. Identify the metal and the halogen ions present in the solution.

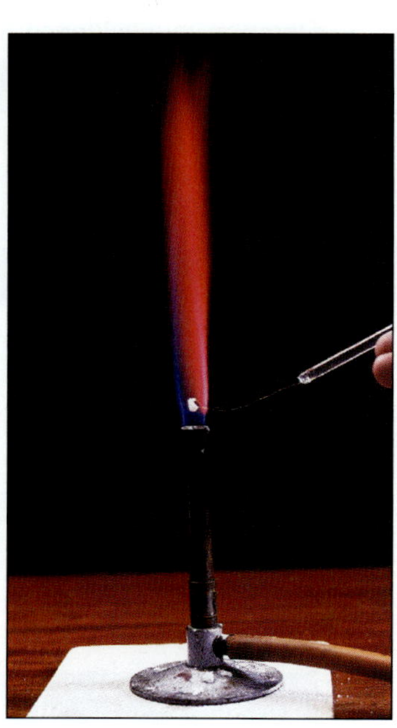

a Metal ion present ______________________

b Halogen ion present ______________________

8 If you mixed a magnesium solution with a solution of ammonium carbonate, what observation would you expect? Justify your answer.

____

____

____

____

▲ *Oat flakes contain some of the minerals that are essential for life.*

Lab 7

Eat Your Wheaties
Nutritional Minerals in Cereal

Nutrients are the chemical components of food that are required for growth and proper biological function. More than 96% of the human body is composed of carbon, oxygen, nitrogen, and hydrogen. These nonmetal elements are most commonly derived from carbohydrates, lipids, proteins, and other organic nutrients. More than 20 other elements also are essential for life, and these are broadly classified as **minerals**.

Minerals generally are absorbed by the body as ions and carry out their biological function in the same form. For example, the sodium ion (Na^+) is critical for conducting nerve impulses and maintaining proper electrolyte concentrations in body fluids. Similarly, calcium (Ca^{2+}) ions are necessary for bone formation and muscle contraction, and zinc (Zn^{2+}) ions are required for proper functioning of certain enzymes responsible for metabolism.

In this experiment, you will test various cereals for the presence of six mineral ions: zinc, iron(III), copper(II), calcium, sodium, and potassium. Most of the organic material in the cereal samples will be removed through a process called **ashing,** which causes the organic molecules to vaporize or combust to form gaseous carbon dioxide and water. The inorganic minerals remain in the ash, mostly as oxide compounds (e.g., Fe_2O_3, ZnO, Na_2O, etc.).

Treatment of the ash with aqueous hydrochloric acid will then produce the corresponding chloride salts (e.g., $FeCl_3$, $ZnCl_2$, $NaCl$, etc.), which are soluble in water. Filtering out the solid material will yield a solution of mineral ions derived from the cereal sample. The tests for each of the six mineral ions are described below. Note that each test is **qualitative**, not **quantitative.** That is, the tests indicate the *identity* of the ions present, but they do not indicate *how much* of the ion is present.

Zinc Test: Any $ZnCl_2$ in your sample will react with an organic molecule called dithizone to produce a compound that is red in color. The dithizone is dissolved in hexane, which is not soluble in water. As a result, the hexane solution will form a layer on top of the aqueous cereal extract. A red color in this layer is a positive result for the presence of zinc ions in the sample (Fig. 7.1A).

Objectives

After completing this lab you should be able to:

- Describe the differences between qualitative and quantitative analyses.
- Explain the difference between the mineral and organic components of food.
- Describe a method for removing the organic material in a food sample without removing the minerals.
- Describe methods for detecting six ions in a food sample.
- Experimentally determine the presence of six mineral ions in a sample of cereal.

Iron Test: Any $FeCl_3$ in your sample will react with an aqueous solution of potassium ferrocyanide to give a blue-colored precipitate. The production of this blue substance is a positive result for the presence of iron(III) ions in the sample (Fig. 7.1B).

Calcium Test: Any $CaCl_2$ in your sample will react with ammonium oxalate to produce a solid white precipitate. The production of a cloudy white solution upon addition of the ammonium oxalate is a positive result for the presence of calcium ions in the sample (Fig. 7.1C).

Copper Test: Any $CuCl_2$ in your sample will react with ammonium hydroxide to produce a deep blue-colored complex ion that remains dissolved in solution. The production of this blue color is a positive result for the presence of copper(II) ions in the sample (Fig. 7.1D).

Sodium Test: The sodium (Na^+) ions in your sample produce an orange-yellow color when heated in a flame (Fig. 7.1E).

Potassium Test: The potassium (K^+) ions in your sample produce a violet color when heated in a flame (Fig. 7.1F). Sodium ions produce an intense orange-yellow color that would swamp the violet produced by K^+ if both ions are present. The potassium test is therefore carried out by observing the flame through a piece of cobalt glass, which filters out any orange color from the presence of sodium ions.

If your instructor provides multiple cereal samples, you may be instructed to gather data for three different cereal samples. Note that you will prepare only one cereal sample for testing. You will obtain data for the other two prepared cereal samples by observing the results of your lab mates' analyses.

In addition to testing the cereal samples, you will perform each mineral analysis on a sample of distilled water (which should contain none of the ions) and a test solution that is known to contain the ion of interest. The water and test solution serve as negative and positive **controls**, respectively. An experimental control is a well-defined sample on which the experiment is performed to gauge the validity of the methodology or as a baseline for comparison when actual samples are analyzed. In this case, performing the tests on distilled water should familiarize you with the visual outcomes of negative results while testing the ion solutions will allow you to recognize the outcomes when the ions are present.

7.1 Positive test for: (A) Zn^{2+}; (B) Fe^{3+}; (C) Ca^{2+}; (D) Cu^{2+}; (E) Na^+; (F) K^+.

Procedure

What Am I Doing?

Part A: Preparing a sample of cereal for mineral analysis by ashing and extracting the minerals into an acidic solution.

Part B: Analyzing the prepared cereal sample and two controls for the presence of Zn^{2+}, Fe^{3+}, Ca^{2+}, Cu^{2+}, Na^{+}, and K^{+} ions in separate tests. You may also gather data for two additional cereal samples by observing results from your lab mates.

Materials

- ❑ Mortar and pestle
- ❑ Crucible and lid
- ❑ Clay triangle
- ❑ Ring stand
- ❑ Iron ring
- ❑ Bunsen burner
- ❑ Tongs
- ❑ Scoopula
- ❑ 50 mL beaker
- ❑ 50 mL Erlenmeyer flask
- ❑ Glass stirring rod
- ❑ Funnel
- ❑ Filter paper
- ❑ Test tubes (16)
- ❑ Cobalt glass filter
- ❑ Flame test wire
- ❑ Striker

7

Part A Preparing a Cereal Sample

1. Obtain approximately 3.0 g of a cereal sample and record its identity on the data sheet, page 97.
2. Describe the appearance of the cereal on the data sheet.
3. Crush the cereal to a fine powder using a mortar and pestle.

SAFETY NOTE

Set up your apparatus and perform Steps 4–6 in a fume hood or appropriately ventilated area. The cereal may produce smoke and flames during the ashing process!

4. Add the powdered cereal to a dry crucible, and place the crucible on a clay triangle and ring stand as shown in Figure 7.2.
5. Heat the crucible in the hottest part of the flame from a Bunsen burner. The crucible should begin to glow orange, and you may observe fumes coming from the top of the crucible. If the fumes ignite, simply allow them to burn off. Continue this vigorous heating for at least 20 minutes.

SAFETY NOTE

Keep tongs and the crucible lid near the reaction apparatus. A fire can be extinguished by depriving the reaction of oxygen. If the flames become dangerously high, use the tongs to place the lid back on your crucible in order to smother the fire.

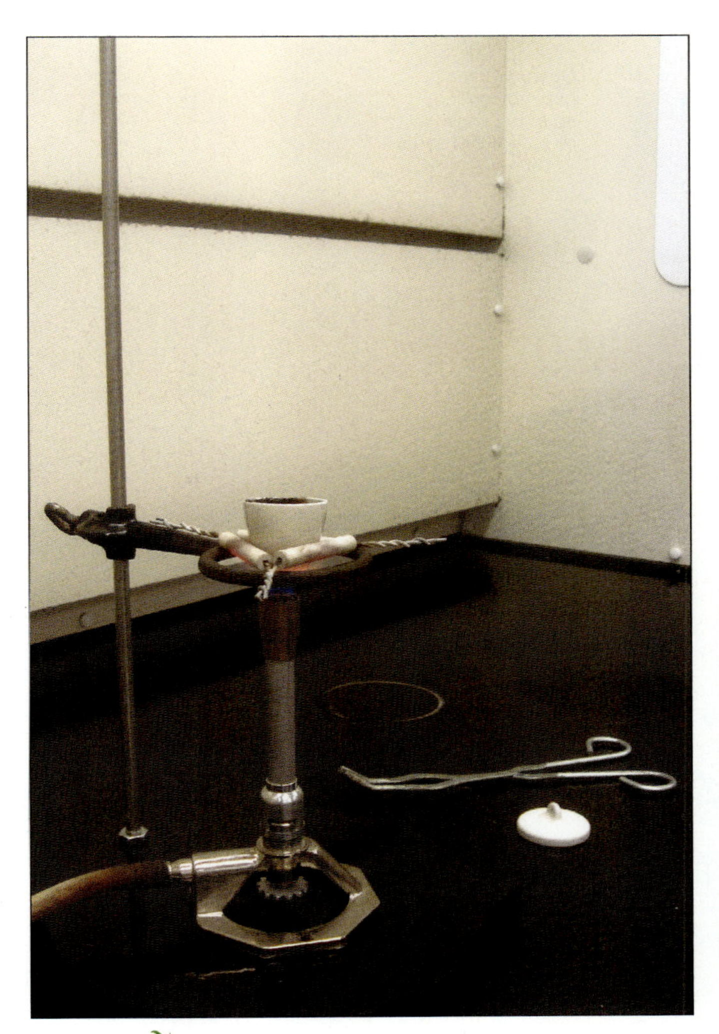

7.2 Apparatus for ashing cereal sample.

6. After the heating period is complete, turn off the Bunsen burner flame and allow the crucible to cool to room temperature. The cereal should now be a black or gray ash. On the data sheet, describe the appearance of the sample.

Hydrochloric acid (HCl) is a corrosive strong acid. ***Handle with caution!***

7. Use a metal scoopula to break up the ash, and transfer the material to a 50 mL beaker.
8. Slowly add 2 mL of 6 M HCl (hydrochloric acid) to the ash. If the material begins to bubble or fizz, slow the addition of HCl to prevent loss of the material by the solution overflowing from the beaker.
9. Add 5 mL of distilled water to the beaker, and use a glass stirring rod to mix the contents and break up any remaining large chunks of ash.
10. Using a fluted piece of filter paper (see Technique Tip 7.1) in a funnel, gravity filter the ash mixture and collect the aqueous filtrate in 50 mL Erlenmeyer flask. This is your cereal sample solution. Dispose of the filter paper with the remaining solid ash as directed by your instructor.

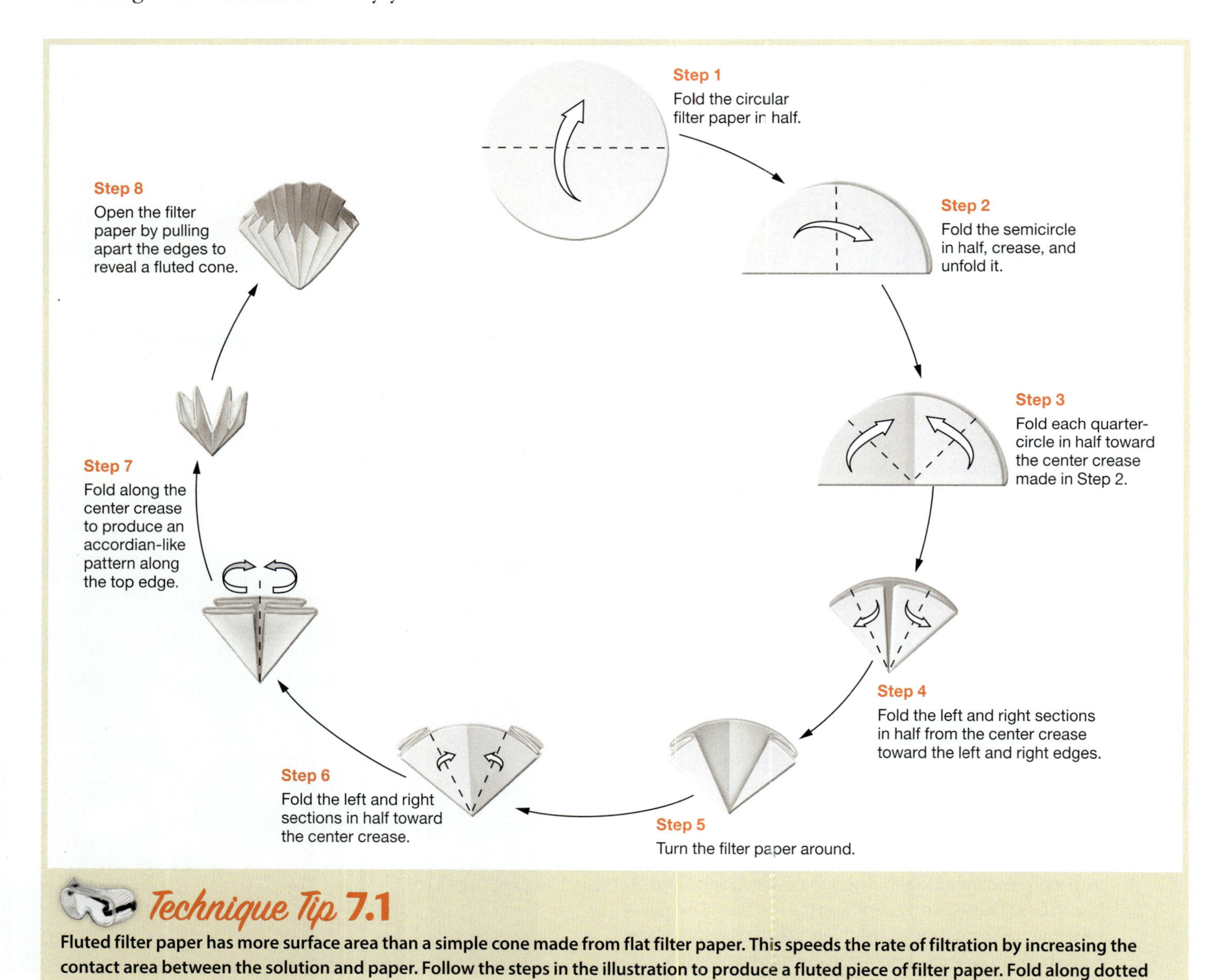

Technique Tip 7.1

Fluted filter paper has more surface area than a simple cone made from flat filter paper. This speeds the rate of filtration by increasing the contact area between the solution and paper. Follow the steps in the illustration to produce a fluted piece of filter paper. Fold along dotted lines in direction indicated by arrows.

Part B Mineral Analyses

You will prepare three sets of test tubes as summarized in Table 7.1. In each mineral test, you will use one test tube from each set. If your class is using more than one cereal sample, observe and record the results from lab mates who used two cereals different from your own.

TABLE 7.1 Summary Description of Three Sets of Test Tubes Prepared for Chemical Analysis

Set Label	Contents	Purpose
N	Distilled water	Negative control
P	Cation solutions	Positive control
C	Cereal sample	Experimental analysis

Sample Preparation

1. Prepare the Set N test tubes.
   - a Label five test tubes N1–N5.
   - b Add 10 drops of distilled water to each test tube.
2. Prepare the Set P test tubes.
   - a Label six test tubes P1–P6.
   - b To each test tube, add 10 drops of the appropriate sample solutions as indicated below:
     - Test Tube P1: $ZnCl_2$ solution
     - Test Tube P2: $FeCl_3$ solution
     - Test Tube P3: $CaCl_2$ solution
     - Test Tube P4: $CuCl_2$ solution
     - Test Tube P5: NaCl solution
     - Test Tube P6: KCl solution
3. Prepare the Set C test tubes.
   - a Label five test tubes C1–C5.
   - b Add 10 drops of the cereal sample solution to each test tube.

Test for Zn^{2+}

1. To Test Tube 1 from each set (N1, P1, and C1):
   - a Add 2 drops of 6 M sodium hydroxide
   - b Add 10 drops of dithizone solution
2. Agitate the test tubes and then allow the mixture to settle for about 1 minute.
3. On the data sheet, page 97, record your observations of the test tubes and interpret the results to indicate the presence (+) or absence (−) of the ion in each sample.

SAFETY NOTE

6 M Sodium hydroxide is caustic. Wear goggles and ***handle it with care!***

Test for Fe^{3+}

1. To Test Tube 2 from each set (N2, P2, and C2), add 3 drops of potassium ferrocyanide solution.
2. Swirl the test tubes to mix the contents.
3. On the data sheet, page 97, record your observations of the test tubes and interpret the results to indicate the presence (+) or absence (−) of the ion in each sample.

Test for Ca^{2+}

≡7

1. To Test Tube 3 from each set (N3, P3, and C3), add 4 drops of ammonium oxalate [$(NH_4)_2C_2O_4$] solution.
2. Swirl the test tubes to mix the contents, and then allow the mixture to settle for 2 minutes.
3. On the data sheet, page 98, record your observations of the test tubes and interpret the results to indicate the presence (+) or absence (−) of the ion in each sample.

Tests for Cu^{2+}

1. To Test Tube 4 from each set (N4, P4, and C4), add 6 drops of concentrated ammonium hydroxide (NH_4OH).
2. On the data sheet, page 98, record your observations of the test tubes and interpret the results to indicate the presence (+) or absence (−) of the ion in each sample.

SAFETY NOTE

Due to the strong odor of ammonium hydroxide, perform this test in a fume hood or appropriately ventilated area.

Tests for Na^+ and K^+

1. Clean a flame test wire loop by dipping it in a small amount of 6 M HCl and then holding it in a Bunsen burner flame until the wire glows. Repeat the cleaning process 2 times.
2. Dip the cleaned flame test wire loop into Test Tube N5, and then hold it to a Bunsen burner flame to heat the sample. On the data sheet, page 98, record the color of the flame produced as the sample is heated and interpret the result to indicate the presence (+) or absence (−) of the Na^+ ion in the sample.
3. Repeat Step 2, but record the color of the flame as observed through a cobalt glass filter on the data sheet. Interpret the result to indicate the presence (+) or absence (−) of the K^+ ion in the sample.
4. Repeat Steps 1–3, on samples P5, P6, and C5.

Nutrition Facts Labels

In order to compare your experimental results to data reported by the manufacturer(s) of your cereal sample, record on the data sheet, page 99, whether the nutrition facts labels on the corresponding cereal boxes indicate that each mineral ion is present (+) or absent (−) in the cereal. Note that all nutrition labels will include information on sodium (Na^+), calcium (Ca^{2+}), and iron (Fe^{3+}). Labels may or may not contain information regarding the zinc (Zn^{2+}), copper (Cu^{2+}), or potassium (K^+) content of the cereal. If no information is available for a certain mineral, record N/A in the table on the data sheet.

Name ______________________

Lab Partner ______________________

Lab Section ______________________ Date ______________

Lab 7 Pre-Laboratory Exercise

1 What is the purpose of ashing the cereal sample?

2 Are the tests used to determine the presence of minerals in the cereal based on the chemical or physical properties of the minerals? Explain your answer.

3 The sample preparation method used in Part A ultimately generates the soluble chloride salts of certain minerals found in a cereal sample. Does this preparation method limit the scope of mineral analysis that may follow? Specifically, are there some types of minerals that cannot form ionic compounds with chloride? Explain your answer. ***Hint:*** What is the common characteristic of all the minerals examined in this experiment?

4 Explain how distilled water is used as a control in this experiment.

5 Explain how the test solutions containing the ions of interest ($ZnCl_2$, $FeCl_3$, NaCl, etc.) are used as controls in this experiment.

Name ______________________________

Lab Partner ______________________________

Lab Section ____________________ Date ____________

Lab 7 DATA SHEET

Data and Observations

Part A Preparing a Cereal Sample

Data Description	Information/Observation
Identity of cereal sample	
Appearance of cereal before ashing	
Appearance of cereal after ashing	

Part B Mineral Analyses

Mineral Analysis—Zinc Ion

Test Tube	Sample Identity	Zn^{2+} Test Observation	Zn^{2+} Test Interpretation (+ or −)
N1	Distilled water		
P1	Zn^{2+}		
C1			
Lab mate's C1			
Lab mate's C1			

Mineral Analysis—Iron(III) Ion

Test Tube	Sample Identity	Fe^{3+} Test Observation	Fe^{3+} Test Interpretation (+ or −)
N2	Distilled water		
P2	Fe^{3+}		
C2			
Lab mate's C2			
Lab mate's C2			

Mineral Analysis—Calcium Ion

Test Tube	Sample Identity	Ca^{2+} Test Observation	Ca^{2+} Test Interpretation (+ or −)
N3	Distilled water		
P3	Ca^{2+}		
C3			
Lab mate's C3			
Lab mate's C3			

Mineral Analysis—Copper(II) Ion

Test Tube	Sample Identity	Cu^{2+} Test Observation	Cu^{2+} Test Interpretation (+ or −)
N4	Distilled water		
P4	Cu^{2+}		
C4			
Lab mate's C4			
Lab mate's C4			

Mineral Analysis—Sodium and Potassium Ions

Test Tube	Sample Identity	Na^{+} Test Observation	Na^{+} Test Interpretation (+ or −)	K^{+} Test Observation	K^{+} Test Interpretation (+ or −)
N5	Distilled water				
P5	Na^{+}				
P6	K^{+}				
C5					
Lab mate's C5					
Lab mate's C5					

Name ____________________

Lab Partner ____________________

Lab Section ____________________ Date ____________________

Lab 7
DATA SHEET
(continued)

Nutrition Facts Labels

Present (+), Absent (−), or Information Not Available (N/A)						
Cereal	Zn^{2+}	Fe^{3+}	Ca^{2+}	Cu^{2+}	Na^{+}	K^{+}

Reflective Exercises

1 In the table below, make a claim regarding the mineral contents of your cereal sample. Describe your evidence and provide a rationale for your claim.

Claim	
Evidence	**Rationale**

2 When performing this experiment, why is it important that you use distilled water instead of tap water during the preparation of your cereal sample?

3 Do you think you could use the sample preparation procedure described in Part A to prepare samples of fresh fruits, such as apples or oranges, for analysis? If not, how could samples of these types of foods be prepared? Explain your answer.

4 Compare the data obtained from the nutritional labels on the cereal boxes to your experimental data. Did any of your cereal samples give negative experimental results for minerals that should be present according to the label? If so, offer a possible explanation for this disparity.

5 The detection tests you used in this lab were *qualitative* in nature. Do you think any of these test procedures could be modified to produce a *quantitative* result? If so, which ones and what would be measured?

6 Another method of preparing food samples for mineral testing involves dissolving the organic material with acid instead of burning it off through the ashing process. If you had used the acid procedure, do you think your results for the inorganic mineral tests would have been any different? Justify your answer.

Name

Lab Partner

Lab Section   Date

(continued)

7 Salts often are used as additives or preservatives in canned foods. The liquid from a can of beans was analyzed for minerals using the tests described in this lab. The results were as follows:

Zn^{2+}	Fe^{3+}	Ca^{2+}	Cu^{2+}	Na^{+}	K^{+}
No color in upper layer	Deep blue precipitate	Cloudy white	No color change	Orange-yellow flame	Violet flame

Based on these data alone, circle all of the compounds below that are possible additives or preservatives used in the can of beans.

a Potassium iodide

b Calcium carbonate

c Sodium chloride

d Zinc oxide

e Iron(II) sulfate

f Silicon dioxide

g Potassium chloride

h Sodium bisulfite

i Zinc citrate

j Sodium bicarbonate

k Calcium sulfate

l Sodium phosphate

m Copper(II) oxide

n Elemental iron

▲ *Copper(II) sulfate is an example of an ionic compound.*

Lab 8

Puzzled?

Ionic and Covalent Bonding Patterns

There are two major classes of compounds: ionic and covalent. **Ionic compounds** are those in which the particles are ions, and the bond is a result of electrostatic attraction between oppositely charged species. The particles in a **covalent compound** are neutral **molecules,** discrete groups of atoms connected to each other in a specific arrangement (Fig. 8.1).

A **chemical formula** expresses the composition of a compound using a combination of element symbols and subscripts, which denote the relative number of each atom or ion present (Fig. 8.1). For example, the chemical formulas of table salt (sodium chloride) and propane are NaCl and C_3H_8, respectively.

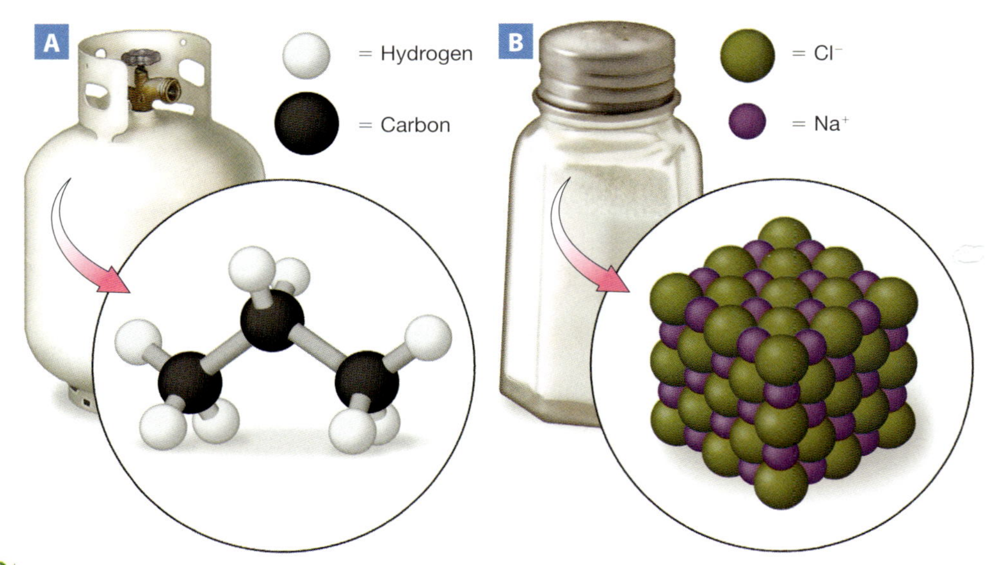

8.1 (**A**) In a covalent compound like propane (C_3H_8), the formula indicates the actual number of each atom present in a molecule of the substance. (**B**) In an ionic compound like sodium chloride (NaCl), the formula indicates the ratios of the ions present in a crystal. Each Na^+ ion interacts with many nearby Cl^- ions.

Objectives

After completing this lab you should be able to:

- Identify the formula of a simple binary ionic compound from its constituent ions.
- Write the formula of a simple binary ionic compound from its name.
- Name a simple binary ionic compound from its constituent ions or formula.
- Recognize the typical covalent bonding patterns of second period nonmetals.
- Identify the relationships between the position of a nonmetal on the periodic table, the number of valence electrons it possesses, and the number of covalent bonds it typically forms.
- Apply your knowledge of bonding patterns for nonmetals to quickly generate reasonable Lewis structures from a molecular formula.

Ionic Bonding

The simplest **ionic compounds** result from bonding between a positively charged metal **cation** and a negatively charged **anion** consisting of one or more nonmetals. (Nonmetals also may form cations, but these are less common and are not considered in this lab).

Ionic compounds have a neutral overall charge because the constituent ions combine in a ratio that balances the positive charge from the cation with the negative charge from the anion. This ratio is indicated by the subscripts in a chemical formula. Ionic compounds are named by combining the name of the cation with the name of the anion.

Consider a typical ionic compound such as $MgCl_2$. The cation (Mg^{2+}) carries a net charge of 2+ and is called the *magnesium ion*. All main group metal cations are named by simply adding *ion* to the corresponding element name. The anion (Cl^-) carries a net 1– charge and is called *chloride*. Nonmetal anions are named by combining the first syllable of the corresponding element name with the suffix *-ide*.

When the magnesium and chloride ions combine to produce a compound, two chloride anions are required to balance the 2+ charge of a single magnesium ion. Thus, the cation and anion combine in a 1:2 ratio, and the formula for the compound is $MgCl_2$. The name of the compound is derived simply by combining the two ion names and dropping the word *ion* from the cation name: *magnesium chloride*.

≡8 **Polyatomic ions** are charged species composed of more than one atom. The names of some common polyatomic ions are provided on the inside back cover of this text. The *nitrate* anion (NO_3^-) bears a 1– charge. Like chloride, it combines in a 1:2 ratio with the magnesium ion, and the result is a compound with the formula $Mg(NO_3)_2$ called *magnesium nitrate*.

Note that whenever more than one polyatomic ion is required in a compound's formula, the polyatomic ion is placed in parentheses to denote that the subscript applies to the entire ion. Just imagine: If parentheses were omitted from this example and the formula for the compound were written $MgNO_{32}$, then it would appear that the compound includes only 1 nitrogen and 32 oxygen atoms! Of course, parentheses are unnecessary if only one polyatomic ion is present in the formula.

Unlike the main group metals, many transition metals may form more than one cation. Cobalt, for example, may form a cation with a 2+ charge (Co^{2+}) or a cation with a 3+ charge (Co^{3+}). The compounds formed by these two cations are necessarily different because different cation-to-anion ratios are required to generate a neutral species. For example, a neutral compound would be formed by combining one Co^{2+} ion with *two* chloride anions to produce $CoCl_2$, but Co^{3+} would combine with *three* chloride anions to produce neutral $CoCl_3$.

If there were no system to distinguish between the cobalt ions, these two different compounds would receive the same name (*cobalt chloride*). To prevent this ambiguity, transition metal cation names include the ionic charge, indicated by a Roman numeral in parentheses after the metal name. For example, Co^{2+} is called the *cobalt(II) ion* while Co^{3+} is called the *cobalt(III) ion*.

Ionic compounds that include transition metals can thus be named without ambiguity by following the same rule used for all other ionic compounds: Combine the two ion names and drop *ion* from the cation name. (Just don't drop the Roman numeral!) The compound $CoCl_2$ would therefore be named *cobalt(II) chloride*, while $CoCl_3$ would receive a different name, *cobalt(III) chloride*. Note that a few transition metals (e.g., silver and zinc) are exceptions to the Roman numeral requirement because they form only one ion.

In Part A of this laboratory exercise, you will explore ionic bonding and naming with puzzle pieces that conceptually represent the charges of ions. You will practice forming neutral ionic compounds and deriving their formulas and names.

Covalent Compounds

Covalent compounds generally result from the sharing of **valence electrons** (those in the outer shell) between nonmetal atoms (Fig. 8.2). Each pair of electrons shared between two atoms constitutes a **covalent bond** and may be referred to as a **bonding pair.** When two atoms share only one bonding pair, the bond is called a **single bond.** The joining of two atoms by two or three bonding pairs is called a **double** or **triple bond**, respectively. Any valence electron pairs not involved in bonding are called **lone pairs**.

The formation of elementary compounds is governed largely by the **octet rule**, which states that most atoms react to achieve a configuration with eight electrons in their outer shell. One exception to this rule is hydrogen, which can only accommodate two valence electrons and reacts to achieve such a configuration.

We may conceive of a covalent bond as being formed when each atom donates one electron to the bonding pair. Thus, the *total number* of bonds in which an atom will typically participate is equal to the number of electrons that the atom requires to fulfill the octet rule. For example, nitrogen has five valence electrons, and it generally forms three covalent bonds in order to "access" the three electrons required to reach eight (Fig. 8.3).

While atoms generally form a characteristic *number* of bonds, the *type* of bonds formed is more variable. Nitrogen, for example, may form three bonds by any of the following combinations: (1) three single bonds; (2) one double bond and one single bond; and (3) one triple bond. The specific arrangement of atoms and the bonding patterns in covalent compounds are visually represented using **Lewis structures**, in which the chemical symbols are surrounded by dots representing nonbonding electrons and dashes representing bonding pairs (Fig. 8.3).

In Part B of this laboratory exercise, you will explore covalent bonding patterns with puzzle pieces that conceptually represent the typical bonding patterns of nonmetal atoms. The puzzle pieces will help you form reasonable connections between atoms to generate covalent compounds and draw the corresponding Lewis structures. With practice, you will be able to recognize how the typical bonding patterns of the elements within a covalent compound can help you quickly derive a valid Lewis structure.

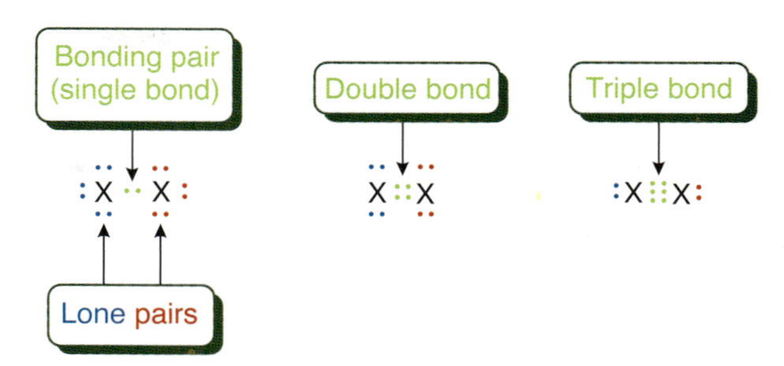

8.2 Covalent compounds are formed by atoms sharing electron pairs. Two generic atoms (X) can share one, two, or three electron pairs to form a single, double, or triple bond. Any unshared electron pairs present on an atom are called lone pairs.

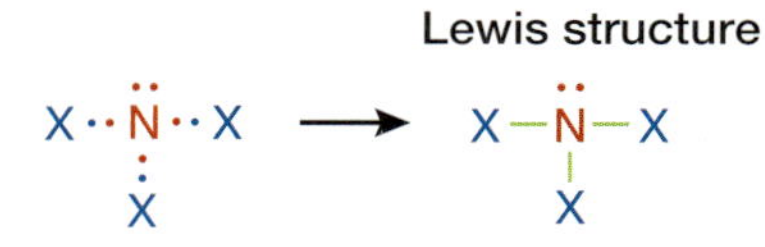

8.3 Nitrogen atom possesses five valence electrons (red). It typically will bond a total of three times because another atom will bring one additional electron to each bond. As a result, the nitrogen atom will be surrounded by eight total valence electrons, satisfying the octet rule. Compounds are usually drawn using Lewis structures in which bonding pairs are represented with lines rather than dots.

Procedure

Materials

- ❑ Puzzle pieces, provided

What Am I Doing?

Part A: Using puzzle pieces to model the 16 neutral compounds that may be formed by combining four different cations with each of four different anions.

Part B: Using puzzle pieces to model compounds listed on the data sheet to gain familiarity with the common bonding patterns associated with carbon, nitrogen, oxygen, hydrogen, and halogen atoms.

Part A The Puzzle of Ionic Compounds

Your task is to form each of the 16 possible compounds from the puzzle pieces provided at the end of this book (Puzzle Pieces pages 1–3). Fill in the table on the data sheet for each compound, pages 111–114. The rules for the ionic bonding puzzle are:

1. The charges on the cation puzzle pieces are represented conceptually using peaks. The number of peaks on a cation piece matches the charge of the cation.
2. The charges on the anion puzzle pieces are represented conceptually using valleys. The number of valleys on an anion piece matches the charge of the anion.

3 Neutral ionic compounds are formed by matching the number of positive charges on a given cation with an equal number of negative charges on an anion.

A compound is modeled with the puzzle pieces by combining the appropriate number of pieces of a given cation with the appropriate number of pieces of a given anion such that all the peaks are matched with a corresponding valley. An example with corresponding data is given in Figure 8.4. Note that for the purposes of this exercise, you should not form compounds that contain more than one type of cation or more than one type of anion. Once a compound is formed, the puzzle may be broken down so that the pieces may be reused to form other compounds.

A

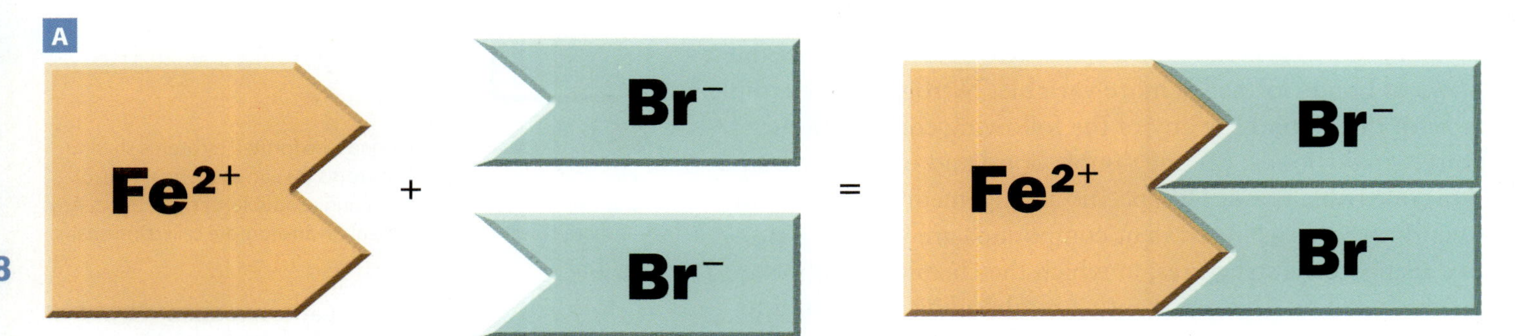

8

	Example Compound	
	Cation Piece: Fe^{2+}	Anion Piece: Br^-
Ion Charge	2^+	1^-
Ion Name	Iron(II)	Bromide
Number of Ions Required	1	2
Compound Formula	$FeBr_2$	
Compound Name	Iron(II) bromide	

B

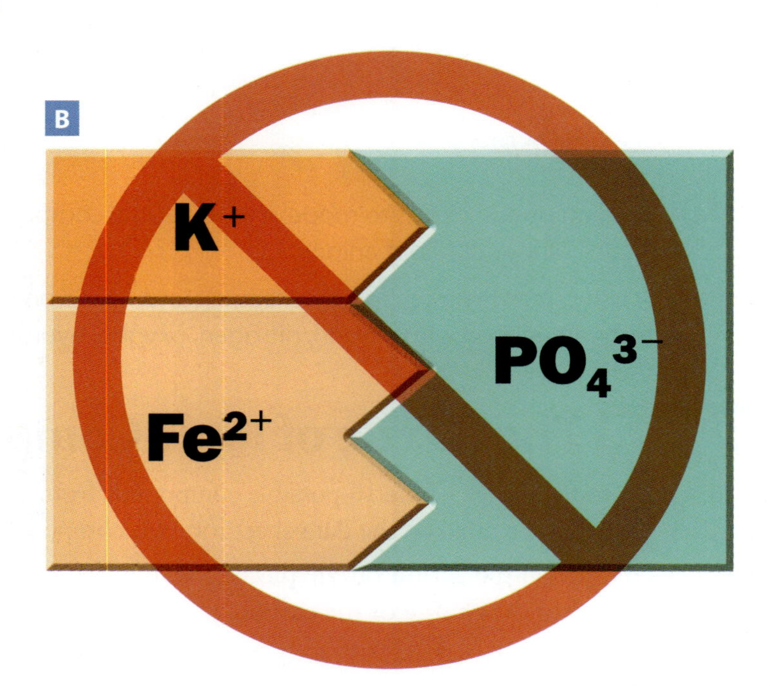

8.4 (**A**) Using puzzle pieces to make a compound. In this example: Iron(II) ions (Fe^{2+}) have a 2+ charge and therefore have two peaks each. Bromide anions (Br^-) have a 1− charge and therefore have one valley each. It takes two bromide puzzle pieces to match with one iron(II) puzzle piece, and the ratio Fe^{2+}: Br^- is 1:2. The corresponding formula for iron(II) bromide would be $FeBr_2$. (**B**) Do not form compounds like this one, which contains two different types of cations.

Part B The Covalent Bonding Puzzle

Your task is to use the puzzle pieces (Puzzle Pieces pages 4–8) to help you form each compound given on the data sheet, pages 115–117. For each, you must draw the Lewis structure of the compound in the corresponding box. The compounds are divided into groups so that you can focus on the bonding pattern for a particular atom. To help you understand the similarities and differences in the bonding patterns of each atom, answer the questions that accompany each group of compounds. The rules for using the covalent bonding puzzle pieces are:

1. The dots on the covalent compound puzzle pieces correspond to the number of valence electrons possessed by that atom. To form a compound, the octet of each atom must be satisfied.
2. When covalent compounds form, one or more pairs of electrons are shared between two atoms. The possible bonding patterns (i.e., the number of single, double, or triple bonds) are represented conceptually on your puzzle pieces by the arrangement of dots around an atom. For example, if a single bond is formed between two atoms, each will donate one electron to the bonding pair that connects them. If a double bond is formed, each atom will donate two electrons, forming two bonding pairs between them. To correctly form a compound, all of the donated electrons on a puzzle piece must be paired with electrons on another piece. The puzzle pieces should fit together smoothly with no "jagged" (or unmatched) edges remaining.
3. Examples of how the puzzles pieces fit together to form bonds are given in Figure 8.5.

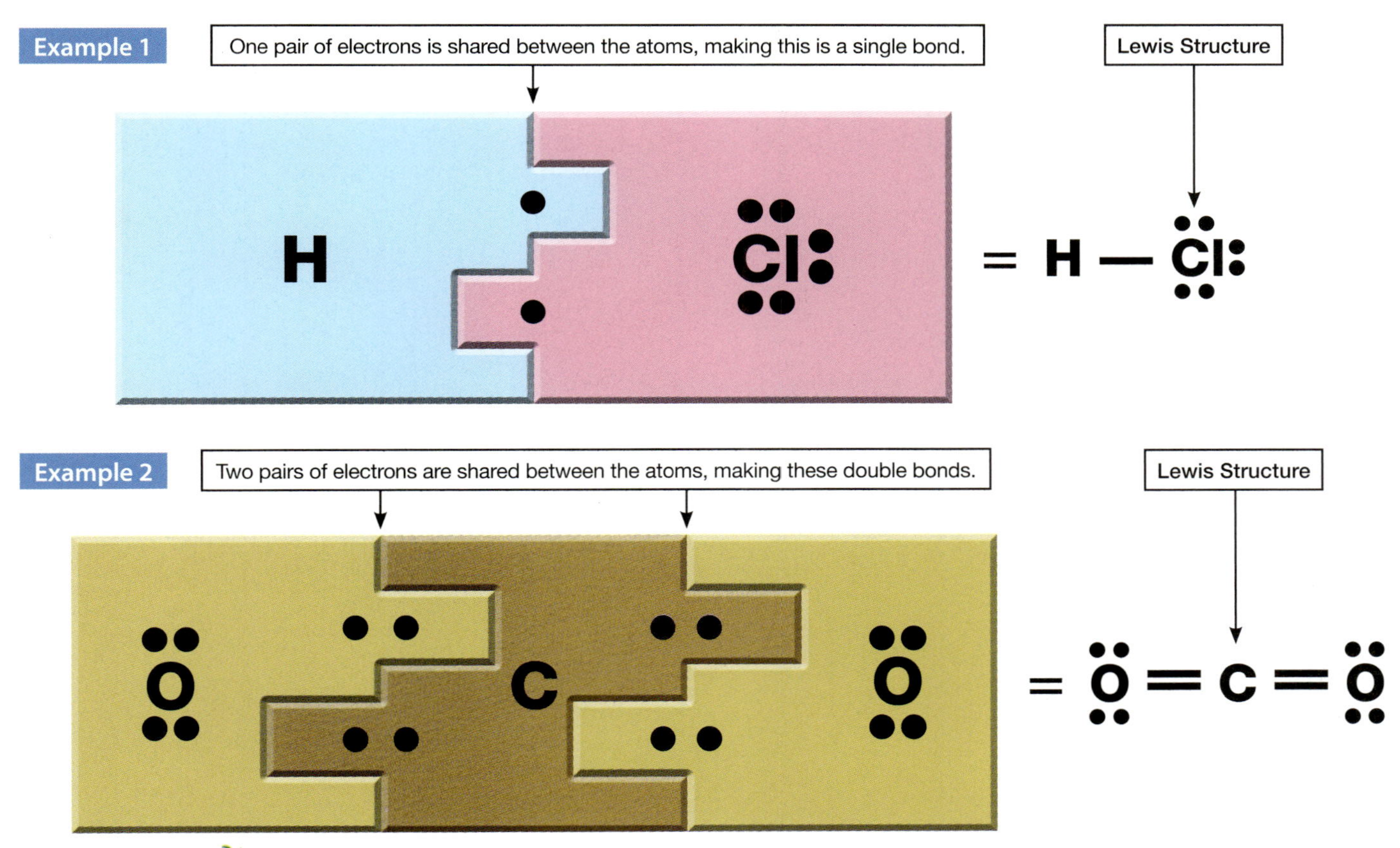

8.5 Correctly matched covalent bonding puzzle pieces and their corresponding Lewis structures.

Name ______________________

Lab Partner ______________________

Lab Section ______________ Date ______________

Lab 8 Pre-Laboratory Exercise

1 Provide a term that matches each description below.

a Generic term for a negatively charged ion. ______________

b Type of compound formed by the attraction between cations and anions. ______________

c A pair of electrons shared between two atoms. ______________

d States that atoms react to achieve a configuration with eight electrons in their valence shell. ______________

e Generic term for a positively charged ion. ______________

f Type of compound formed by electrons being shared between atoms. ______________

g Refers to the sharing of two electron pairs between two atoms. ______________

h Representation of a molecule that shows the bonding electrons between atoms and any nonbonding electrons on an atom. ______________

i Ion composed of more than one atom. ______________

2 Provide the name of each ion below.

a K^{+} ______________

b Cu^{+} ______________

c Cu^{2+} ______________

d Al^{3+} ______________

e Br^{-} ______________

f NO_3^{-} ______________

g O^{2-} ______________

h PO_4^{3-} ______________

3 Consider the calcium ion (Ca^{2+}) and the iodide anion (I^{-}).

a How many iodide ions are necessary to balance the charge on one calcium ion? ______________

b Given your answer to 3a, what should be the formula for the compound formed between Ca^{2+} and I^{-}? ______________

c What should be the name of the compound formed between Ca^{2+} and I^{-}? ______________

4 Consider the sodium ion (Na^+) and the sulfate anion (SO_4^{2-}).

a How many sodium ions are necessary to balance the charge on one sulfate anion? ______

b Given your answer to 4a, what should be the formula for the compound formed between Na^+ and SO_4^{2-}? ______

c What should be the name of the compound formed between Na^+ and SO_4^{2-}? ______

5 Use the electron dot structure of the compound below to answer the following questions.

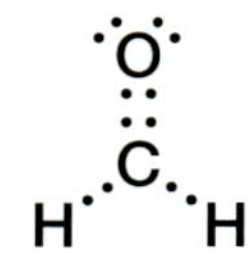

a Draw a Lewis structure of this compound using lines for bonding electrons.

b Is the octet of carbon satisfied in this structure? ______

c How many total bonds does the carbon possess? ______

d Is the bond between the carbon and hydrogen identical to the bond between the carbon and the oxygen? If not, what is the difference?

______

______

e How many of each bond type does carbon possess?

______

Name ______________________________

Lab Partner ______________________________

Lab Section ____________________ Date ____________

Lab 8 DATA SHEET

Data and Observations

Part A The Puzzle of Ionic Compounds

Ionic Compounds

	Compound 1	
	Cation Piece:	Anion Piece:
Ion Charge		
Ion Name		
Number of Ions Required		
Compound Formula		
Compound Name		

	Compound 2	
	Cation Piece:	Anion Piece:
Ion Charge		
Ion Name		
Number of Ions Required		
Compound Formula		
Compound Name		

	Compound 3	
	Cation Piece:	Anion Piece:
Ion Charge		
Ion Name		
Number of Ions Required		
Compound Formula		
Compound Name		

	Compound 4	
	Cation Piece:	Anion Piece:
Ion Charge		
Ion Name		
Number of Ions Required		
Compound Formula		
Compound Name		

	Compound 5	
	Cation Piece:	Anion Piece:
Ion Charge		
Ion Name		
Number of Ions Required		
Compound Formula		
Compound Name		

	Compound 6	
	Cation Piece:	Anion Piece:
Ion Charge		
Ion Name		
Number of Ions Required		
Compound Formula		
Compound Name		

	Compound 7	
	Cation Piece:	Anion Piece:
Ion Charge		
Ion Name		
Number of Ions Required		
Compound Formula		
Compound Name		

	Compound 8	
	Cation Piece:	Anion Piece:
Ion Charge		
Ion Name		
Number of Ions Required		
Compound Formula		
Compound Name		

Name ______________________________

Lab Partner ______________________________

Lab Section ____________________ Date ____________

Lab 8
DATA SHEET
(continued)

	Compound 9	
	Cation Piece:	Anion Piece:
Ion Charge		
Ion Name		
Number of Ions Required		
Compound Formula		
Compound Name		

	Compound 10	
	Cation Piece:	Anion Piece:
Ion Charge		
Ion Name		
Number of Ions Required		
Compound Formula		
Compound Name		

	Compound 11	
	Cation Piece:	Anion Piece:
Ion Charge		
Ion Name		
Number of Ions Required		
Compound Formula		
Compound Name		

	Compound 12	
	Cation Piece:	Anion Piece:
Ion Charge		
Ion Name		
Number of Ions Required		
Compound Formula		
Compound Name		

	Compound 13	
	Cation Piece:	Anion Piece:
Ion Charge		
Ion Name		
Number of Ions Required		
Compound Formula		
Compound Name		

	Compound 14	
	Cation Piece:	Anion Piece:
Ion Charge		
Ion Name		
Number of Ions Required		
Compound Formula		
Compound Name		

	Compound 15	
	Cation Piece:	Anion Piece:
Ion Charge		
Ion Name		
Number of Ions Required		
Compound Formula		
Compound Name		

	Compound 16	
	Cation Piece:	Anion Piece:
Ion Charge		
Ion Name		
Number of Ions Required		
Compound Formula		
Compound Name		

Name ______________________

Lab Partner ______________________

Lab Section ______________ Date ______________

Lab 8
DATA SHEET
(continued)

Part B The Covalent Bonding Puzzle

Carbon Compounds

Compound	Structure
CH_4	
CH_2ClBr	
C_2H_4	
$C_2H_2Cl_2$	

Compound	Structure
C_2H_6	
CH_3OH	
HCN	
HCO_2H	

1 How many bonds does carbon typically form? Does this make sense based on the number of valence electrons in a carbon atom? Explain your answer.

____

____

____

2 Circle the structures below that represent a typical bonding pattern for carbon.

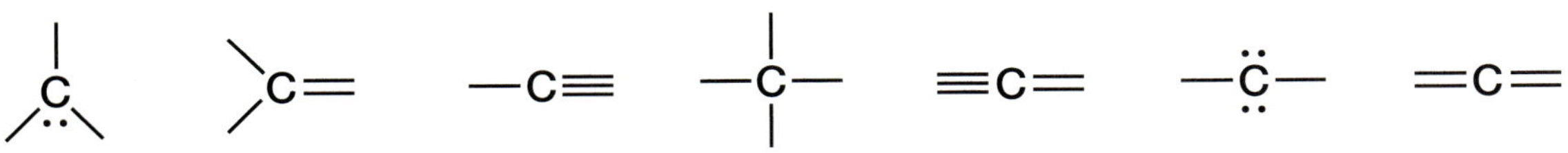

Nitrogen and Related Compounds

Compound	Structure
NH_3	
NH_2OH	
N_2H_4	

Compound	Structure
N_2H_2	
N_2	
PH_3	

1 How many bonds does nitrogen typically form? Does this make sense based on the number of valence electrons in a nitrogen atom? Explain your answer.

2 Does phosphorus typically form the same number of bonds as nitrogen? Use the periodic table to explain why your answer makes sense.

3 Circle the structures below that represent a typical bonding pattern for nitrogen.

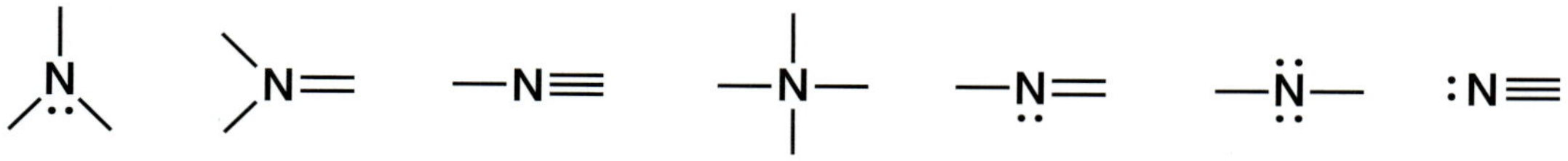

Name ______________________

Lab Partner ______________________

Lab Section ______________________ Date ______________________

Oxygen and Related Compounds

Compound	Structure
H_2O	
CH_3OCH_3	
OCl_2	

Compound	Structure
O_2	
CH_2O	
H_2S	

1 How many bonds does oxygen typically form? Does this make sense based on the number of valence electrons in an oxygen atom? Explain your answer.

2 Does sulfur typically form the same number of bonds as oxygen? Use the periodic table to explain why your answer makes sense.

3 Circle the structures below that represent a typical bonding pattern for oxygen.

Reflective Exercises

1 Provide the formula and name of the compound formed by each pair of ions.

	Ion pair	Formula	Name
a.	Li^+ and NO_3^-		
b.	Pb^{2+} and S^{2-}		
c.	Mg^{2+} and P^{3-}		
d.	K^+ and PO_4^{3-}		
e.	Ti^{4+} and O^{2-}		

2 What is the charge on the cation in each compound below?

a $PdCl_2$ __________

b $Al(OH)_3$ __________

c $FeCO_3$ __________

d $Pb(SO_4)_2$ __________

e Bi_2O_3 __________

3 Provide the correct formula for each compound named below.

a sodium sulfate __________

b magnesium nitride __________

c calcium sulfide __________

d copper(I) hydroxide __________

e lead(IV) oxide __________

4 For each pair below, decide whether the name and formula are correctly written and match. Provide a corrected version for any mismatch or if either the name or formula is incorrect. If there is no mistake in the pair, write *none*.

a NiO_2, nickel dioxide __________

b Li_3P, lithium phosphide __________

c Ti_4O, titanium(IV) oxide __________

d $NaCO_3$, sodium carbonate __________

e NiO, nickel(I) oxide __________

f $SrCl_2$, strontium(II) chloride __________

Name

Lab Partner

Lab Section Date

Lab 8 DATA SHEET

(continued)

5 Explain why oxygen forms two bonds to hydrogen to make a water molecule (H_2O), while nitrogen forms three bonds to hydrogen to make a molecule of ammonia (NH_3).

6 Consider the molecular formula C_2H_2. Would you expect this compound to contain any double or triple bonds? Use your knowledge of typical bonding patterns to fully support your answer.

7 Consider the molecular formula CH_5N. Would you expect this compound to contain any double or triple bonds? Use your knowledge of typical bonding patterns to fully support your answer.

8 How many covalent bonds would you expect silicon (Si) to form? Explain your answer.

9 Consider the molecular formula $COCl_2$. Based on your knowledge of bonding patterns, does it makes sense for the chlorine atoms to be bound to the carbon or to the oxygen atom? Explain your answer.

10 Draw a Lewis structure for each compound below.

a C_2Cl_4

b $CHCl_3$

c PBr_3

d H_2CNH

▲ Fluorite is the mineral form of calcium fluoride, whose formula is CaF_2.

Lab 9

Tried and True
Formula of a Compound

Objectives

After completing this lab you should be able to:

- Describe how the empirical formula of a compound could be determined from experimental data.
- Explain the difference between empirical and molecular formulas.
- Write balanced chemical equations to describe the reaction of two elements to produce an ionic compound.
- Determine the empirical formula of magnesium oxide using experimental data from the combustion of magnesium metal.
- Calculate the formula of a simple ionic compound from experimental data.

Chemical formulas may be expressed in several different ways. Among these, an empirical formula represents the *relative number* of nuclei in a compound while a molecular formula represents the *actual number* of nuclei within a single molecule of the compound. We generally choose the formula type used to represent a compound in order to convey or highlight specific information.

For example, the molecular formula of blood sugar (glucose) is $C_6H_{12}O_6$. This means that one molecule of glucose is made up of 6 carbon atoms bonded to 12 hydrogen atoms and 6 oxygen atoms. The empirical formula for glucose (CH_2O) is less informative because it indicates *only* that the compound contains twice as many hydrogen atoms as carbon or oxygen atoms.

In general, covalent compounds are represented using molecular formulas because we can easily identify the actual number of each atom type in a single molecule of the compound (Fig. 9.1A). In an ionic crystal, however, it is not possible to identify individual bonding interactions between the various ions. As a result, empirical formulas are used to represent ionic compounds (Fig. 9.1B).

We often use the concept of charge balance to generate the theoretical empirical formula for a compound formed by the interaction of two ions. For example, we claim that magnesium cations (Mg^{2+}) and chloride anions (Cl^-) combine in a 1:2 ratio to form magnesium chloride ($MgCl_2$). To verify such a formula, we must demonstrate with experimental evidence that the nuclei actually combine in the ratio indicated by the formula.

This can be done by experimentally determining a compound's **mole ratio,** which is simply the ratio of nuclei expressed in mole units. The mole ratio is convenient because molar quantities are readily calculated from easily measured mass data by employing an element's molar mass as a conversion factor. The following example illustrates how such experimental data may be used.

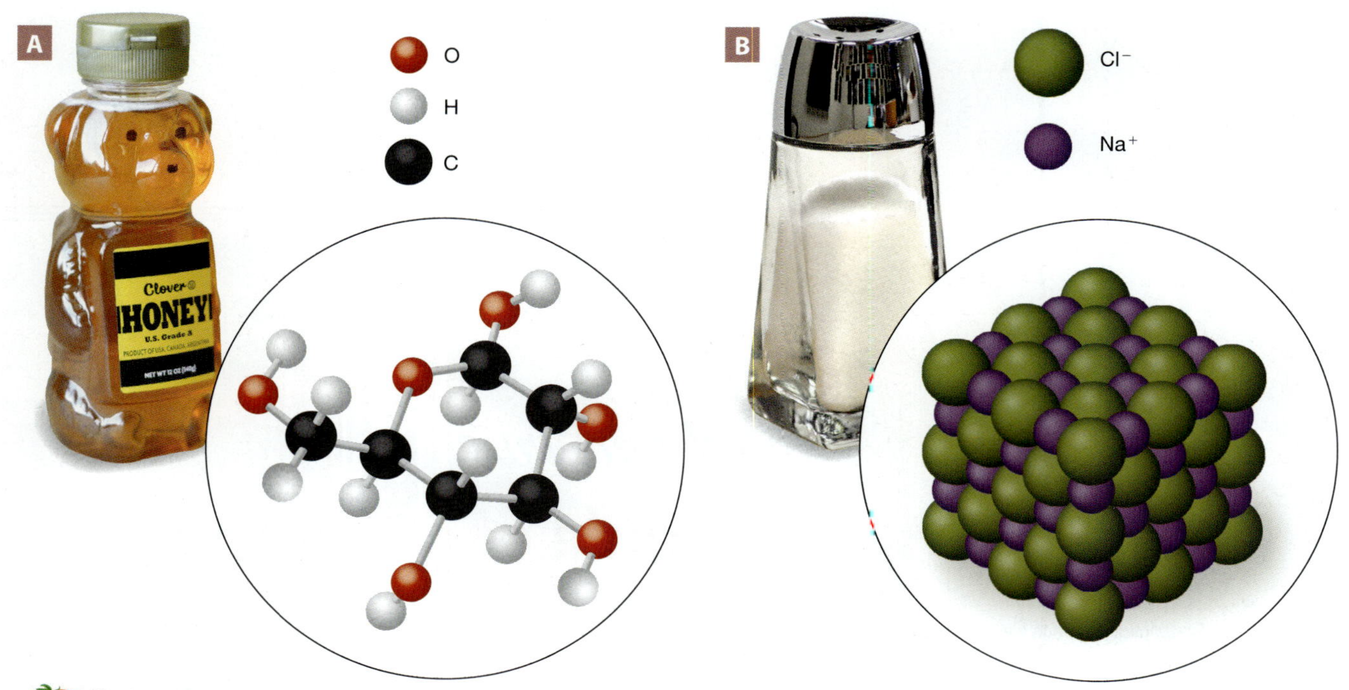

9.1 (A) For covalent compounds like glucose ($C_6H_{12}O_6$), molecular formulas are used to indicate the actual number of each atom present in a molecule of the substance. (B) For an ionic compound like sodium chloride (NaCl), the empirical formula is employed to indicate the ratios of the ions present in a crystal. Each Na^+ ion interacts electrostatically with many nearby Cl^- ions and vice versa.

Suppose a student found that 3.764 g of a new substance was produced when 3.343 g of pure copper reacted with oxygen in the air. Applying the law of conservation of matter, which states that "matter is neither created nor destroyed in a chemical reaction," we know that the number of copper atoms present is the same before and after the reaction. We are therefore able to reason that the mass of copper before and after the reaction must be equal.

Since the student obtained 3.764 g of product, the additional 0.421 g (3.764 g product–3.343 g copper) in the product must be the mass of oxygen incorporated into the new substance. To determine the mole ratio of copper to oxygen in the product compound, these measured mass values must first be converted to mole units. The requisite mass-to-moles conversions are accomplished by employing the molar masses of each element:

$$\text{Moles of copper in product} = 3.343\ \cancel{\text{g Cu}} \times \frac{1\ \text{mol Cu}}{63.55\ \cancel{\text{g Cu}}} = 0.05260\ \text{mol Cu} \qquad \textbf{[Eq. 1]}$$

$$\text{Moles of oxygen in product} = 0.421\ \cancel{\text{g O}} \times \frac{1\ \text{mol O}}{16.00\ \cancel{\text{g O}}} = 0.0263\ \text{mol O} \qquad \textbf{[Eq. 2]}$$

Next, the ratios of the two nuclei can be determined. A simple way to do this is to divide each mole value by the smaller of the two values:

$$\text{Copper mole ratio} = \frac{0.05260\ \text{mol}}{0.0263\ \text{mol}} = 2 \qquad \textbf{[Eq. 3]}$$

$$\text{Oxygen mole ratio} = \frac{0.0263\ \text{mol}}{0.0263\ \text{mol}} = 1 \qquad \textbf{[Eq. 4]}$$

These calculations indicate that there are twice as many copper nuclei as oxygen nuclei. Because the ratio of the copper to oxygen nuclei in the product is 2 to 1, the empirical formula for the compound must be Cu_2O.

In this example, the data gave us exact whole numbers. In reality, small errors generally yield experimental data that do not produce exact whole number ratios for the nuclei in a compound. In such cases, we may round the ratio values to the nearest whole numbers as long as this does not cause a significant deviation from the experimental values.

In this lab, you will experimentally determine the empirical formula of the ionic compound produced by reacting magnesium with oxygen from the air. The *unbalanced* chemical equation with an *unsolved* empirical formula for the magnesium oxide product (Mg_xO_y) is given in Figure 9.2. Your task is to use your experimental data to confirm the theoretical formula for magnesium oxide by determining the values of x and y.

Unfortunately, oxygen is not the only substance in air that will react with magnesium under the reaction conditions. Gaseous nitrogen (N_2) also will react with magnesium, producing magnesium nitride (Mg_3N_2) as an undesirable side product (Fig. 9.3). If the formation of the Mg_3N_2 remains uncorrected, your calculations will be invalid because the measured mass of the product would correspond to the mixture of products rather than to pure magnesium oxide.

Fortunately, there is a simple solution. Heating the initial product mixture with water will convert any Mg_3N_2 into the desired magnesium oxide. As a result, all of the magnesium in your sample ultimately will be transformed into magnesium oxide, and the change in sample mass during the reaction will reflect the transformation of magnesium metal into the magnesium oxide compound.

$$Mg\,(s) + O_2\,(g) \longrightarrow Mg_xO_y\,(s)$$

9.2 Unbalanced equation for the determination of the magnesium oxide empirical formula.

$$Mg\,(s) + O_2\,(g) + N_2\,(g) \longrightarrow Mg_xO_y\,(s) + Mg_3N_2\,(s)$$

Undesirable side product

Both reactants present in air

H_2O

Side product converted to Mg_xO_y by heating with water

9.3 When magnesium is heated, two substances in air (O_2 and N_2) react with it to produce two different products. The magnesium nitride side product (Mg_3N_2) can be completely converted into the target magnesium oxide compound by heating the initial product mixture with distilled water.

Procedure

What Am I Doing?

Steps 1–6: Cleaning the crucible until it displays a constant mass.

Steps 7–13: Heating a sample of magnesium in order to produce magnesium oxide.

Steps 14–17: Heating the initial reaction product with water to convert any magnesium nitride to magnesium oxide.

Steps 18–20: Heating the final product until the mass is constant, indicating that the reaction is complete.

Steps 21–22: Calculating the empirical formula of magnesium oxide from the experimental data.

Materials

- ❑ Ring stand
- ❑ Bunsen burner
- ❑ Crucible and cover
- ❑ Clay triangle
- ❑ Tongs
- ❑ Striker
- ❑ Forceps
- ❑ Glass stirring rod
- ❑ Steel wool
- ❑ Heat-resistant pad
- ❑ Iron ring

Determining Empirical Formula

1. Clean a crucible by filling it with 6 M HCl and allowing it to stand for 5 minutes. Discard the HCl and rinse the crucible with several milliliters of distilled water.

Hydrochloric acid (HCl) is a strong acid.
Handle with caution!

2. Place the empty crucible with its cover slightly ajar in a clay triangle supported on a ring stand as illustrated in Figure 9.4.
3. Heat the crucible with an intense flame from a Bunsen burner for about 5 minutes. The blue tip of the flame's inner cone should just touch the bottom of the crucible.

SAFETY NOTE

The apparatus will get very hot! Use caution to avoid burns or injuries! Remember: The apparatus looks the same whether it is hot or cold.

4. After the heating period, turn off the flame and allow the crucible to cool to room temperature. *For the remainder of the experiment, handle the crucible with tongs only!*

9.4 Apparatus for the reaction of magnesium with oxygen.

5. Using tongs, move the crucible and lid from the clay triangle to a balance and record their combined mass on the data sheet, page 131. During the transfer, you may wish to hold a heat-resistant pad or other rigid object beneath the crucible. This will help steady it in case the tongs slip.
6. Repeat Steps 3–5. If the mass of the crucible and lid does not change by more than 0.005 g between the two heating periods, move on to Step 7. If the mass of the crucible changes by more than 0.005 g, repeat Steps 3–5 until two consecutive mass measurements of the crucible are within 0.005 g of each other.
7. To avoid contaminating the metal surface, use forceps to obtain a piece of magnesium ribbon with a mass of approximately 0.3 g. To facilitate the reaction, clean the surface of the ribbon with steel wool. Fold the ribbon into a loosely packed ball and place it in the crucible.
8. Reweigh the crucible and lid with the magnesium ball and record the mass on the data sheet.
9. Place the crucible back on the clay triangle and add the cover so that it is slightly ajar.

10. Use a Bunsen burner to heat the crucible. Adjust the height of the ring so that the inner blue cone of the flame sits just below the bottom of the crucible. The magnesium should smoke and begin to burn after a few minutes. When this happens, use tongs to cover the crucible entirely with the lid.

Do not look directly at the bright flame of the burning magnesium.

11. Briefly check the reaction by slightly lifting the lid with tongs every 3 to 4 minutes. When there are no more embers or smoke, the ribbon should have changed appearance to a dull, gray solid. At this point, half-cover the crucible with the lid and continue heating it for an additional 5 minutes.
12. After the heating period, turn off the flame and allow the crucible to cool to room temperature.
13. When the crucible reaches room temperature, use a glass rod to break up the solids.
14. Add enough distilled water to cover the solids in the crucible. To prevent the loss of any product, you also may need to use the distilled water to wash solid residue off the glass rod back into the crucible.
15. Handling the lid with tongs, half-cover the crucible and *gently* heat it with the Bunsen burner flame until all of the water has evaporated (about 5 minutes). Be careful to avoid heating too intensely, which may cause you to lose material due to splattering as the water boils.
16. When the water has evaporated, heat the crucible with an intense flame for an additional 10 minutes.
17. After the heating period, turn off the burner and allow the crucible to cool to room temperature.
18. Use tongs to transfer the crucible and lid to a balance and record the combined mass of the crucible, lid, and product on the data sheet, page 131.
19. Using tongs, place the crucible with lid back on the clay triangle and heat it intensely again for 5 minutes.
20. Cool the crucible to room temperature and then reweigh the crucible, lid, and product. Record this value on the data sheet. If the mass has not changed by more than 0.005 g, move on to Step 21. If the mass changed by more than 0.005 g, repeat Steps 19 and 20 until two consecutive mass measurements are within 0.005 g of each other.
21. Fill in the calculations table on the data sheet. Using the final mass measurement of your crucible, calculate the mass of magnesium oxide produced. Determine the mass of oxygen in your sample by subtracting the mass of magnesium you used from the mass of the product you prepared.
22. Determine the empirical formula of magnesium oxide and record it on your data sheet.

Name ____________________

Lab Partner ____________________

Lab Section ____________________ Date ____________________

Lab 9
Pre-Laboratory Exercise

1 Oxygen (O_2) was first isolated by the decomposition of mercury(II) oxide (HgO) into mercury and oxygen. If 121.3 g of mercury(II) oxide were completely decomposed to generate 112.3 g of mercury, how many grams of oxygen should have been produced?

2 A student has 15.85 g of iron metal. Answer the following questions and show your work!

a How many moles of iron does the student have?

b Rust has the formula Fe_2O_3. What is the mole ratio of oxygen to iron in this compound?

c Given your answer to 2a and 2b, how many moles of oxygen would be required to convert all of the student's iron to rust?

d Given your answer to 2c, how many grams of oxygen would be required?

e What would be the mass of rust produced if all of the iron were converted to rust?

3 Aluminum metal can react with oxygen in the air (O_2) to produce aluminum oxide, a white powder. Write a balanced chemical equation for this reaction.

Name

Lab Partner

Lab Section Date

4 The black "tarnish" that can develop on silver objects is a result of the metal reacting with sulfur-containing compounds in the air to produce silver sulfide. A sample of silver sulfide was found to contain 90.64 g of silver and 13.47 g of sulfur. Use these data to calculate the empirical formula of silver sulfide.

Name ______________________________

Lab Partner ______________________________

Lab Section ____________________ Date ____________

Lab 9 DATA SHEET

Data and Observations

Determining Empirical Formula

Mass Data

Procedure Reference	Quantity	Mass (g)
Step 5	Empty crucible + lid (1st heating)	
Step 6	Empty crucible + lid (2nd heating)	
Step 6, repeated as necessary	Empty crucible + lid (additional heatings)	
Step 8	Crucible + lid + magnesium	
Step 18	Crucible + lid + product (1st heating)	
Step 20	Crucible + lid + product (2nd heating)	
Step 20, repeated as necessary	Crucible + lid + product (additional heatings)	

Calculations

Quantity	Show Calculation	Result
Mass of magnesium sample (g)	=	
Moles of magnesium in sample (mol)	=	
Mass of product obtained (g)	=	
Mass of oxygen in product (g)	=	
Moles of oxygen in product (mol)	=	
Mole ratio of magnesium to oxygen in product (mol Mg: mol O)	=	:
Experimental empirical formula for product		

Reflective Exercises

1 Using your experimental data, make a claim regarding the empirical formula of magnesium oxide. Describe your evidence and provide a rationale for your claim.

Claim	
Evidence	**Rationale**

2 Using the concept of charge balance, what is the theoretical empirical formula for magnesium oxide? Describe how well your experimental results match the theoretical formula. Identify any experimental errors that may have influenced the accuracy of your data.

3 Write a balanced chemical equation for the reaction of magnesium with atmospheric oxygen (O_2).

Name ______________________

Lab Partner ______________________

Lab Section ______________ Date ______________

Lab 9 DATA SHEET

(continued)

4 How would each of the following experimental errors affect your data and the empirical formula of the product?

a You did not handle the crucible with tongs, as directed in Step 4.

b You skipped the steps that involved adding water and heat to the initial product.

c You added the water in Step 14, but heated the crucible too strongly and caused spattering during Step 15.

5 Why is it important to ensure that there is little difference between the mass measurements of the product after two successive heatings, as indicated in Step 20 of the procedure? If you failed to do this, what compound(s) other than magnesium oxide might be in your crucible when you take the final mass?

6 Write the empirical formula for each combination of elements below. Show any work necessary to determine the formulas.

a 0.360 mol carbon (C) and 1.44 mol chlorine (Cl)

b 0.600 mol potassium (K), 0.300 mol carbon (C), and 0.900 mol oxygen (O)

c 4.50 g nitrogen (N) and 122 g iodine (I)

7 A student reacted 11.54 g of zinc metal with iodine and obtained 56.34 g of zinc iodide. What is the empirical formula of the zinc iodide? Show your work.

Experimental empirical formula of zinc iodide ________________

▲ *Copper metal reacts with a silver nitrate solution to form elemental silver and a blue solution of copper(II) nitrate.*

Lab 10

Don't Lose Your Balance

Verification of Reaction Stoichiometry

Objectives

After completing this lab you should be able to:

- Describe how coefficients for a balanced chemical equation could be determined from experimental data.
- Calculate the coefficients for the chemical equation describing the decomposition of potassium chlorate using experimental data.
- Balance a chemical equation using the law of conservation of mass and/or experimental data.

The concept of balancing a chemical equation is based on the **law of conservation of matter**, which states that atoms are neither created nor destroyed during a chemical change. As a result, reactants and products will only be consumed or generated in proportions that maintain the *total quantity* of each atom during the transformation. These proportions are reflected by the coefficients in a balanced equation.

Consider the reaction of iron metal with oxygen gas (O_2) to produce iron(III) oxide (Fe_2O_3), which you are probably familiar with as rust. The law of conservation of matter requires the number of iron atoms in the reactants to equal the number of iron atoms in the product. Similarly, the number of oxygen atoms cannot change during the reaction.

The chemical equation representing this transformation is balanced by adding coefficients before each chemical formula in order to equalize the number of each atom present in the reactants and products (Fig. 10.1). This balanced equation indicates that the reaction requires four iron atoms to react with three oxygen molecules in order to produce two formula units of iron(III) oxide.

Of course, most reactions involve many more than a handful of atoms, and we may think of the coefficients in a balanced equation as providing the required *ratios* through which reactants interact to form products. For example, the equation in Figure 10.1 suggests that 4 *dozen* iron atoms would react with 3 *dozen* oxygen molecules to yield 2 *dozen* units of iron(III) oxide. Similarly, 4 *moles* of iron would react with 3 *moles* of oxygen to produce 2 *moles* of iron(III) oxide.

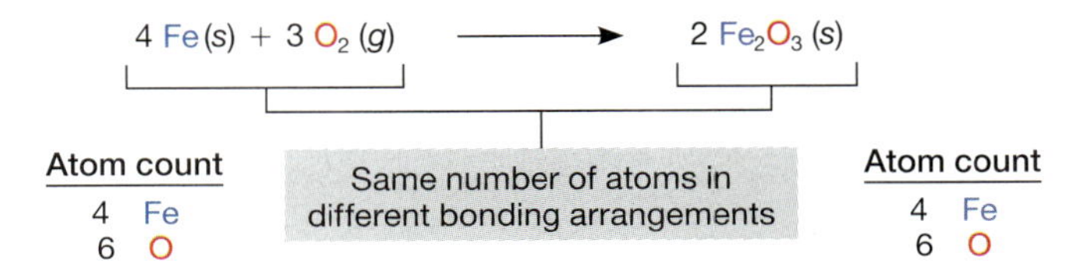

10.1 The balanced chemical equation for the reaction of iron with oxygen includes coefficients that indicate the ratios of chemical species involved in the reaction. Equations are "balanced" because the total quantity of each atom in the reactants equals the quantity in the products.

Keep in mind that the coefficients are used to balance the *total number* of each atom on either side of the chemical equation and therefore do not necessarily correspond to the mass (i.e., gram) ratios of the various species involved in the reaction. For example, it is *not* the case that reacting 4 *grams* of iron with 3 *grams* of oxygen would produce 2 *grams* of iron(III) oxide. The coefficient ratios apply to the number (i.e., moles) of each species involved in the reaction, and we must use their individual molar masses to convert the mole values into mass values.

Is it possible to verify the coefficients of a balanced chemical equation with experimental evidence? Of course! After all, experimental evidence is the basis for the law of conservation of matter. If we follow the equation in Figure 10.1 exactly, we could measure out 4 moles of iron and then investigate how much O_2 it takes to completely convert the iron to Fe_2O_3. If the reaction proceeds according to the balanced equation, then our experimental results should demonstrate that 3 moles of O_2 are consumed and 2 moles of Fe_2O_3 are produced.

In reality, this experiment is not very practical because it would require us to use over 223 grams (4 mol × 55.85 g/mol) of iron! Fortunately, it is not necessary to use exactly 4 moles of iron because we are only interested in whether the *ratios* match those represented by the equation. In fact, we could use *any* amount of iron in our experiment, and the ratios between the moles of iron used, O_2 consumed, and Fe_2O_3 produced should always be 4 Fe : 3 O_2 : 2 Fe_2O_3.

For example, suppose a student found that 3.45 g of iron reacted with 1.47 g of O_2 to produce 4.95 g Fe_2O_3. We may convert these mass values to molar quantities and then examine whether they fit the expected 4 Fe : 3 O_2 : 2 Fe_2O_3 ratios given by the reaction coefficients. This process is detailed in the steps that follow.

Keep in mind that experimental data are never perfect (due to small amounts of experimental error) and that it may be necessary to round some of the numbers. For our purposes, rounding is valid when our calculations produce coefficient values that do not deviate significantly from whole numbers.

≡10

Step 1: Convert measured mass values to moles using the molar mass of each substance in the reaction.

Moles of iron (Fe) consumed:

$$3.45\,\cancel{g} = \frac{1\text{ mol Fe}}{55.85\,\cancel{g}} = 0.0618\text{ mol} \qquad \text{[Eq. 1]}$$

Moles of oxygen (O_2) consumed:

$$1.47\,\cancel{g} = \frac{1\text{ mol }O_2}{32.00\,\cancel{g}} = 0.0459\text{ mol} \qquad \text{[Eq. 2]}$$

Moles of iron(III) oxide (Fe_2O_3) produced:

$$4.95\,\cancel{g} = \frac{1\text{ mol }Fe_2O_3}{159.70\,\cancel{g}} = 0.0310\text{ mol} \qquad \text{[Eq. 3]}$$

Next, the mole-to-mole ratios must be generated. To do this, divide the molar quantity of each substance by the smallest number of moles of any substance that appears in the reaction. In this case, the smallest mole value is that of Fe_2O_3 (0.0310 mol).

Step 2: Calculate the mole-to-mole ratios of the substances in the reaction.

Mole ratio for Fe:

$$\frac{0.0618\text{ mol}}{0.0310\text{ mol}} = 1.99 \qquad \text{[Eq. 4]}$$

Mole ratio for O_2:

$$\frac{0.0459\text{ mol}}{0.0310\text{ mol}} = 1.48 \qquad \text{[Eq. 5]}$$

Mole ratio for Fe_2O_3:

$$\frac{0.0310 \text{ mol}}{0.0310 \text{ mol}} = 1.00 \quad \text{[Eq. 6]}$$

Finally, recall that the coefficients in a balanced equation must be the *smallest whole number ratios* of the reaction components. Notice that our calculations do not give us whole number ratios. The mole ratio for O_2 is 1.48. Multiplying by 2 would convert this number to a value that is approximately a whole number ($2 \times 1.48 = 2.96$, which rounds to 3). Therefore, each of the calculated mole ratios must be multiplied by a factor of 2 in order to generate the whole number ratios that would correspond to the coefficients in our balanced equation.

Step 3: Convert the mole ratios to reaction coefficients by multiplying each mole ratio by the same integer. The chosen integer should produce whole numbers (within rounding error) from all fractional values upon carrying out the operation.

Experimental reaction coefficient for Fe:

$$1.99 \times 2 = 3.98 \rightarrow 4 \quad \text{[Eq. 7]}$$

Experimental reaction coefficient for O_2:

$$1.48 \times 2 = 2.96 \rightarrow 3 \quad \text{[Eq. 8]}$$

Experimental reaction coefficient for Fe_2O_3:

$$1.00 \times 2 = 2.00 \rightarrow 2 \quad \text{[Eq. 9]}$$

The student's experimental data therefore indicate that the ratios of Fe : O_2 : Fe_2O_3 are 4 : 3 : 2, just as predicted by our balanced equation!

In this lab, you will use experimental data to verify the coefficients in the balanced equation for the decomposition reaction of potassium chlorate ($KClO_3$), which is converted to potassium chloride and oxygen gas when heated to a temperature greater than 356°C. The unbalanced equation that describes this transformation is given in Figure 10.2.

You will need to measure the initial mass of $KClO_3$ and the final mass of KCl produced by the reaction. It will not be possible to measure directly the quantity of O_2 released into the air, but this value must be equal to the difference in the masses of solids present at the beginning and end of the reaction.

You will use your three experimentally determined mass values to calculate the whole number ratios of $KClO_3$: KCl : O_2. These ratios should provide you with coefficients that match the theoretical values generated by applying the law of conservation of matter to balance the equation.

The data you generate for the $KClO_3$ decomposition equation are valid only if the potassium perchlorate undergoes the expected reaction. To verify the conversion of $KClO_3$ to KCl, you will test the product of your reaction to ensure that it is indeed potassium chloride. When KCl is dissolved in water, the chloride anions react with aqueous silver nitrate ($AgNO_3$) to produce silver chloride (AgCl), an insoluble white solid. The equation for this reaction is given in Figure 10.3. Your verification test will involve dissolving your product residue in water and treating it with $AgNO_3$. The precipitation of white AgCl solid indicates that you successfully produced KCl from your initial sample of $KClO_3$.

$$KClO_3(s) \longrightarrow KCl(s) + O_2(g)$$

10.2 Unbalanced equation describing the decomposition of $KClO_3$.

$$KCl(aq) + AgNO_3(aq) \longrightarrow KNO_3(aq) + AgCl(s)$$

10.3 Reaction of potassium chloride with silver nitrate.

Procedure

What Am I Doing?

Steps 1–6: Cleaning the crucible until it displays a constant mass.

Steps 7–12: Heating a sample of $KClO_3$ in order to carry out the decomposition reaction.

Steps 13–15: Heating the reaction product until the mass is constant, indicating that the reaction is complete.

Steps 16–18: Testing the reaction product and authentic samples of $KClO_3$ and KCl with $AgNO_3$ to verify identity.

Step 19: Repeating the experiment, if instructed to do so.

Steps 20–21: Calculating reaction coefficients for the decomposition of $KClO_3$ using experimental data.

Materials

- ❑ Ring stand
- ❑ Bunsen burner
- ❑ Crucible and cover
- ❑ Clay triangle
- ❑ Iron ring
- ❑ Tongs
- ❑ Scoopula
- ❑ Heat-resistant pad
- ❑ Test tubes (3)
- ❑ Striker
- ❑ Microspatula

Using Experimental Data to Verify Coefficients

≡10

SAFETY NOTE

Hydrochloric acid (HCl) is a strong acid. ***Handle with caution!*** Potassium **chlorate is harmful if swallowed or inhaled.**

1. Clean a crucible by filling it with 6 M HCl and allowing it to stand for 5 minutes. Discard the HCl and rinse the crucible with several milliliters of distilled water.
2. Place the empty crucible with its cover slightly ajar in a clay triangle supported on a ring stand as illustrated in Figure 10.4.
3. Heat the crucible with an intense flame from a Bunsen burner for about 5 minutes. The blue tip of the flame's inner cone should just touch the bottom of the crucible.

SAFETY NOTE

The apparatus will get very hot! Use caution to avoid burns or injuries! Remember: The apparatus looks the same whether it is hot or cold.

10.4 Apparatus for the decomposition of $KClO_3$.

4. After the heating period, turn off the flame and allow the crucible to cool to room temperature. *For the remainder of the experiment, handle the crucible with tongs only!*

5. Using tongs, move the crucible and cover from the clay triangle to a balance and record their mass on the data sheet, page 143. During the transfer, you may wish to hold a heat-resistant pad or other rigid object beneath the crucible. This will help steady it in case the tongs slip.

6. Repeat Steps 3–5. If the mass of the crucible does not change by more than 0.005 g between the two heating periods, move on to Step 7. If the mass of the crucible changes by more than 0.005 g, repeat Steps 3–5 until two consecutive mass measurements of the crucible are within 0.005 g of each other.

7. Obtain approximately 2.0 g of potassium chlorate ($KClO_3$) and add it to the crucible.

8. Reweigh the crucible and cover with the added $KClO_3$ and record the mass on the data sheet.

9. Place the crucible back on the clay triangle and add the cover so that it is slightly ajar.

10. Use a Bunsen burner to heat the $KClO_3$ in the crucible with a mild flame for about 10 minutes. The inner blue cone of the flame should be a few inches below the bottom of the crucible. The potassium chlorate should melt and produce some bubbles. If this does not occur after a few minutes, move the crucible closer to the flame.

SAFETY NOTE

While monitoring the reaction progress, do not place your face or hands directly above the crucible. Wear safety goggles and examine the reaction from the side of crucible.

11. After any intense bubbling has ceased, move the crucible closer to the inner flame of the burner. Heat it intensely for about 15 minutes.

12. After the heating period, turn off the flame and allow the crucible to cool to room temperature.

13. Use tongs to transfer the crucible and cover to a balance, and record the mass of the crucible and contents on the data sheet, page 143.

14. Using tongs, place the crucible back on the clay triangle and heat it intensely again for 5 minutes.

15. Cool the crucible to room temperature and then reweigh the crucible and lid with the contents. Record this value on the data sheet. If the mass of the crucible has not changed by more than 0.01 g, move on to Step 16. If the mass of the crucible changed by more than 0.01 g, repeat Steps 14 and 15 until two consecutive crucible mass measurements are within 0.01 g of each other.

16. Label three test tubes 1 through 3. Prepare to verify that the reaction product is potassium chloride by adding a rice grain-sized quantity of the appropriate solid to each of the test tubes as indicated below:
    - Test Tube 1: $KClO_3$
    - Test Tube 2: KCl
    - Test Tube 3: Product from your crucible

17. Add about 2 mL of distilled water to each test tube and mix to dissolve the contents.

18. To each test tube, add 5 drops of 0.1 M silver nitrate ($AgNO_3$). The formation of a white precipitate indicates that the chloride ion is present. The absence of a white precipitate indicates that no chloride ion is present.

19. Repeat Steps 1–18 to obtain a second set of data for the chemical reaction if directed to do so by your instructor.

20. Fill in the calculations table on the data sheet, page 144. Using the final mass measurement of your crucible, calculate the mass of potassium chloride produced by your reactions. Determine the mass of oxygen released by subtracting the mass of KCl produced from the initial mass of $KClO_3$ in the crucible.

21. Use your Trial 1 (and Trial 2, if applicable) calculations to report experimental reaction coefficients for the decomposition of $KClO_3$.

Name ______________________________

Lab Partner ______________________________

Lab Section ______________________ Date ______________

Lab 10 Pre-Laboratory Exercise

1 Apply the law of conservation of matter to determine the coefficients necessary to balance the chemical equation for the decomposition of potassium chlorate:

$$\_\_\_\_ KClO_3(s) \longrightarrow \_\_\_\_ KCl(s) + \_\_\_\_ O_2(g)$$

2 "Oxygen candles" release breathable oxygen (O_2) through the chemical decomposition of potassium chlorate or related compounds. These devices are used to provide emergency oxygen sources to aircraft passengers, firefighters, miners, and astronauts. Use your balanced equation from Question 1 to calculate the mass of $KClO_3$ needed for an oxygen candle to provide a three-hour supply of oxygen if the average adult consumes 114 g of O_2 over this time period.

3 In Step 4, the procedure states: "For the remainder of the experiment, handle the crucible with tongs only!" In addition to a safety measure, this instruction helps ensure that your data are as accurate as possible. Explain how touching the crucible with your hands might affect the data.

4 Explain the test you will perform in this lab to confirm that your reaction was successful.

5 The equation for the combustion of propane is given below.

a Use the law of conservation of mass to balance the equation, and place the appropriate coefficients in front of each species in the equation.

b For each line in the table below, provide the number of moles of the other reactants or products necessary for complete reaction to take place with the given molar quantity.

$$\_\_\_\_\ C_3H_8\,(g) + \_\_\_\_\ O_2(g) \longrightarrow \_\_\_\_\ CO_2(g) + \_\_\_\_\ H_2O\,(g)$$

mol C_3H_8	mol O_2	mol CO_2	mol H_2O
5			
		18	
			0.500

Name ____________________

Lab Partner ____________________

Lab Section ____________________ Date ____________________

Lab 10 DATA SHEET

Data and Observations

Using Experimental Data to Verify Coefficients

Mass Measurements & Product Testing Data

Procedure Reference	Description	Trial 1	Trial 2
Step 5	Empty crucible + lid (1st heating)		
Step 6	Empty crucible + lid (2nd heating)		
Step 6, repeated as necessary	Empty crucible + lid (additional heatings)		
Step 8	Crucible + lid + $KClO_3$		
Step 13	Crucible + lid + product (1st heating)		
Step 15	Crucible + lid + product (2nd heating)		
Step 15, repeated as necessary	Crucible + lid + product (additional heatings)		
Step 18	Observation: $KClO_3 + H_2O + AgNO_3$		
Step 18	Observation: $KCl + H_2O + AgNO_3$		
Step 18	Observation: product + $H_2O + AgNO_3$		

Calculations

Trial 1						
	Show Calculations	$KClO_3$	Show Calculations	KCl	Show Calculations	O_2
Mass (g)	=		=		=	
Moles (mol)	=		=		=	
Mole-to-mole ratios	=		=		=	
Reaction coefficients (whole number ratios)	=		=		=	

Trial 2						
	Show Calculations	$KClO_3$	Show Calculations	KCl	Show Calculations	O_2
Mass (g)	=		=		=	
Moles (mol)	=		=		=	
Mole-to-mole ratios	=		=		=	
Reaction coefficients (whole number ratios)	=		=		=	

Name ______________________

Lab Partner ______________________

Lab Section ______________________ Date ______________

Lab 10 DATA SHEET

Reflective Exercises

1 Make a claim regarding the experimental stoichiometry of the reaction by writing a chemical reaction with the appropriate coefficients. List your evidence and provide a rationale for your claim.

Claim	
Evidence	**Rationale**

2 Did your experimental data verify the theoretical reaction coefficients for the decomposition of $KClO_3$? Fully support your answer by comparing your results to the theoretical values.

3 If you carried out two trials, did one of your data sets give you results closer to the coefficients of the balanced equation? If so, can you identify any experimental errors that might explain the difference between the two trials?

4 Why is it important to ensure that there is little difference between the mass measurements of the product after two successive heatings, as indicated in Step 15 of the procedure? What would it indicate if there was a significant change in mass from one heating to another?

5 Identify some other substances (besides KCl) that might give a positive test for chloride upon addition of $AgNO_3$. Based on the reactant used in your experiment, do you think it is reasonable to exclude these types of substances as contaminants that would give a false positive when you tested your reaction product to verify that it is KCl?

6 Aluminum metal can be produced by the decomposition of ores containing aluminum oxide. The unbalanced equation for this reaction is given below. Balance the equation using the law of conservation of matter and then verify that the experimental data in the accompanying table match the predicted reaction coefficients. Show your calculations.

$$\_\_\_\_ Al_2O_3\ (s) \rightarrow \_\_\_\_ Al\ (s) + \_\_\_\_ O_2\ (g)$$

Reaction	Show Calculations	Al_2O_3	Show Calculations	Al	Show Calculations	O_2
Mass (g)		16.31 g		8.63 g		7.68 g
Moles (mol)	=		=		=	
Mole-to-mole ratios	=		=		=	
Reaction coefficients (whole number ratios)	=		=		=	

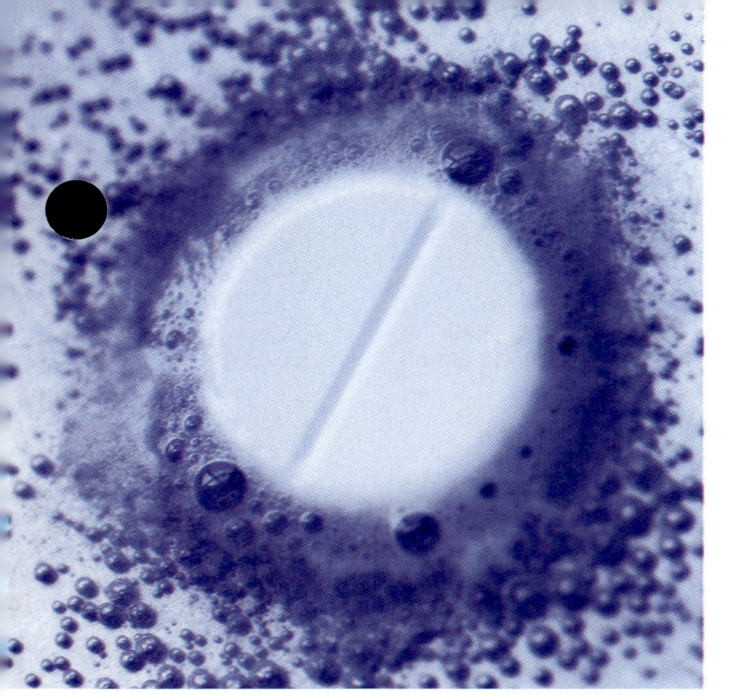

▲ *Aspirin, a common pain reliever, was first synthesized in 1899.*

Lab 11

Lab Is Such a Headache

Synthesis of Aspirin

A functional group is a set of atoms that are bound together in a specific pattern and exhibit characteristic chemical and physical properties. A table of common functional groups is given on the inside back cover of this book. Refer to this table to help you identify the reactive centers described in this lab.

The ester is an important functional group found in many biologically active molecules (Fig. 11.1). For example, esters are found in various fragrances and flavoring agents, including the compound responsible for the smell of fresh raspberries (2-methylpropyl formate). In some cases, a molecule's ester group may participate in a chemical reaction that leads to physiological benefits. Aspirin, the world's most widely used pain reliever, derives its ability to reduce inflammation, pain, and fever from a chemical reaction that takes place between the ester group and an enzyme found in the human body.

In this lab, you will use two common methods to prepare compounds that contain ester groups. In Part A, you will synthesize a sample of aspirin. You will analyze your product to determine the success of your reaction by two measures: the amount of aspirin produced (i.e., the yield) and the quality of aspirin produced (i.e., the purity). In Part B, you will use the second ester-preparation method to generate small solutions of several esters that have different aromas. You will gauge the success of the reactions by detecting a characteristic smell from the presence of the new ester-containing compound.

Objectives

After completing this lab you should be able to:

- Identify the ester functional group in an organic compound.
- Prepare and isolate a sample of aspirin from simple laboratory chemicals.
- Calculate the theoretical and percent yield of a chemical reaction.
- Identify alcohol and carboxylic acid structural components necessary to produce a given ester.

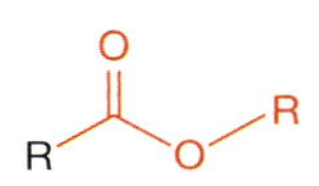

Generic ester functional group

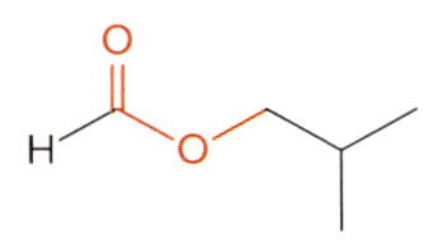

2-methylpropyl formate (raspberry fragrance)

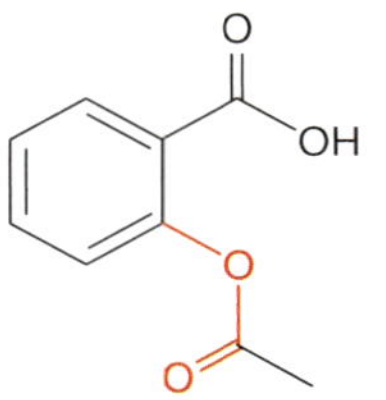

Acetylsalicylic acid (aspirin)

11.1 Generic structure of the ester functional group and some compounds that contain esters.

The method you will use to synthesize aspirin involves the reaction of an alcohol functional group in salicylic acid with acetic anhydride (Fig. 11.2). Salicylic acid is a natural analgesic found in the bark of the willow tree. The fifth century Greek physician Hippocrates suggested chewing willow bark to alleviate pain, and Native Americans used infusions of willow bark for medicinal purposes.

Salicylic acid (138.12 g/mol) + Acetic anhydride (102.09 g/mol) $\xrightarrow{H_3PO_4}$ Aspirin (180.16 g/mol) + Acetic acid (60.05 g/mol)

11.2 Synthesis of aspirin.

However, pure salicylic acid causes severe stomach irritation. In 1899, chemists at Bayer & Co. prepared a pure sample of acetylsalicylic acid, which was found to retain the analgesic properties of salicylic acid but cause less irritation. They soon began marketing the medicine as aspirin.

Notice that the balanced reaction for the synthesis of aspirin requires one mole of acetic anhydride for each mole of salicylic acid. In addition, a small amount of strong acid (H_3PO_4) is necessary to catalyze the reaction. You will use an excess of acetic anhydride to ensure that all of your salicylic acid is converted to aspirin. Thus, the quantity of aspirin you can potentially produce (the **theoretical yield**) depends only upon the amount of salicylic acid you use. For example, if a student uses 1.145 g of salicylic acid ($C_7H_6O_3$), the theoretical yield of aspirin ($C_9H_8O_4$) would be calculated as shown in Equation 1.

$$\text{Theoretical yield} = 1.145\text{ g } \cancel{C_7H_6O_3} \times \frac{1\ \cancel{\text{mol } C_7H_6O_3}}{138.12\text{ g } \cancel{C_7H_6O_3}} \times \frac{1\ \cancel{\text{mol } C_9H_8O_4}}{1\ \cancel{\text{mol } C_7H_6O_3}} \times \frac{180.16\text{ g } C_9H_8O_4}{1\ \cancel{\text{mol } C_9H_8O_4}} = 1.494\text{ g } C_9H_8O_4 \qquad \textbf{[Eq. 1]}$$

11

In general, the amount of product actually obtained from a chemical reaction is less than the theoretical yield for a number of reasons. Incomplete reactions may leave behind unreacted started material, or undesirable side reactions may use some of the reactants without producing any of the desired product. Additionally, small amounts of product may be left behind or otherwise lost during isolation and purification steps. One important measure of success for a chemical reaction is the quantity of product actually obtained relative to the theoretical yield. Chemists express this value as the **percent yield**, which is calculated according to Equation 2.

$$\text{Percent yield} = \frac{\text{Mass of product obtained}}{\text{Theoretical yield}} \times 100 \qquad \textbf{[Eq. 2]}$$

If the student from our example obtained 0.987 g of aspirin from the reaction using 1.145 g of salicylic acid, the percent yield would be 66.1% (Eq. 3).

$$\frac{0.987\text{ g}}{1.494\text{ g}} \times 100 = 66.1\% \qquad \textbf{[Eq. 3]}$$

Another measure of the success of a chemical reaction is the purity of isolated product. Recall that the melting point of a solid is a characteristic physical property. As such, measuring the melting point of a solid product provides a general measure of its purity and evidence of its identity. Although melting points are often reported as single values, they are actually measured as a temperature range over which a substance is in equilibrium between the solid and liquid state.

According to the chemical literature, aspirin melts between 136°C and 140°C. If your product melts within this range, it is likely to be relatively pure aspirin. Any impurities present in a sample cause a lowering and broadening of a substance's melting point. Thus, it is possible to assess the purity of your aspirin sample by comparing the melting point range of your sample to the 136°C–140°C reference value. If your value is lower and broader, your sample likely contains some impurities.

You may be wondering how you can be certain that you actually made *any* aspirin if your measured melting point is not a close match to the reference value due to the presence of impurities. The fact is: you *cannot* be certain! To "prove" the structure of a compound, chemists must provide multiple pieces of evidence to support their conclusion and exclude other possibilities.

In addition to the melting point, you will provide one additional source of evidence that you successfully produced aspirin. A yellow solution of iron(III) chloride reacts with the alcohol functional group of aromatic compounds such as salicylic acid to produce dark-colored complexes. This functional group is no longer present in aspirin, so the lack of a color change upon addition of aqueous $FeCl_3$ to a solution of your product indicates the successful conversion of salicylic acid to aspirin.

You will use a different method, exemplified in Figure 11.3, to prepare a series of fragrant esters. In this type of reaction, the ester group is formed by a **condensation reaction** between the carboxylic acid functional group of one molecule and the alcohol functional group of another. A condensation reaction involves the combination of two or more small molecules to produce a larger molecule and a different small molecule (usually water). This reaction also requires a small amount of acid catalyst (H_2SO_4).

Figure 11.3 illustrates the synthesis of raspberry flavor (2-methylpropyl formate) by the reaction of formic acid with 2-methylpropan-1-ol. You will use analogous reactions to prepare six different esters by combining different carboxylic acids and alcohols. Use Figure 11.3 as a template to predict the structures of the esters you prepare, and confirm the success of your reactions by detecting the new fragrance produced by each ester.

Formic acid + 2-methylpropan-1-ol $\xrightarrow{H_2SO_4}$ 2-methylpropyl formate (raspberry fragrance) + H_2O

11.3 Example synthesis of a fragrant ester.

Procedure

Part A: Mixing salicylic acid, acetic anhydride, and phosphoric acid to synthesize aspirin; chilling the reaction mixture to precipitate the aspirin product; isolating the aspirin by vacuum filtration; testing the identity and purity of the product by melting point and $FeCl_3$ analysis.

Part B: Reacting various alcohols and carboxylic acids in test tubes to produce ester compounds with characteristic fragrances.

Part A Synthesis of Aspirin

1. Assemble the apparatus illustrated in Figure 11.4 under a fume hood or appropriately ventilated space.
   - **a** Fill a 400 mL beaker approximately two-thirds full of water, and then place it on a hot plate.
   - **b** Clamp a thermometer to a ring stand and lower it into the water bath.
   - **c** Turn on the hot plate to begin heating the water bath to a temperature between 70 and 80°C.
   - **d** You will add the Erlenmeyer flask after filling it with the reactants.

Materials

- ❑ 400 mL beaker (2)
- ❑ Thermometer
- ❑ 50 mL Erlenmeyer flasks (2)
- ❑ Boiling stick
- ❑ Glass stirring rod
- ❑ Watch glass
- ❑ Filter paper
- ❑ Büchner funnel with rubber adapter
- ❑ Filter flask with tubing
- ❑ Test tubes (6)
- ❑ Hot plate
- ❑ Ring stand
- ❑ Clamps (2)
- ❑ 10 mL graduated cylinder
- ❑ Dropper
- ❑ Tongs
- ❑ Test-tube holder
- ❑ Melting point apparatus
- ❑ Melting point capillary
- ❑ Scoopula

2. Obtain approximately 1.0 g of salicylic acid. Record the exact mass of the salicylic acid on the data sheet, page 155, and then add it to the 50 mL Erlenmeyer flask.

3. Add approximately 3 mL of acetic anhydride to the flask.

4. Add 3 drops of concentrated phosphoric acid to the Erlenmeyer flask.

SAFETY NOTE

Acetic anhydride is a flammable liquid and vapor. Acetic anhydride and phosphoric acid are harmful if inhaled or swallowed and may cause severe skin burns. Wear gloves and handle these chemicals in a fume hood or appropriately ventilated space!

5. Add a boiling stick to the Erlenmeyer flask. Clamp it to the ring stand and lower the flask into the hot-water bath so the liquid level in the flask is just beneath the surface of the hot-water bath.

6. Allow the reaction flask to remain in the hot-water bath for 15 minutes after the bath reaches a temperature above 70°C.

7. While the reaction is heating, chill about 20 mL of distilled water in an ice-water bath.

8. At the end of the 15-minute heating period, add about 1 mL of the chilled distilled water to the Erlenmeyer flask. This step will consume any unused acetic anhydride.

9. Remove the flask from the hot-water bath and add approximately 9 mL of the cold distilled water. You may now remove the flask from the fume hood if you wish. **Keep the hot-water bath on the hot plate, because you will use it again in Step 14.**

10. Allow the flask to cool to room temperature. As the cooling proceeds, crystalline aspirin should appear in the flask. If no solid appears once the flask has cooled to room temperature, crystallization can usually be induced by using a glass stirring rod to gently scratch the bottom and sides of the flask. Once the solution reaches room temperature and solid has appeared, cool the flask in an ice-water bath in order to maximize the formation of crystals.

11.4 Apparatus for the synthesis of aspirin.

11. While the solution is cooling, set up the vacuum filtration apparatus (Fig. 11.5):
    - **a** Clamp a filter flask to a ring stand.
    - **b** Fit the filter flask with a rubber adapter and a Büchner funnel.
    - **c** Attach a vacuum hose from the vacuum source to the port on the vacuum flask.
    - **d** Place an appropriately sized piece of filter paper flat in the funnel. Ensure that all the perforations are covered and that the paper does not curl up the sides.

12. When the product crystals are completely cooled, isolate them by vacuum filtration:
    - **a** Use a small amount of the chilled distilled water to wet the filter paper.
    - **b** Turn on the vacuum source. This should seal the filter paper to the bottom of the funnel.
    - **c** Slowly pour the mixture of aspirin crystals into the funnel. The liquid will be pulled through the funnel, leaving the solids on the filter paper.
    - **d** Wash the solids with the remaining portion of the chilled distilled water.

13. Allow the solids to dry in the funnel with the vacuum on for at least 10 minutes. While the crystals are drying, weigh a large watch glass and record the mass on the data sheet, page 155. Also, turn up the heat on the hot plate so that the hot-water bath boils gently.
14. After drying the product under vacuum, break the filter vacuum by pulling the hose off of the vacuum flask. Carefully scrape all of your crystals onto the pre-weighed watch glass.
15. Place the watch glass on top of the boiling-water bath in order to further dry the crystals. After about 10 minutes, the crystals should be fully dried.

Steam from the boiling-water bath can cause severe burns. Exercise caution, and use tongs or a hot mitt to remove the watch glass!

16. Use tongs or a hot mitt to remove the watch glass from the heat and dry the bottom with a paper towel.
17. On the data sheet, record the mass of the watch glass with the dried product.

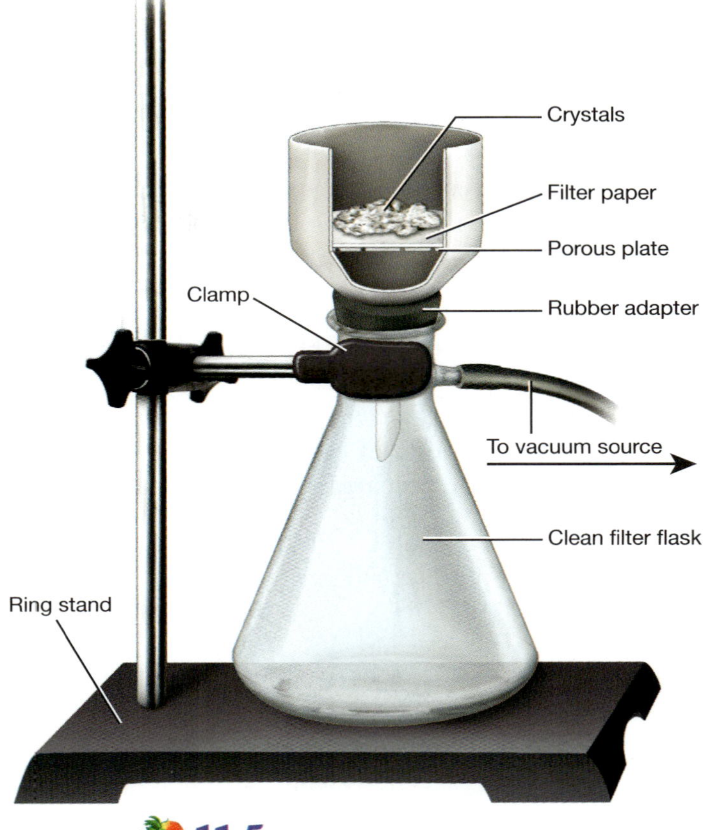

11.5 Apparatus for vacuum filtration.

18. Obtain the melting point of your aspirin as directed by your instructor. Record your result on the data sheet.
19. Perform the $FeCl_3$ test by mixing the appropriate components as indicated below. Record your observations on the data sheet.
    - Test Tube 1: A few crystals of salicylic acid, 1 mL ethanol, 2 drops 0.1 M $FeCl_3$
    - Test Tube 2: A few crystals of authentic aspirin, 1 mL ethanol, 2 drops 0.1 M $FeCl_3$
    - Test Tube 3: A few crystals of your product, 1 mL ethanol, 2 drops 0.1 M $FeCl_3$
20. Dispose of the test tube contents, your product, and the liquid in the filter flask as directed by your instructor.
21. Complete the calculations table, and record your results on the data sheet.

Part B Syntheses of Fragrant Esters

1. Ensure that your hot-water bath is at a gentle boil.
2. Label six test tubes 1–6.
3. To Test Tubes 1–5, add 6 drops of the appropriate alcohol followed by 2 drops of the appropriate carboxylic acid as indicated below:
    - Test Tube 1: propan-1-ol + acetic acid
    - Test Tube 2: isoamyl alcohol + acetic acid
    - Test Tube 3: octan-1-ol + acetic acid
    - Test Tube 4: methanol + butyric acid
    - Test Tube 5: ethanol + butyric acid

Butyric acid has a strong odor and should be handled under a fume hood.

4. In Test Tube 6, combine a few crystals of salicylic acid and 6 drops of methanol.
5. Add 1 drop of concentrated sulfuric acid to each test tube.
6. Gently agitate each test tube to ensure good mixing, and then place them in a boiling-water bath for 5 minutes (see Technique Tip 11.1).

Concentrated sulfuric acid can cause severe skin burns. Wear gloves and ***handle it with caution!***

7. After the heating period is complete, remove each test tube from the boiling-water bath and add about 2 mL of distilled water. Gently swirl the test tube to mix the contents.
8. Test the odor of the solution by wafting the air above the test tube toward your nose (see Technique Tip 11.2).
9. On the data sheet, page 156, record the odor of each solution. Use Figure 11.3, to help you identify the structure of each ester you prepared.

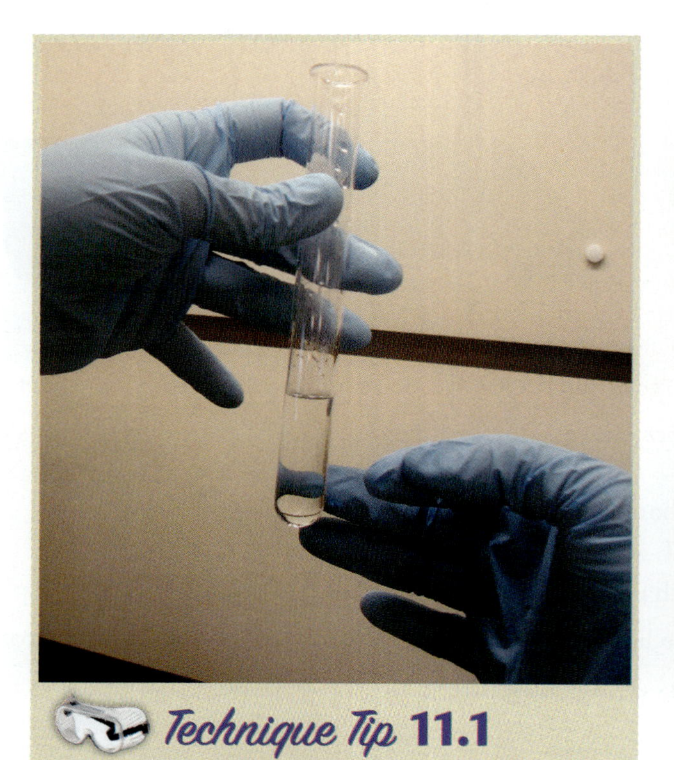

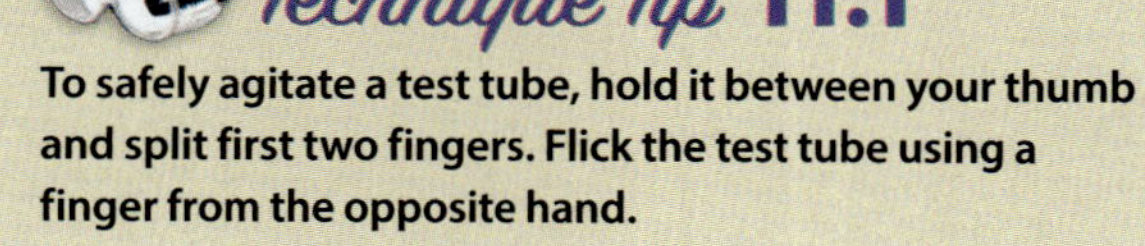

To safely agitate a test tube, hold it between your thumb and split first two fingers. Flick the test tube using a finger from the opposite hand.

Technique Tip 11.2

To safely test the odor of a sample, waft the air above the test tube toward your nose. Do not put your nose directly above the test tube to smell the odor.

Name ______________________

Lab Partner ______________________

Lab Section ______________ Date ______________

Lab 11 Pre-Laboratory Exercise

1 Name three physiological properties of aspirin.

a ______________________

b ______________________

c ______________________

2 Naproxen is a potent painkiller found in the drug Aleve. Based on the structure of naproxen, would you expect its analgesic properties to result from the same chemical reaction with the human body that makes aspirin a good pain reliever? Explain your answer.

____

____

____

3 A student trying to prepare aspirin begins with 2.52 g of salicylic acid and excess acetic anhydride.

a What is the student's theoretical yield of aspirin? Show your work.

b If the student obtains 1.71 g of aspirin, what is her percent yield? Show your work.

4 In your synthesis of aspirin, should you spend time measuring out exactly 1.00 g of salicylic acid or should you measure out approximately 1.0 g and record the exact mass of salicylic acid that you use? Explain your answer.

5 In the condensation reaction used in Part B, the water molecule is produced when –OH from the carboxylic acid functional group in one molecule bonds to the hydrogen of an alcohol functional group in the other molecule. Consider the two molecules below and answer the questions that follow.

OH + HO(C=O) $\xrightarrow{H_2SO_4}$

a Circle the –OH of the carboxylic acid group and the hydrogen of the alcohol that combine to form H_2O.

b Recall that oxygen generally participates in two bonds and carbon participates in four. Draw a dotted line between the two atoms in the organic molecules that must bind to each other in order to maintain the typical bonding number for all atoms if the circled elements of the H_2O molecule separate from the reactants.

c In the box provided, draw the product formed by bonding the atoms connected by your dotted line.

Name ______________________

Lab Partner ______________________

Lab Section ______________ Date ______________

Lab 11 DATA SHEET

Data and Observations

Part A Synthesis of Aspirin

Data Table

Procedure Reference	Description	Value/Observations
Step 2	Mass: salicylic acid (g)	
Step 13	Mass: watch glass (g)	
Step 17	Mass: watch glass + product (g)	
Step 18	Melting range: product (°C)	
Step 19	Test Tube 1: salicylic acid + $FeCl_3$	
Step 19	Test Tube 2: aspirin + $FeCl_3$	
Step 19	Test Tube 3: product + $FeCl_3$	

Calculations Table

Procedure Reference	Description	Calculation	Value
Step 20	Calculated mass: product (g)	=	
Step 20	Theoretical yield (g)	=	
Step 20	Percent yield	=	

Part B Syntheses of Fragrant Esters

Fragrant Ester Synthesis

Test Tube	Carboxylic Acid	Alcohol	Ester Structure	Observed Fragrance
1	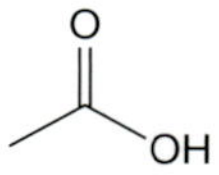 acetic acid	 propan-1-ol		
2	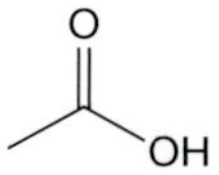 acetic acid	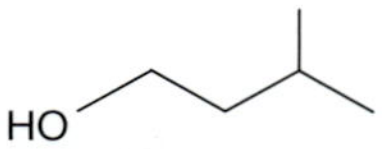 isoamyl alcohol		
3	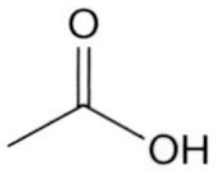 acetic acid	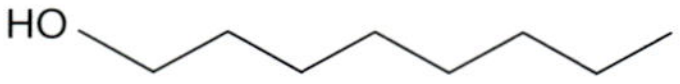 octan-1-ol		
4	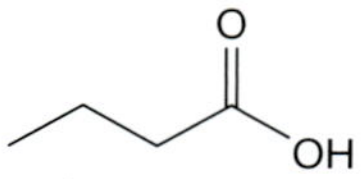 butyric acid	CH_3OH methanol		
5	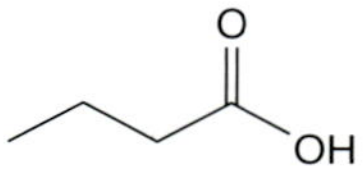 butyric acid	CH_3CH_2OH ethanol		
6	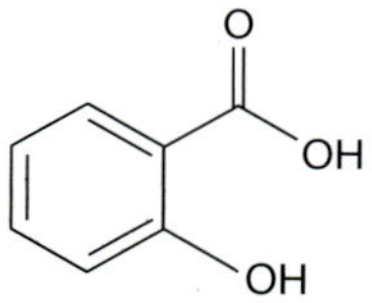 salicylic acid	CH_3OH methanol		

Reflective Exercises

1 What data did you collect to indicate that the identity of your product was aspirin? What did your results indicate about the purity of the product you obtained? Explain your answers.

Name ______________________________

Lab Partner ______________________________

Lab Section ____________________ Date ______________

Lab 11
DATA SHEET
(continued)

2 Aspirin and salicylic acid have a very similar appearance.

a Given that the melting point of salicylic acid is 159°C, can you be certain that the product you isolated was not *pure* salicylic acid that was unchanged during the reaction? Justify your answer using your data.

b Can you be certain that the product you isolated was not *impure* salicylic acid? Justify your answer using your data.

3 Was your percent yield of aspirin exactly 100%? If not, give reasons why your yield was greater or less than 100%.

4 Why did the procedure direct you to isolate your aspirin crystals using vacuum filtration rather than simply using gravity filtration?

5 Acetaminophen is the analgesic sold under the brand name Tylenol. Acetaminophen contains an amide functional group that can be prepared through a reaction similar to the aspirin synthesis you performed:

H_2N / OH + O O / O —H_3PO_4→ H N O / OH + O HO

p-aminophenol **Acetic anhydride** **Acetaminophen (Tylenol)** **Acetic acid**

a If a student starts with 1.88 g of p-aminophenol and excess acetic anhydride, what is the theoretical yield of acetaminophen? Show your calculation.

b If 1.85 g of acetaminophen is obtained from the reaction, what is the percent yield? Show your calculation.

c Would the $FeCl_3$ test be useful as evidence that you successfully produced acetaminophen by this reaction? Explain your answer.

Name ______________________________

Lab Partner ______________________________

Lab Section ____________________ Date ____________

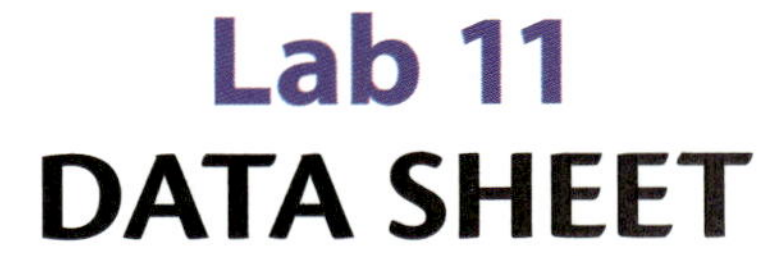

(continued)

6 The smell of peaches is due to benzyl acetate (A), and the smell of rum results from 2-methylpropyl propionate (B). In the boxes below, draw the structures of the carboxylic acids and alcohols that could be used to synthesize each of these esters.

A

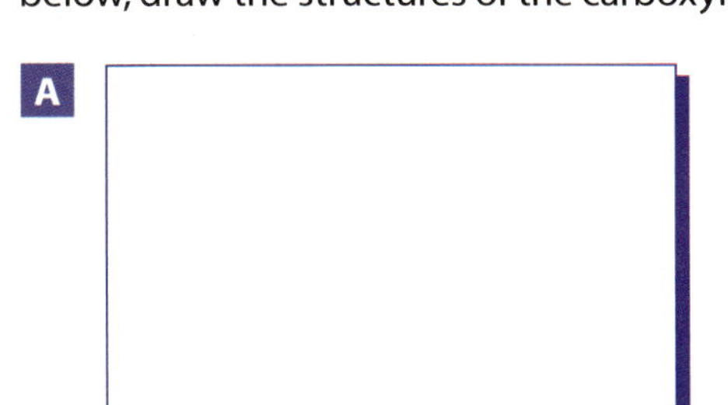

\+

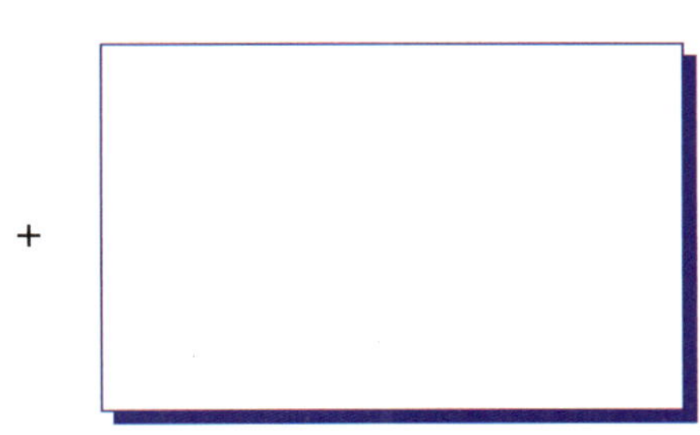

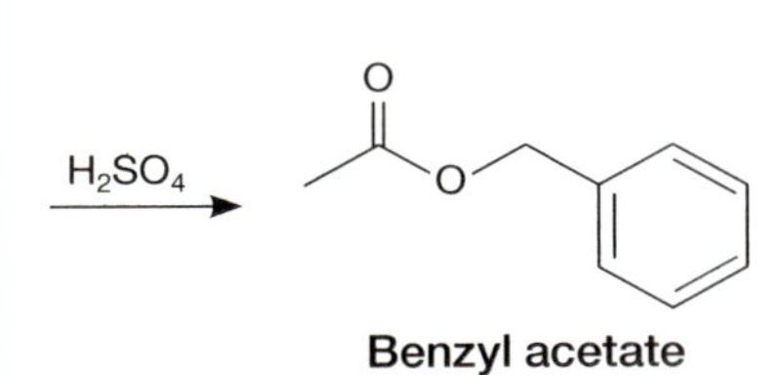

Benzyl acetate

B

\+

H_2SO_4

O

O

2-methylpropyl propionate

▲ *Petroleum is a hydrocarbon that is refined for use as gasoline and other combustible fuel.*

Lab 12

Skeletons and Suffixes

Hydrocarbon Isomerism and Bond Reactivity

Organic molecules are composed of a skeletal framework of carbon and hydrogen atoms. Recall that carbon typically forms a total of four bonds through some combination of single, double, and triple bonding with other atoms. For example, octane, a primary component of gasoline, has eight carbon atoms singly bound to each other in a linear chain (Fig. 12.1). Each carbon atom also is bound to the number of hydrogen atoms necessary to complete its octet.

```
    H   H   H   H   H   H   H   H
    |   |   |   |   |   |   |   |
H — C — C — C — C — C — C — C — C — H
    |   |   |   |   |   |   |   |
    H   H   H   H   H   H   H   H
```

12.1 Lewis structure of octane.

Hydrocarbon Structures and Nomenclature

Octane is an example of a **hydrocarbon**, which is a molecule composed only of carbon and hydrogen atoms. The properties of these molecules are defined by the bonding patterns between their carbon atoms. Hydrocarbons are therefore classified by this feature as summarized in Table 12.1.

Alk*anes*, such as octane and hexane, are hydrocarbons that contain only single bonds. They often are referred to as **saturated hydrocarbons** because each carbon atom has the maximum number of possible bonds to hydrogen atoms. **Alk*enes*** and **alk*ynes*** are hydrocarbons that contain one or more double and triple bonds, respectively. These two classes are collectively referred to as **unsaturated hydrocarbons** because not all of the carbon atoms are "saturated" with the maximum number of possible bonds to hydrogen.

For example, the carbons involved in the hexene double bond are each "missing" a potential bond to hydrogen as a result of the presence of the double bond. Alkanes or alkenes may occur as cyclic compounds, but the geometric restrictions of the triple bond prevent alkynes from occurring in cyclic compounds unless the ring is very large.

Objectives

After completing this lab you should be able to:

- Classify a hydrocarbon as an alkane, alkene, alkyne, or aromatic based on its chemical structure.
- Define the terms *saturated hydrocarbon*, *unsaturated hydrocarbon*, and *isomer*.
- Draw and recognize various isomers of simple alkanes and alkenes.
- Name simple, branched alkanes.
- Describe the different chemical reactivities of alkanes, alkenes, and aromatics toward bromine, potassium permanganate, and chloroform/aluminum chloride.
- Identify a hydrocarbon using a combination of physical and chemical properties.
- Use molecular models to help you determine the relationship between different structural representations of the same molecular formula.
- Translate the three-dimensional structure of a molecule into a two-dimensional representation.
- Identify structural representations as constitutional isomers, stereoisomers, or identical molecules.

TABLE **12.1** ■ Four Major Classes of Hydrocarbons

Hydrocarbon Class	Typical Bonding Pattern	Examples
Alkane	—C—C—	hexane; cyclopentane
Alkene	C=C	hex-3-ene; cyclopentene
Alkyne	—C≡C—	hex-3-yne
Aromatic	benzene (other patterns possible)	toluene; naphthalene

Typical **aromatic hydrocarbons** are recognizable by the presence of one or more benzene rings, although other bonding patterns also are possible. The skeletal structure of benzene displays an alternating pattern of single and double bonds within a six-membered carbon ring. This structural pattern imparts special stability and reactivity to aromatic compounds. Note that aromatic compounds include those with fused benzene rings such as naphthalene, the active ingredient in some brands of mothballs.

A molecule's carbon "backbone" is not necessarily linear. In fact, there often are many different ways in which the atoms of a given molecular formula may combine to form a molecule. Compounds that share a molecular formula but have different structures are called **isomers**. We separate isomers into two major classes based on the structural differences between them (Fig. 12.2).

Consider the three drawings of alkanes in Figure 12.3. All of the structures have a molecular formula of C_4H_{10}, but the atoms that make up the carbon skeletons of molecules A and B are connected in a different pattern. In compound A, the carbon atoms are connected in a linear fashion. In compound B, three carbon atoms are connected linearly, but the fourth carbon extends from the central carbon atom in the same way a branch radiates from the trunk of a tree.

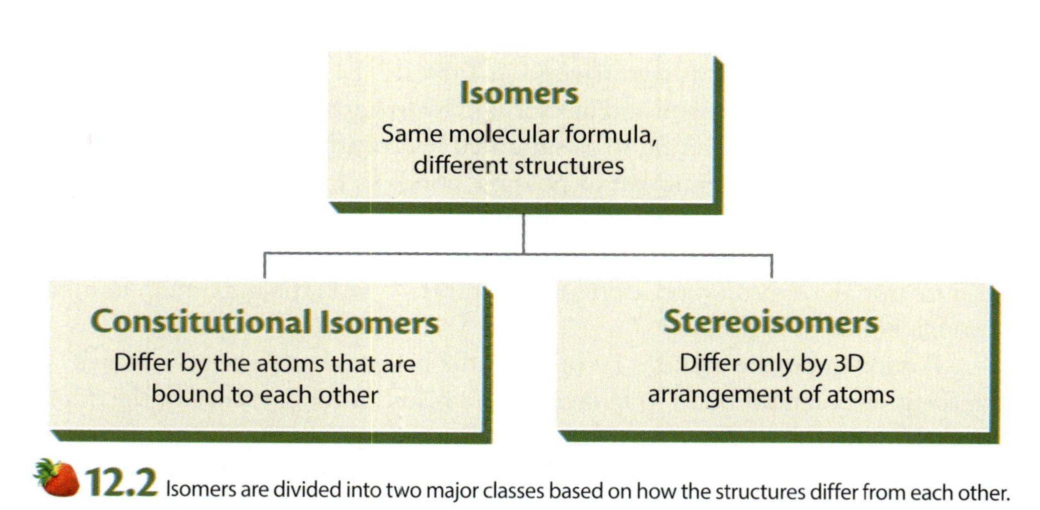

12.2 Isomers are divided into two major classes based on how the structures differ from each other.

Molecules A and B are an example of **constitutional isomers**, which differ in the way atoms are bound to each other. Constitutional isomers have different skeletons and are therefore entirely different compounds that exhibit different physical and chemical properties. Structure C at first may appear different from both A and B, but careful examination reveals that all the atoms are bound together in the same bonding pattern as in compound A. Thus, structures A and C are simply two different ways to draw the same molecule. The two representations are considered **stereoisomers.**

Two molecules are stereoisomers if their atoms are bound in an identical pattern but arranged differently in three-dimensional space. Consider the molecule CHBrClF, which is illustrated in Figure 12.4. The only reasonable bonding pattern has the carbon atom singly bound to each of the other atoms. Constitutional isomerism is not possible. However, the four atoms may be arranged around the central carbon atom in two distinct spatial arrangements. The fact that these structures are not identical is evidenced by the inability to superimpose the two. Thus, there are two stereoisomers that match the formula CHBrClF.

Given the diversity of structures that may match any given chemical formula, it is important to have a method for specifying the compound of interest in words as well as illustrations. The International Union of Pure and Applied Chemistry (IUPAC) has defined a series of systematic naming rules that provide a unique name for each unique compound. As a result, constitutional isomers (e.g., Fig. 12.3A and B) will always have different names.

By contrast, drawings that are simply different representations of the same compound (e.g., Fig. 12.3A and C) would receive the same name. Although a detailed discussion of IUPAC nomenclature is not presented here, an introduction to a few key features may be instructive when considering constitutional isomerism in alkanes and related compounds.

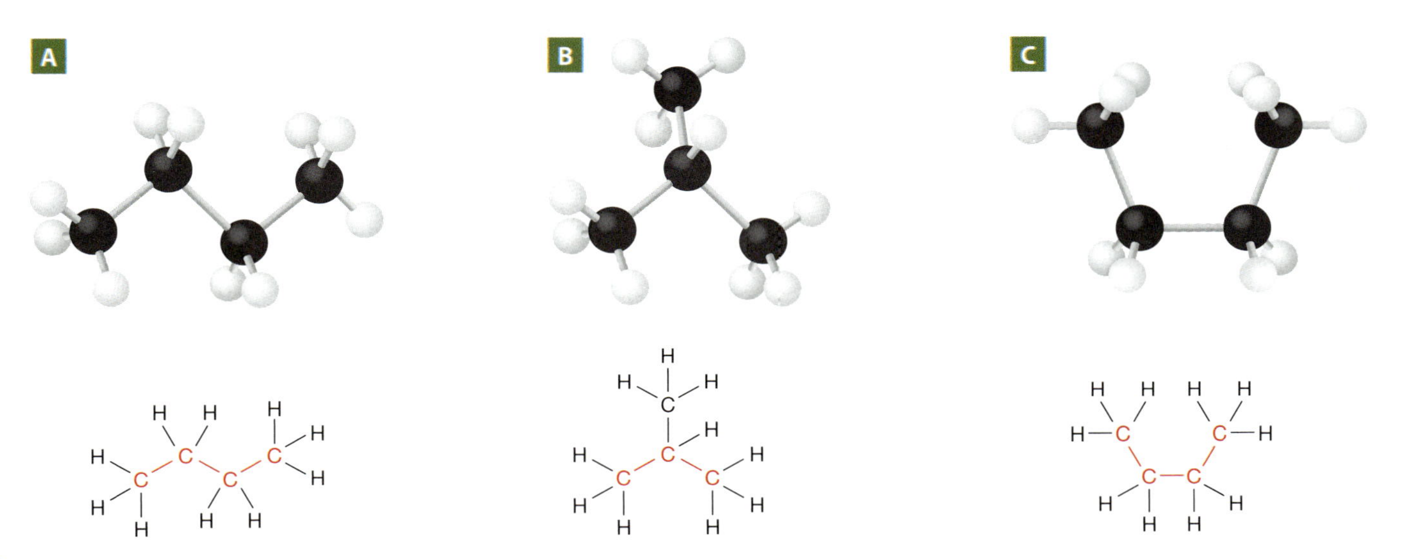

12.3 Constitutional isomerism in compounds with the molecular formula C_4H_{10}. Drawings (**A**) and (**C**) are different representations of the same molecule. Each has four carbons connected linearly (highlighted in red). Compound (**B**) has a different skeleton because only three carbons are connected linearly (red), while the fourth branches from the central carbon atom (black).

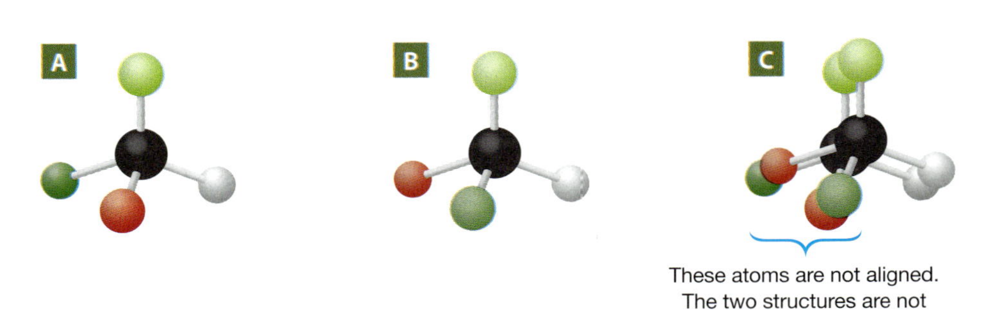

12.4 (**A** and **B**) There are two possible stereoisomers of CHBrClF. The difference in spatial arrangement is demonstrated by attempting to superimpose the two molecules, which is not possible. (**C**) Two atoms are not aligned. The two structures are not superimposable.

All alkanes are named based on the *longest continuous carbon chain* in the molecule. Table 12.2 provides the names of straight-chain alkanes with skeletons of one to ten carbon atoms. Any **substituent**, or group that is not part of the main chain, is denoted using a prefix that indicates its *identity and location* on the main chain. If a substituent is a shorter alkane chain, the prefix identifier is derived from the alkane name with the corresponding number of carbons by removing *–ane* and replacing it with *–yl*. The location of the substituent is indicated by numbering the carbons in the main chain.

TABLE **12.2** ■ Names and Formulas of the Straight-Chain Alkanes with One to Ten Carbons

Carbons in Skeleton	Name	Condensed Formula
1	Methane	CH_4
2	Ethane	CH_3CH_3
3	Propane	$CH_3CH_2CH_3$
4	Butane	$CH_3CH_2CH_2CH_3$
5	Pentane	$CH_3CH_2CH_2CH_2CH_3$
6	Hexane	$CH_3CH_2CH_2CH_2CH_2CH_3$
7	Heptane	$CH_3CH_2CH_2CH_2CH_2CH_2CH_3$
8	Octane	$CH_3CH_2CH_2CH_2CH_2CH_2CH_2CH_3$
9	Nonane	$CH_3CH_2CH_2CH_2CH_2CH_2CH_2CH_2CH_3$
10	Decane	$CH_3CH_2CH_2CH_2CH_2CH_2CH_2CH_2CH_2CH_3$

For example, the name 2-methylpropane indicates that the compound is composed of a three-carbon main chain (propane) with a one-carbon substituent (methyl) attached to the second carbon of the main chain (2-). Note that the location number in the prefix is separated from the rest of the name using a hyphen (Fig. 12.5).

The situation is somewhat more complex if the parent chain is longer and/or possesses more than one substituent. To avoid ambiguity in naming these types of compounds, we must consider three more basic procedures:

≡12

1. Number the carbon skeleton at the end nearer the substituent(s).
2. Explicitly include in the name the carbon number on which each substituent is located.
3. Denote two or more identical substituents by adding a Greek prefix (*di*, *tri*, *tetra*, etc.) before the substituent name.

Figure 12.6 illustrates the basic process for applying these additional procedures to name an alkane.

It can be difficult to envision the shape of a molecule's skeleton from a two-dimensional drawing. Molecular models, such as the one shown for methane (CH_4) in Figure 12.7, are useful tools for exploring the three-dimensional nature of a compound. Manipulating a molecular model will help you become more comfortable with the idea of isomerism and determining whether two molecules are superimposable (i.e., identical).

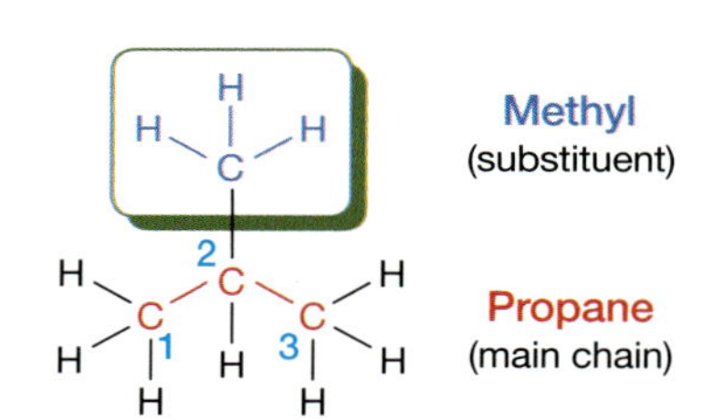

12.5 The IUPAC name 2-methylpropane denotes a three-carbon chain bearing a one-carbon substituent bound to carbon 2.

The three-dimensional representation provided by a model can be translated onto a two-dimensional page using wedge and dash bonds, as illustrated in Figure 12.7. According to this convention, the solid lines lie in the same plane as the paper. The wedge represents a bond coming out of the paper toward the reader, and a dash represents a bond going away from the reader (behind the paper).

Name this alkane:

Step 1: Identify the longest continous chain and its name.

Hints:

1. You should be able to trace this chain by moving in a single direction without lifting a pencil.
2. It does not matter which direction (up & down/right & left) the chain travels in a drawing.
3. When tracing the chain, take the longer path at every junction.

Heptane

Step 2: Number the chain, starting at the end nearest a substituent.

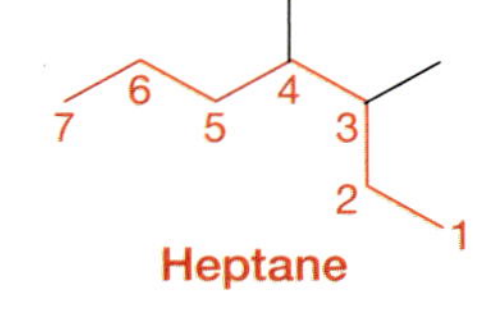

Step 3: Identify substituents by prefix name and location number.

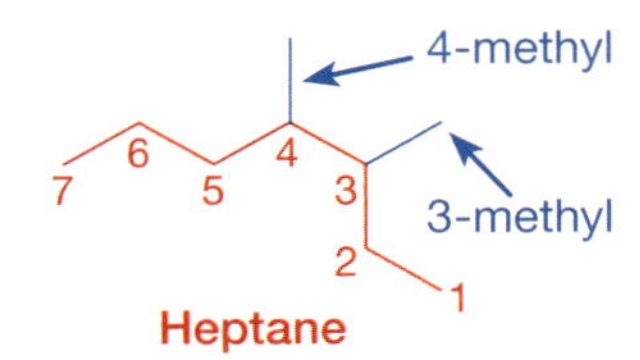

Step 4: Combine the substituent prefixes with the name of the main chain.

3,4-dimethylheptane

In this case, there are two identical substituents. Rather than name them individually, the Greek prefix *di-* indicates the presence of two methyl groups. The numbers 3 and 4 indicate the locations of the individual methyl substituents. Note that numbers are separated by commas. A hyphen always separates numbers and letters. If there are two nonidentical substituents, the prefixes are listed in alphabetical order.

12.6 Basic procedure for naming a substituted alkane.

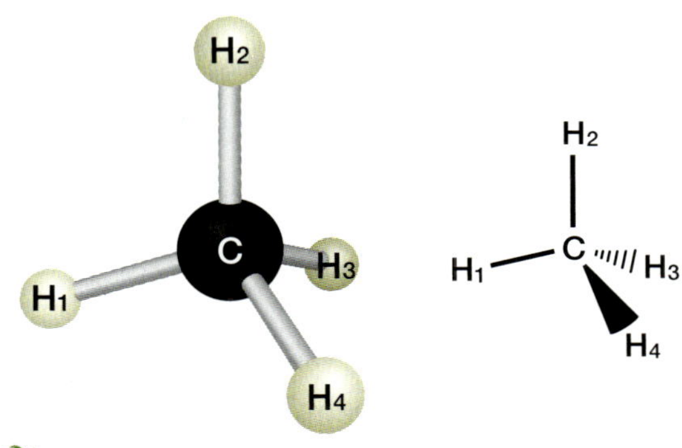

12.7 Molecular model of methane (CH_4) and the translation into a two-dimensional drawing using the wedge-and-dash convention.

12

Hydrocarbon Properties and Reactivity

All hydrocarbons exhibit similar physical properties, which makes it difficult to use these properties to distinguish between the different classes of hydrocarbons. For example, almost all hydrocarbons are nonpolar compounds that are less dense than water. As a result, none of them mix well with water, and they all tend to form a separate layer that floats on the water surface.

Even so, certain distinctive physical properties, such as boiling point, may provide useful information about the *possible* identity of a substance. Care must be taken when drawing conclusions from a boiling point analysis, however, because hydrocarbons of similar size (molar mass) tend to have very similar boiling points. Table 12.3, which lists the boiling points for selected hydrocarbons, illustrates this point. A hydrocarbon sample with a measured boiling point of 82°C, might be cyclohexane (an alkane), cyclohexene (an alkene), or benzene (an aromatic), all of which have a boiling point similar to the sample's measured value.

TABLE **12.3** ■ Possible Identities of the Mystery Sample Organized by Hydrocarbon Class and Boiling Point

Alkanes	Boiling Point	Alkenes	Boiling Point	Aromatics	Boiling Point
Pentane	36°C	2-pentene	36°C	—	—
Hexane	69°C	1-hexene	63°C	—	—
Cyclohexane	81°C	Cyclohexene	83°C	Benzene	80°C
Heptane	98°C	1-heptene	94°C	—	—
2-methylheptane	118°C	Cycloheptene	115°C	Toluene	111°C
Ethylcyclohexane	132°C	4-vinylcyclohexene	130°C	Ethylbenzene	136°C

≡12 Chemical properties can provide more definitive information about the structural features of a particular hydrocarbon because a hydrocarbon's chemical reactivity depends on the types of bonds present in the compound. The discussion that follows explains the structural information that can be obtained by subjecting a hydrocarbon sample to a series of chemical tests.

Unsaturated hydrocarbons, such as alkenes, react quickly with bromine (Br_2) to produce addition products, in which the two reactants combine to form a single product (Fig. 12.8A). The bromine reagent is a reddish color, but the addition product is colorless. Thus, the reaction can be monitored by observing the loss of color as bromine is consumed during the reaction.

Alkanes and aromatic hydrocarbons do not immediately react with bromine in this manner (Fig. 12.8B–C). The chemical reaction therefore provides a method of distinguishing between alkenes and the other two classes of hydrocarbons. If the red color persists after exposing a hydrocarbon to Br_2, then it cannot be an alkene (Fig. 12.8D).

To avoid handling a hazardous solution of elemental bromine, you will test your hydrocarbon samples using a safer combination of reagents that generates Br_2 in situ (i.e., within the reaction mixture). The hydrocarbon sample will be treated with aqueous sodium bromide (NaBr), bleach, and nitric acid (HNO_3). This mixture forms two layers because the nonpolar hydrocarbons are immiscible with the aqueous solution.

As nonpolar Br_2 is formed in the aqueous solution, it migrates into the less dense, nonpolar hydrocarbon layer. If the hydrocarbon is an alkene, it will immediately react with the Br_2 and the upper layer will remain colorless. If the hydrocarbon is not an alkene, then no reaction occurs and the upper layer will become red-orange in color due to the presence of Br_2 (Fig. 12.8D).

A second chemical test also distinguishes between alkenes and other classes of hydrocarbons. Unsaturated hydrocarbons are oxidized by potassium permanganate ($KMnO_4$) to produce a type of compound called a diol. As the reaction occurs, the purple color of $KMnO_4$ in solution changes into the brown color of the MnO_2 precipitate produced as a side product of the reaction (Fig. 12.9). Alkanes and aromatic compounds do not undergo reaction with $KMnO_4$, and the purple color of the $KMnO_4$ persists when these hydrocarbons are mixed with the reagent.

Reaction with both bromine and potassium permanganate provides strong evidence that a hydrocarbon belongs to the alkene class. If a hydrocarbon does not react with either of these reagents, it may be either an alkane or an aromatic. A third chemical reaction is needed to distinguish between these two classes of hydrocarbons.

One such test exploits the reactivity of aromatic compounds, which form vibrant colored cations upon exposure to a mixture of $AlCl_3$ and chloroform ($CHCl_3$). The color generated depends upon the structure of the aromatic compound and ranges from orange-red to purple (Fig. 12.10). Neither alkanes nor alkenes react in this manner, and no color change is observed upon their treatment with $CHCl_3$ and $AlCl_3$.

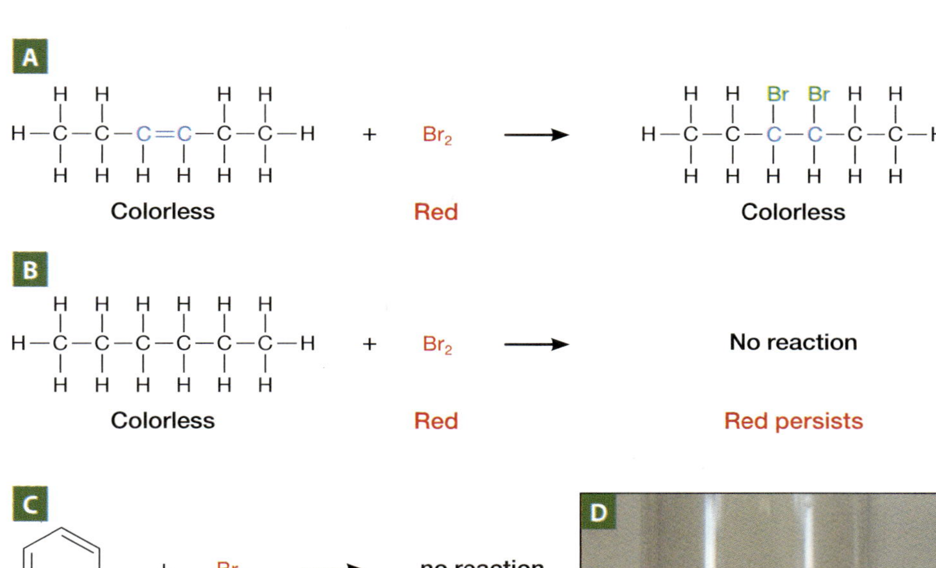

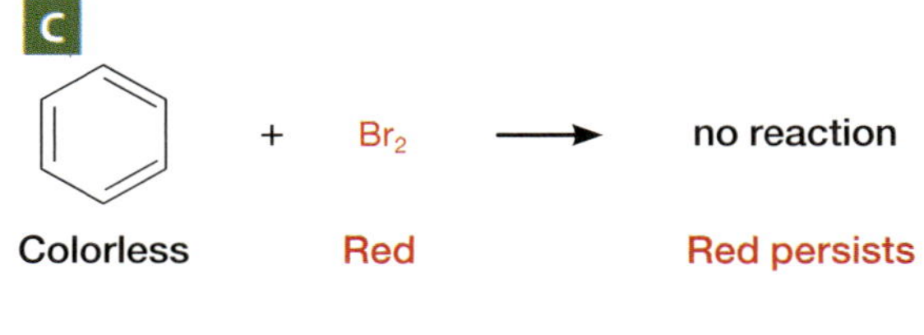

12.8 Reaction of Br_2 with (**A**) an alkene; (**B**) an alkane; and (**C**) an aromatic hydrocarbon. In the alkene reaction, the red color of Br_2 disappears as it is consumed to produce the addition product. (**D**) There is no reaction with the other hydrocarbon classes, so the bromine is unchanged and the red color persists in the reaction mixture.

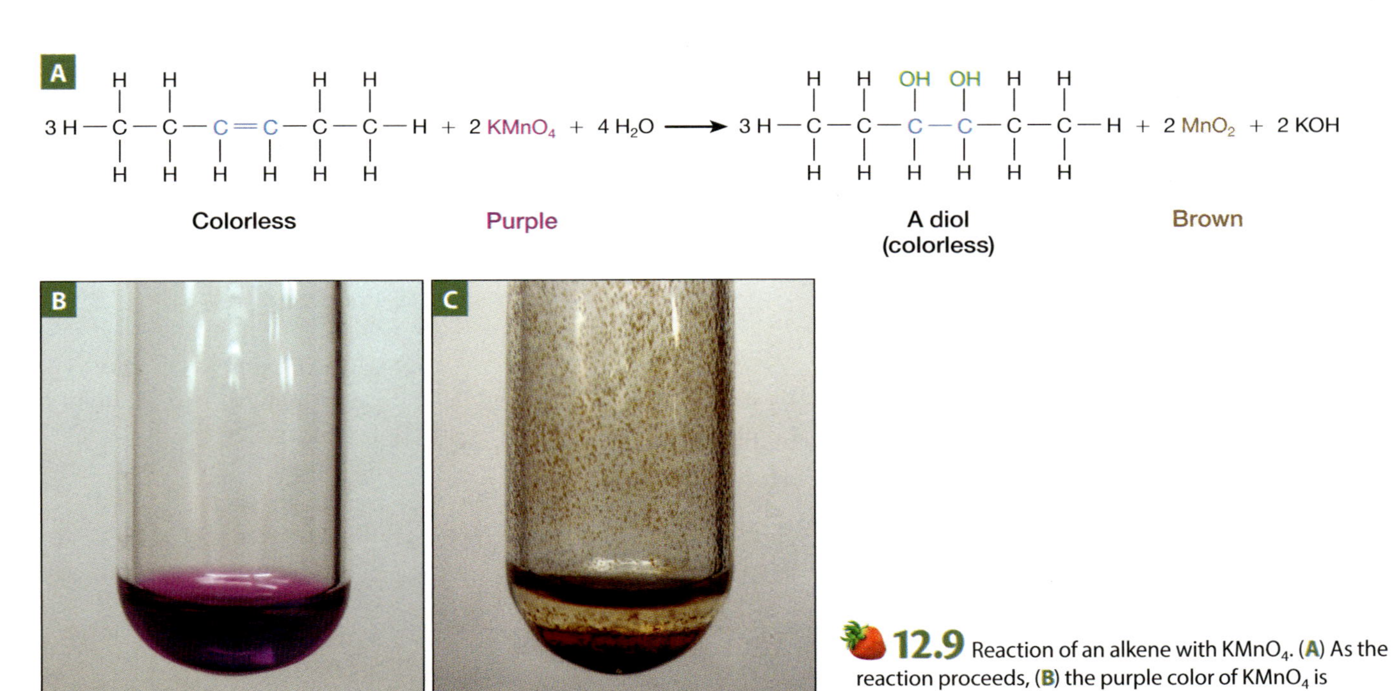

12.9 Reaction of an alkene with $KMnO_4$. (**A**) As the reaction proceeds, (**B**) the purple color of $KMnO_4$ is (**C**) replaced by the brown color of MnO_2.

$$C_6H_6 + AlCl_3 + CHCl_3 \longrightarrow (C_6H_5)_3C^+ + AlCl_4^- + \text{Other products}$$

Vibrant color (varies)

12.10 Reaction of an aromatic with $AlCl_3$ and $CHCl_3$ produces a vibrant colored species and a complex mixture of other products.

In Part A of this lab, you will use molecular models to explore certain types of isomerism found in some hydrocarbons, and you will practice naming some simple, branched alkanes.

In Part B of this lab, you will practice taking a boiling point measurement on a sample of hexane and then measure the boiling point of a mystery hydrocarbon sample. The mystery sample is one of those listed in Table 12.3, and the boiling point measurement should allow you to narrow the possible identities to two or three compounds.

In Part C of this lab, you will explore the chemical reactivity of alkanes, alkenes, and aromatic compounds. You will then use the observed reactivity of your mystery sample to provide a definitive identification of the sample as one of the compounds listed in Table 12.3.

Procedure

What Am I Doing?

Part A: Investigating the concept of isomerism by manipulating molecular models and practicing IUPAC naming of simple alkanes.

Part B: Measuring the boiling point of hexane and a mystery hydrocarbon sample; identifying possible hydrocarbon identities based on the boiling point data.

Part C: Reacting the mystery hydrocarbon with three different reagent combinations to determine the hydrocarbon class to which it belongs.

Part D: Identifying your mystery hydrocarbon.

Materials

- ❑ Hot plate
- ❑ Sand
- ❑ 250 mL beaker
- ❑ Large test tube (25 × 200 mm)
- ❑ Test tubes (12)
- ❑ Boiling chips
- ❑ Aluminum foil
- ❑ Thermometer
- ❑ Ring stand
- ❑ Clamps (2)
- ❑ Microspatula
- ❑ Rubber stopper or cork (4)
- ❑ Molecular model kit

Part A Molecular Models

Constitutional Isomers: C_5H_{12}

1. Make a model of a pentane. Refer to Table 12.2 (p. 164), if necessary.
2. On the data sheet, page 179, draw a Lewis structure that matches your model.
3. There are two additional compounds with the molecular formula C_5H_{12}. Use your model to help you identify these isomers, and draw a Lewis structure of each on the data sheet.
4. Recall that the IUPAC names for a three- or four-carbon chain are propane and butane, respectively (Table 12.2). Also recall that a one-carbon substituent is called a methyl group. Using this information and the procedure in Figure 12.6 (p. 165) as a model, provide a name for each of the C_5H_{12} isomers you constructed.

Constitutional Isomers: C_4H_8

1. Make a model of butane (C_4H_{10}). Refer to Table 12.2 if necessary.
2. Remove two hydrogen atoms from your model such that the formula is C_4H_8. Find a way to alter the structure such that the model represents a legitimate compound with the appropriate number of electrons and a filled valence shell for each atom. Draw a Lewis structure of the new molecule on the data sheet, page 180.
3. Use your model to help you identify four more isomers that have the formula C_4H_8, and draw a Lewis structure of each on the data sheet.

Stereoisomers: Single vs. Double Bonds

1. Construct a molecular model that matches illustration A in Figure 12.11. Note that in the ball-and-stick model the black, white, and green balls represent carbon, hydrogen, and chlorine atoms, respectively.
2. Try to convert your model of illustration A into illustration B (Fig. 12.11), breaking bonds *only* if necessary. On the data sheet, page 181, describe how you completed the transformation. Include in your description whether you could do so without dismantling any bonds in your model.
3. Construct a molecular model that matches illustration C in Figure 12.11.
4. Try to convert your model of illustration C into illustration D (Fig. 12.11), breaking bonds *only* if necessary. On the data sheet, describe how you completed the transformation. Include in your description whether you could do so without dismantling any bonds in your model.
5. Try to convert your model of illustration D into illustration E (Fig. 12.11), breaking bonds *only* if necessary. On the data sheet, describe how you completed the transformation. Include in your description whether you could do so without dismantling any bonds in your model.
6. Answer Questions 1–4 on the data sheet.

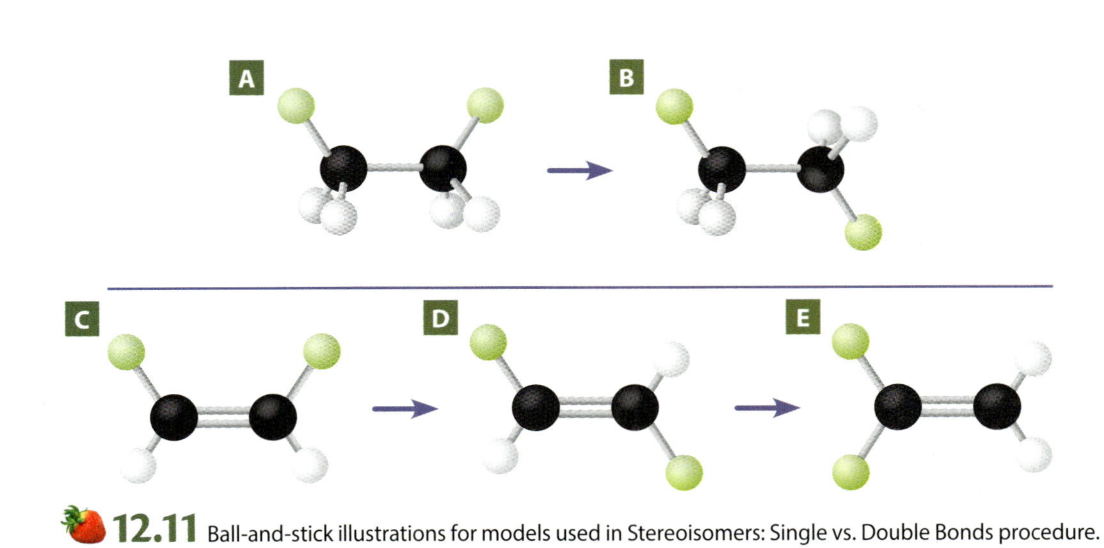

12.11 Ball-and-stick illustrations for models used in Stereoisomers: Single vs. Double Bonds procedure.

Stereoisomers: Ring Structures

1. Make two molecular models of cyclobutane, illustrated in Figure 12.12. Designate them Model A and Model B. Use both models to complete Steps 2–7.
2. Orient the models as shown in Figure 12.12. On the data sheet, page 182, draw accurate three-dimensional representations of the models using lines, wedges, and dashes for bonds. On each model, designate two adjacent carbons as C1 and C2. Label C1 and C2 in your drawings.
3. Replace one of the hydrogen atoms on C1 in your models with a chlorine atom (green).
4. Orient the two models as illustrated in Figure 12.12 (no chlorine atom is illustrated). Rotate Model B as necessary to determine whether it is possible to superimpose the two models. On the data sheet, record your answer and draw accurate three-dimensional representations of the models using lines, wedges, and dashes for bonds. Label C1 and C2 in your drawings.
5. On the C2 atom of each model, designate the hydrogen atom illustrated with a wedge as H1 and the hydrogen atom illustrated with a dash as H2. Label H1 and H2 in the drawings you made for Step 4.
6. In Model A, replace H1 with a chlorine atom. In Model B, replace H2 with a chlorine atom.
7. Orient the two models as illustrated in Figure 12.12 (no chlorine atoms are illustrated). Rotate Model B as necessary to determine whether it is possible to superimpose the two models. On the data sheet, record your answer and draw accurate three-dimensional representations of the models using lines, wedges, and dashes for bonds. In your drawings, label C1, C2, H1, and H2 (where present).
8. Answer Questions 1–3 on the data sheet.

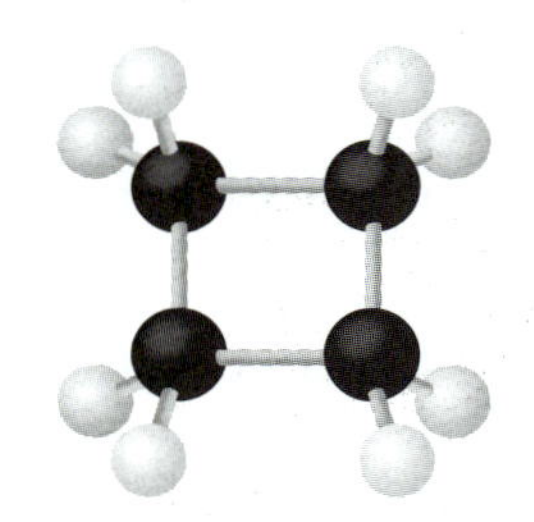

12.12 Ball-and-stick model of cyclobutane.

≡12

Part B Boiling Point Determination

SAFETY NOTE

Hydrocarbons are flammable and volatile. Prolonged exposure to their vapors may be harmful. In addition, many hydrocarbons have an unpleasant odor. Handle all hydrocarbon samples and perform all experiments in a fume hood or appropriately ventilated space.

1. Begin assembling the apparatus for boiling point determination (Fig. 12.13) by heating about 1 inch of sand in a 250 mL beaker to approximately 100°C on a hot plate.
2. Determine the boiling point of hexane using the following steps:
   - **a** Add a boiling chip and about 2 mL of hexane to a large test tube (25 × 200 mm).
   - **b** Cover the tube with aluminum foil and use a sharp implement, such as the narrow end of a microspatula, to cut a slit in the center of the covering.
   - **c** Clamp the test tube to a ring stand above the sand bath.
   - **d** Separately clamp a thermometer above the tube.
   - **e** When the sand bath is heated, lower the test tube into the bath so that sand is even with the surface of the liquid. If vigorous boiling starts immediately, raise the tube so that the bottom just touches the surface of the sand.
   - **f** Lower the thermometer through the slit in the tube cover until the bottom of the bulb is about 1 cm above the liquid surface.

g The liquid should soon begin to boil and you should see evidence of the refluxing vapors. When the thermometer reading stops rising and is stable for at least 1 minute, record the boiling point on the data sheet, page 182.

h Raise the thermometer and remove the test tube from the sand bath. Allow the liquid to cool, and then dispose of it as directed by your instructor.

3 Obtain a mystery hydrocarbon and record its identification code on the data sheet.

4 Determine the boiling point of your mystery hydrocarbon using the following steps:

a Add a boiling chip and about 2 mL of your mystery hydrocarbon to a dry test tube.

b Cover the tube with aluminum foil and use a sharp implement, such as the narrow end of a microspatula, to cut a slit in the center of the covering.

c Clamp the test tube to a ring stand above the sand bath.

d Separately clamp a thermometer above the tube.

e Test the boiling point of the hydrocarbon by lowering the test tube so that the bottom of the tube rests on top of the sand.

- If the liquid boils quickly, it has a relatively low boiling point. If the boiling is not overly vigorous, you may proceed with the analysis by following Steps g–i. If boiling on top of the sand bath is too vigorous, raise the tube, reduce the heat, and allow the sand bath to cool to about 70°C before proceeding with Steps f–i.
- If the liquid does not boil on top of the sand, it has a relatively high boiling point. Proceed with the analysis by following Steps f–i.

f Lower the test tube into the bath so that sand is even with the surface of the liquid.

g Lower the thermometer into position with the bulb about 1 cm above the liquid surface.

h The liquid should soon begin to boil and you should see evidence of the refluxing vapors. When the thermometer reading stops rising and is stable for at least 1 minute, record the boiling point on the data sheet. If the liquid does not begin to boil after a few minutes, you may need to increase the heat until the sand bath reaches a temperature of approximately 140°C.

i Raise the thermometer and remove the test tube from the sand bath. Allow the liquid to cool, and then dispose of it as directed by your instructor.

5 Use the boiling point measurement and Table 12.3 to determine possible identities of your mystery hydrocarbon.

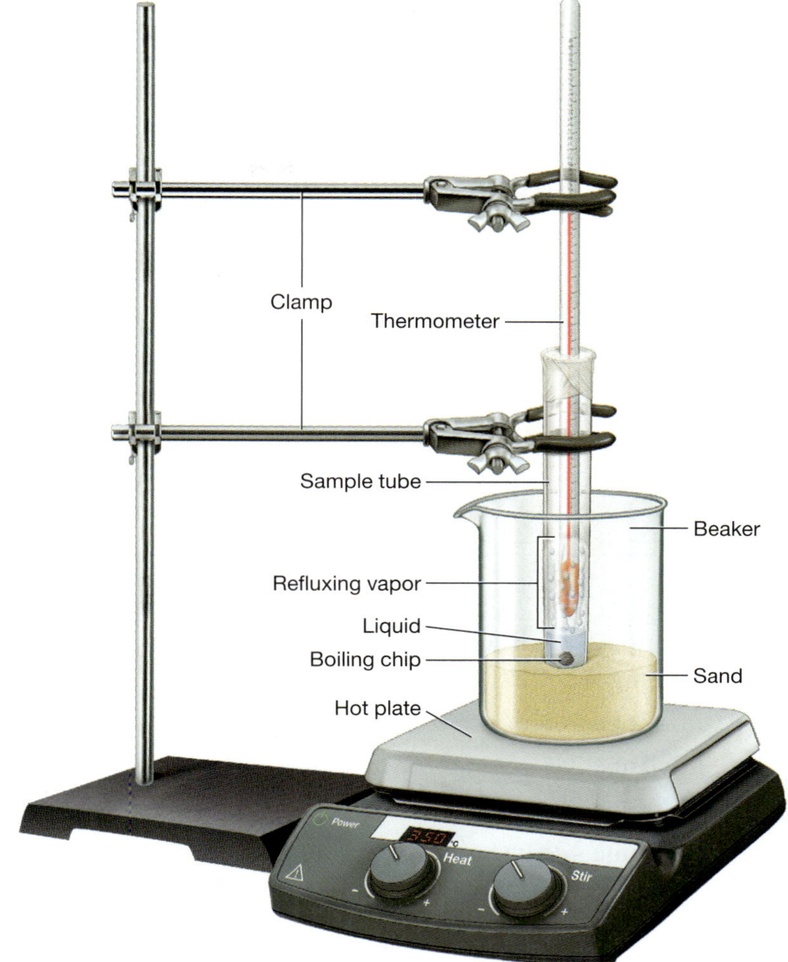

12.13 Apparatus for small scale determination of boiling point.

12

Part C Chemical Reactions of Hydrocarbons

Reaction with Bromine

1. Label four test tubes 1–4.

SAFETY NOTE

Perform all reactions in a fume hood or appropriately ventilated space! Stopper all test tubes when you are not actively using them!

2. To each test tube, add 10 drops of the appropriate hydrocarbon sample as indicated below. Stopper each test tube when not adding reagents.
   - Test Tube 1: Hexane
   - Test Tube 2: Cyclohexene
   - Test Tube 3: Toluene
   - Test Tube 4: Mystery hydrocarbon
3. To each test tube, add 10 drops of 0.1 M NaBr, 5 drops of bleach, and 1 drop of 6 M nitric acid.
4. Gently agitate each test tube, and then record the color of the upper layer on the data sheet, page 183.
5. Dispose of the test contents as directed by your instructor.

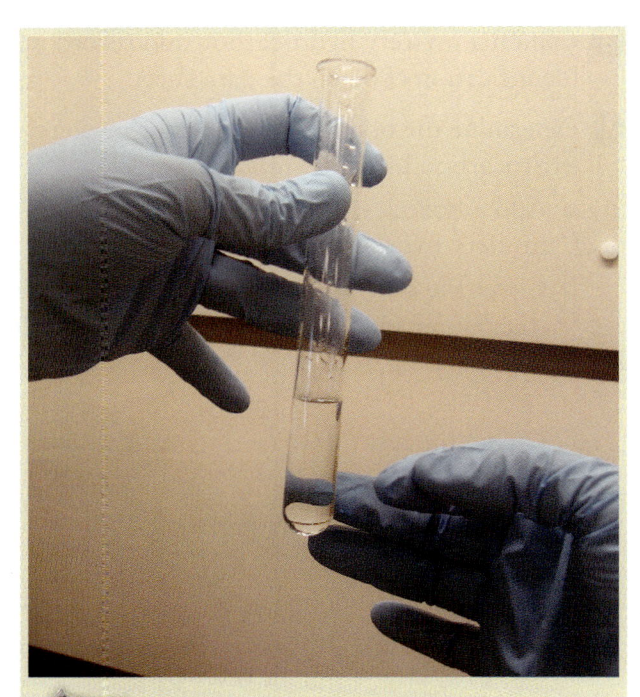

Technique Tip 12.1

To safely shake a test tube, hold it between your thumb and split first two fingers. Flick the test tube using a finger from the opposite hand.

Reaction with Potassium Permanganate ($KMnO_4$)

1. Label four test tubes 1–4.
2. To each test tube, add 10 drops of acetone and 2 drops of the appropriate hydrocarbon sample as indicated below. Stopper each test tube when not adding reagents.
   - Test Tube 1: Hexane
   - Test Tube 2: Cyclohexene
   - Test Tube 3: Toluene
   - Test Tube 4: Mystery hydrocarbon
3. Add 2 drops of 1% $KMnO_4$ to each test tube. Stopper each tube.
4. Agitate each tube carefully for 10 seconds and then allow them to stand for 2 minutes.

5. Record the color of each solution on the data sheet, page 183.
6. Dispose of the test tube contents as directed by your instructor.

Reaction with Chloroform and Aluminum Chloride

1. Label four **dry** test tubes 1–4. **Note that the test tubes must be completely dry for the reaction to function properly.**
2. To each test tube, add 5 drops of the appropriate hydrocarbon sample as indicated below. Stopper each test tube when not adding reagents.
   - Test Tube 1: Hexane
   - Test Tube 2: Cyclohexene
   - Test Tube 3: Toluene
   - Test Tube 4: Mystery hydrocarbon
3. Add 5 drops of chloroform ($CHCl_3$) to each test tube.
4. Use a microspatula to add a few crystals of anhydrous aluminum chloride ($AlCl_3$) to the reaction mixture.
5. Agitate each tube gently.
6. Wait 3 to 4 minutes, and then record the color on the solids in each test tube on the data sheet, page 183.
7. Dispose of the test tube contents as directed by your instructor.
8. Use the data from all three chemical reactions in Part B to determine the class of hydrocarbons to which your mystery sample belongs and record this conclusion in the table on the data sheet.

Part D Identification of the Mystery Hydrocarbon

The mystery hydrocarbon is one of those listed in Table 12.3, page 166. Using your results from Parts B and C, propose an identity for your mystery hydrocarbon sample, and record it on the data sheet, page 183. Check with your instructor to ensure that none of your tests need to be repeated.

Name ______________________________

Lab Partner ______________________________

Lab Section ____________________ Date ______________

Lab 12
Pre-Laboratory Exercise

1 Provide a term that matches each description below.

- **a** Compounds composed only of carbon and hydrogen. ____________________
- **b** Compounds that share a molecular formula but have different structures. ____________________
- **c** Class of hydrocarbons that contain only single bonds. ____________________
- **d** Isomers that differ only in the spatial arrangement of atoms. ____________________
- **e** Hydrocarbons that contain double or triple bonds. ____________________
- **f** Compounds with the same molecular formula that differ in the way the atoms are connected. ____________________
- **g** Class of hydrocarbons that may incorporate one or more benzene rings in their structure. ____________________

2 Circle the longest continuous carbon chain in each skeletal structure below.

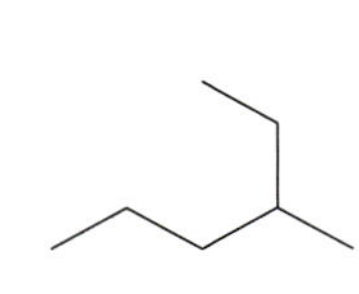

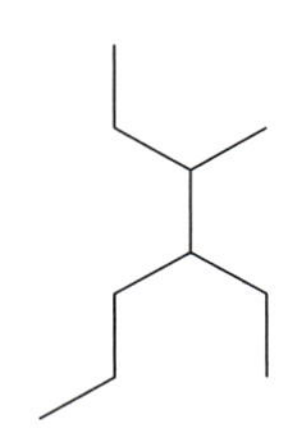

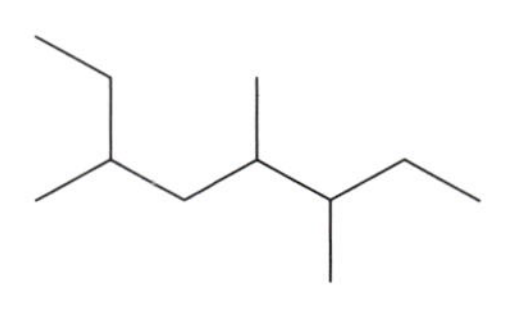

3 Draw a skeletal structure that corresponds to each IUPAC name below.

Name	Structure
3-methylpentane	
2,4-dimethylhexane	
3-ethyl-4-methyloctane	

4 The procedure used to test the reaction between your mystery hydrocarbon and bromine involves the production of two liquid layers in the test tube. In which layer should you judge the color? Explain why this is the correct layer to observe.

5 Describe the visual results you would expect from each of the following experiments.

a An alkene is treated with a mixture of $NaBr/NaOCl/HNO_3$.

b An alkane is treated with $KMnO_4$.

c An aromatic compound is treated with a mixture of $CHCl_3/AlCl_3$.

d An alkane is treated with $NaBr/NaOCl/HNO_3$.

e An alkene is treated with a mixture of $CHCl_3/AlCl_3$.

Name

Lab Partner

Lab Section   Date

Lab 12
Pre-Laboratory Exercise
(continued)

6 Why do you think the reaction of an alkene with Br_2 is considered an example of an *addition* reaction?

7 A student measures the boiling point of a mystery hydrocarbon and finds that it is 111°C, which matches the reported boiling point of toluene exactly. The student reports that the identity of the sample is toluene without performing any further tests. Explain why this student's conclusion is unreliable.

Name ______________________________

Lab Partner ______________________________

Lab Section ____________________ Date ____________

Lab 12 DATA SHEET

Data and Observations

Part A Molecular Models

Constitutional Isomers: C_5H_{12}

Isomers of C_5H_{12}	Name

Constitutional Isomers: C_4H_8

Isomers of C_4H_8

Name

Lab Partner

Lab Section Date

Lab 12 DATA SHEET

(continued)

Stereoisomers: Single vs. Double Bonds

Procedure Reference	Transformation Modeled	Description of Transformation
Step 2	A ⟶ B	
Step 4	C ⟶ D	
Step 5	D ⟶ E	

1 There are many different kinds of stereoisomers. Some are interconvertible without breaking bonds while others are not. If the stereoisomers are interconvertible without breaking bonds, then they simply represent different shapes adopted by the same molecule. That is, the two stereoisomers are actually identical molecules.

a Based on your data, are Molecules A and B identical? Explain your answer.

b Based on your data, are Molecules C and D identical? Explain your answer.

c What structural feature accounts for the difference in your answers to 1a and 1b? What difference did this structural feature impart on your models?

2 Are Compounds A and C isomers? If so, are they constitutional isomers or stereoisomers? Explain your answer.

3 Are Compounds C and D isomers? If so, are they constitutional isomers or stereoisomers? Explain your answer.

4 Are Compounds C and E isomers? If so, are they constitutional isomers or stereoisomers? Explain your answer.

Stereoisomers: Ring Structures

Procedure Reference	Model A	Model B	Superimposable? (Yes/No)
Step 2			
Steps 4–5			
Step 7			

1 Based on your data, were the molecules illustrated in Step 4 identical? If not, how would you classify their relationship? Explain your answer.

2 Based on your data, were the molecules illustrated in Step 7 identical? If not, how would you classify their relationship? Explain your answer.

3 What structural feature accounts for the difference in your answers to Questions 1 and 2? What was the influence of this structural feature on your models? Explain your answer.

Part B Boiling Point Determination

Procedure Reference	Sample	Boiling Point
Step 2	Hexane	
Step 4	Mystery hydrocarbon	

Mystery hydrocarbon identification code: __________

Possible identities: __________

Name ______________________________

Lab Partner ______________________________

Lab Section ____________________ Date ____________________

Lab 12 DATA SHEET

(continued)

Part C Chemical Reactions of Hydrocarbons

Hydrocarbon Chemical Reactions

Test Tube	Hydrocarbon	Reaction with Br_2 ($NaBr$ + bleach + HNO_3)	Reaction with $KMnO_4$	Reaction with $CHCl_3 + AlCl_3$	Hydrocarbon Class
1	Hexane				Alkane
2	Cyclohexene				Alkene
3	Toluene				Aromatic
4	Mystery hydrocarbon				

Part D Identification of the Mystery Hydrocarbon

Proposed identity of your mystery hydrocarbon: ______________________________

Reflective Exercises

1 There are four constitutional isomers that share the molecular formula $C_3H_6Cl_2$. Draw the skeletal structure of each isomer. The prefix for a chlorine substituent is *chloro-*. Provide the IUPAC name of each compound you draw.

Structure	Name

2 What structural feature(s) are possible when two hydrogen atoms are removed from the molecular formula of an alkane? For example, what kinds of bonds are possible for the molecular formula C_4H_8 that are not possible for the molecular formula C_4H_{10}?

Name ______________________

Lab Partner ______________________

Lab Section ______________ Date ______________

Lab 12
DATA SHEET
(continued)

3 For each problem below, choose the best description of the relationship between the molecules represented by the pair of illustrations: constitutional isomers, stereoisomers, or identical molecules.

a ______________ and

b 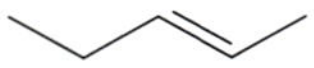 ______________ and

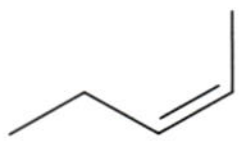

c 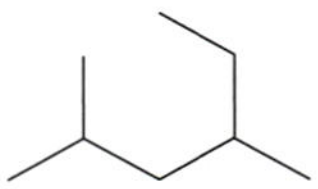 ______________ and

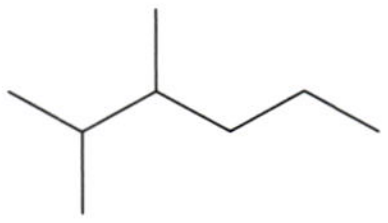

d  ______________ and

e 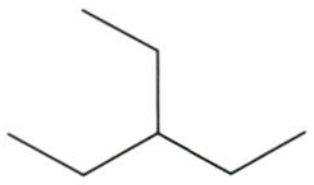 ______________ and

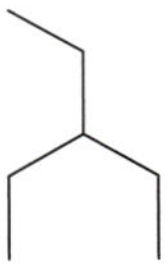

f 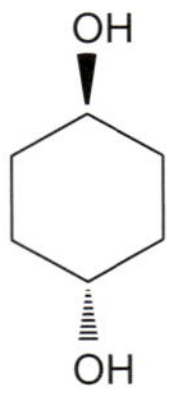 ______________ OH OH and OH OH

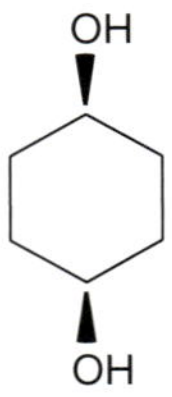

4 In the table below, present a claim regarding the identity of your mystery hydrocarbon. Describe your evidence and provide a rationale to justify your claim.

Claim	
Evidence	**Rationale**

5 Suppose that a mystery hydrocarbon sample is one of the compounds shown here. Using the data presented in the chart, circle the compound that is the best choice for the identity of the mystery substance. Justify your answer by filling in the boxes provided with the structure of the organic compound produced when each test is performed. Write NR if no reaction takes place.

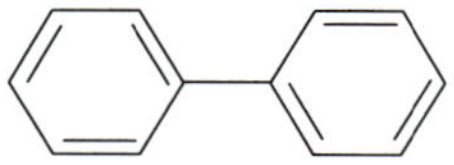

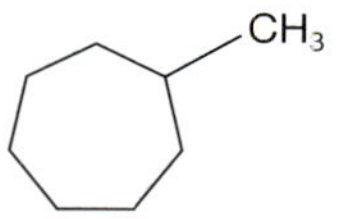

CH_3

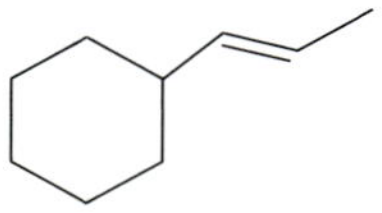

Test	Observations
Reaction with $NaBr/NaOCl/HNO_3$	No color observed in the upper layer
Reaction with $KMnO_4$	Color changed from purple to brown

Mystery Compound + Br_2 $\longrightarrow$

Mystery Compound + $KMnO_4$ $\longrightarrow$

Name ______________________

Lab Partner ______________________

Lab Section ______________ Date ______________

Lab 12 DATA SHEET

(continued)

6 Fill in the data table below with the visual results you would expect for naphthalene in each of the indicated tests.

Naphthalene

Test	Observations
Reaction with $NaBr/NaOCl/HNO_3$	
Reaction with $KMnO_4$	
Reaction with $CHCl_3/AlCl_3$	

7 Consider the structures of Compounds A–C, and answer the following questions.

A B C

a Is it possible to distinguish between Compounds A and B using any of the chemical tests described in this lab? Explain.

____

____

____

b Is it possible to distinguish between Compounds B and C using any of the chemical tests described in this lab? Explain.

____

____

____

8 Suppose a mystery hydrocarbon sample has a boiling point of 36°C.

a Based on the information available in this lab, what are two possible identities of the substance?

b How could you distinguish between these two possibilities? Describe how the results of any tests would allow you to choose the most likely identity.

Amaranth starch is a polysaccharide.

Lab 13

How Sweet It Is
Structures of Carbohydrates

Carbohydrate Structures

The term **carbohydrate** is derived from the outdated belief that sugars, starches, and related molecules are hydrates of carbon with the general formula $C_n(H_2O)_n$. Carbohydrates are now more precisely defined as polyhydroxylated aldehydes, polyhydroxylated ketones, or substances that produce these molecules upon hydrolysis (Fig. 13.1). The **monosaccharides** ("single sugars") D-glucose and D-fructose, illustrated in Figure 13.2, are examples of a polyhydroxylated aldehyde and a polyhydroxylated ketone, respectively.

Note that carbon 1 in D-glucose is an aldehyde functional group while carbons 2–6 all contain hydroxyl (or alcohol) functional groups. In D-fructose, carbon 2 is a ketone, while the remaining carbons are all hydroxylated. D-glucose and other aldehyde-containing sugars are generically referred to as **aldoses** while ketone-bearing sugars like D-fructose are called **ketoses**. As you may know, D-glucose is the primary energy source for human cells and is also known as *blood sugar* because human blood contains dissolved glucose.

Monosaccharides generally exist in equilibrium between the open chain form (e.g., Fig. 13.2) and a cyclic constitutional isomer formed by an intramolecular ("within the molecule") reaction. Various isomeric ring

Alcohol (hydroxyl) | Ketone | Aldehyde

R = carbon skeleton; R' = carbon skeleton or H

13.1 Generic structures of the alcohol, ketone, and aldehyde functional group. The –OH substituent of an alcohol is referred to as a "hydroxyl" group.

D-glucose (an aldose) | D-fructose (a ketose)

13.2 Structures of D-glucose (an aldose) and D-fructose (a ketose).

Objectives

After completing this lab you should be able to:

- Define the terms *monosaccharide*, *disaccharide*, and *polysaccharide*.
- Identify reducing and nonreducing sugars from chemical structures.
- Interpret the results of classic chemical tests for carbohydrates, including Benedict's test, Barfoed's test, Seliwanoff's test, and the iodine test.
- Identify the carbohydrate in a mystery sample using the results of classic chemical tests.
- Predict the products of condensation and hydrolysis reactions involving monosaccharides, disaccharides, and polysaccharides.

structures are possible, but the forms shown in Figure 13.3 dominate in the biological environment. Note that aldoses tend to form a six-membered ring called a **pyranose,** and ketoses tend to form a five-membered ring called a **furanose.** The original aldehyde or ketone carbon atom (1 for aldoses and 2 for ketoses) is the site of ring formation and is referred to as the **anomeric carbon.**

Disaccharides are formed by the condensation of two monosaccharide subunits. The new bond between the monomers is called a **glycosidic bond,** which links the anomeric carbon of one sugar with an alcohol group in another. For example, lactose is the disaccharide produced by the condensation of one D-galactose molecule with one D-glucose molecule as shown in Figure 13.4.

Similarly, **polysaccharides** such as starch are formed by the condensation of many monosaccharide subunits. Note that the reverse of a condensation reaction is a hydrolysis reaction, which involves the breaking of a large molecule into smaller subunits. Both disaccharides and polysaccharides qualify as carbohydrates because they may be hydrolyzed to produce monosaccharides. In fact, the first step in the metabolism of these carbohydrates is digestion, which involves their complete hydrolysis to liberate the constituent monosaccharides.

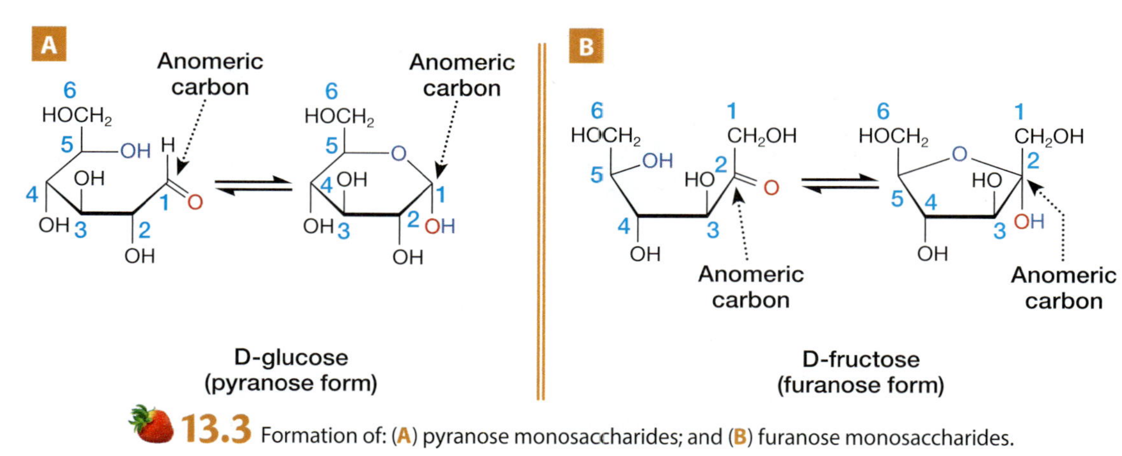

13.3 Formation of: (A) pyranose monosaccharides; and (B) furanose monosaccharides.

Condensation and Hydrolysis of a Disaccharide

D-galactose + D-glucose ⇌ (Condensation / Hydrolysis) Lactose + H_2O

Condensation and Hydrolysis of a Polysaccharide

Chain continues … + D-glucose ⇌ (Hydrolysis / Condensation) Starch + H_2O

13.4 Condensation and hydrolysis reactions of disaccharides and polysaccharides.

Chemical Reactivity Tests

Chemical reactivity tests are a classic way to identify important structural features in a carbohydrate sample. Four common chemical tests for carbohydrates are Benedict's test, Barfoed's test, Seliwanoff's test, and the iodine test. Each of these tests is described in the discussion that follows.

The aldehyde functional group in a carbohydrate is easily oxidized to the corresponding carboxylic acid by aqueous Cu^{2+} ions. An example of this reaction is illustrated in Figure 13.5. During the reaction, the Cu^{2+} ions are reduced to Cu^{+}, and the blue solution, called **Benedict's reagent**, is usually converted to brick red as solid copper(I) oxide (Cu_2O) precipitates. Occasionally, the copper(I) will form other compounds that precipitate and vary in color from yellow to green. As long as a precipitate is formed, the result is considered a "positive" Benedict's test (Fig. 13.6).

Carbohydrates that give a positive Benedict's test are called **reducing sugars** because they are responsible for the reduction of the copper ions and the resulting change in the appearance of the solution. Although monosaccharides like D-glucose exist primarily in the cyclic form that does not possess an aldehyde functional group, the equilibrium with the open chain form allows this reaction to proceed readily.

Under the Benedict's test reaction conditions, even ketoses will give a positive result because the sugar undergoes a rearrangement reaction to generate an intermediate aldehyde, which may then be oxidized. An example is given in Figure 13.7.

Lactose and other disaccharides with hydroxyl-bearing anomeric carbons (often referred to as "free" anomeric carbons) are reducing sugars as well. This structural feature is highlighted in Figure 13.8. Disaccharides that lack this structural component (e.g., sucrose) are not reducing sugars because they are not in equilibrium with an open chain form possessing an aldehyde. As a result, sucrose and related disaccharides give a negative Benedict's test. Polysaccharides also give negative Benedict's tests.

Glucose (pyranose form) ⇌ Glucose (open chain) + Cu^{2+} (aq) $\xrightarrow{H_2O}$ + Cu_2O (s)

The anomeric carbon has been oxidized to a carboxylic acid

Cu^{2+} has been reduced to Cu^{+} in the red preciptate copper(I) oxide

13.5 Example reaction between an aldose and Benedict's reagent.

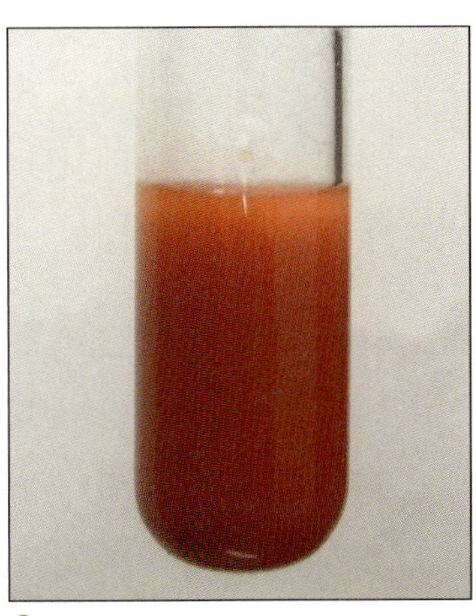

13.6 Visual results for a positive Benedict's test.

D-fructose (open chain) ⇌ (Rearrangement) D-fructose (a ketose) + Cu^{2+} (aq) $\xrightarrow{H_2O}$ + Cu_2O (s)

13.7 Example reaction between a ketose and Benedict's reagent.

13

Free anomeric carbon (reducing sugar)

Lactose

Neither anomeric carbon is free (nonreducing sugar)

Sucrose

13.8 Free anomeric carbons in certain disaccharides make them reducing sugars.

Barfoed's reagent is a modified version of the Benedict's reagent and is used to distinguish between monosaccharide and disaccharide reducing sugars. Like Benedict's test, Barfoed's test relies on the carbohydrate acting as reducing agent toward Cu^{2+}, and a positive test is visualized by the generation of the copper(I) precipitate. However, Barfoed's reagent reacts much more quickly with monosaccharides than disaccharides, allowing the two types of carbohydrates to be distinguished from each other.

Ketoses react very quickly with **Seliwanoff's reagent** to produce a red organic derivative of the original saccharide (Fig. 13.9). By contrast, aldoses react very slowly with the reagent. Thus, ketoses and aldoses are easily distinguished based on differences in the time it takes to form a deep-red solution in the presence of Seliwanoff's reagent.

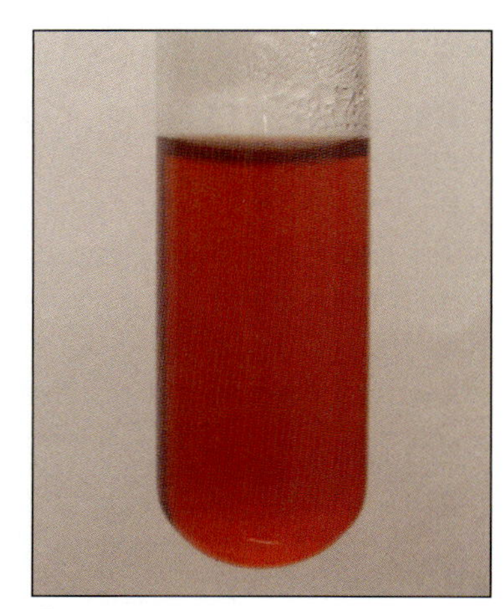

13.9 Positive visual results for the Seliwanoff's test.

Polysaccharides react with a solution of iodine (I_2) to produce complexes that vary in color from blue black to red purple (Fig. 13.10). The large structures of polysaccharides are necessary for the reaction, and simpler carbohydrates do not produce the same dark-colored structures.

In this lab, you will perform the Benedict's, Barfoed's, Seliwanoff's, and iodine tests on various carbohydrates in order to observe the differences in their reactivities. The tests, results, and associated structural features are summarized in Table 13.1. You also will use the chemical tests to deduce the structural features and the identity of a mystery carbohydrate sample. Additionally, you will perform a hydrolysis reaction on a disaccharide (sucrose) and a polysaccharide (starch). You will then confirm the success of the hydrolysis products using relevant chemical tests.

TABLE **13.1** Reagent Test Results

Reagent	Test For	Positive Result	Negative Result
Benedict's	Reducing Sugar	Precipitate (usually red)	No precipitate formed
Barfoed's	Monosaccharide	Precipitate (usually red) formed quickly	Precipitate may form slowly
Seliwanoff's	Ketose	Solution turns red quickly	Solution turns red slowly
Iodine	Polysaccharide	Solution turns dark blue or purple	No color change

13.10 Positive visual results for the iodine test.

Procedure

What Am I Doing?

Materials

- ❑ Test tubes (10)
- ❑ 250 mL beaker
- ❑ Hot plate
- ❑ 10 mL graduated cylinder
- ❑ Blue litmus paper
- ❑ Test-tube holder

Part A: Treating distilled water, five identified carbohydrates, and one mystery sample with Benedict's reagent to determine whether or not each is a reducing sugar.

Part B: Treating distilled water, five identified carbohydrates, and one mystery sample with Barfoed's reagent to determine whether or not each is a monosaccharide.

Part C: Treating distilled water, five identified carbohydrates, and one mystery sample with Seliwanoff's reagent to determine whether or not each is a ketose.

Part D: Treating distilled water, five identified carbohydrates, and one mystery sample with iodine to determine whether or not each is a polysaccharide.

Part E: Determining the identity of your mystery carbohydrate using the data from Parts A–D and the structures in Figure 13.12.

Part F: Hydrolyzing samples of sucrose and starch and performing relevant chemical tests to determine whether the reactions were successful.

In Parts A–D, you will examine the reactivities of the carbohydrates in Figure 13.11 and a mystery sample. Choose one of the available solutions containing a mystery carbohydrate and record the sample number on the data sheet, page 201. Use the same mystery sample for Parts A–D.

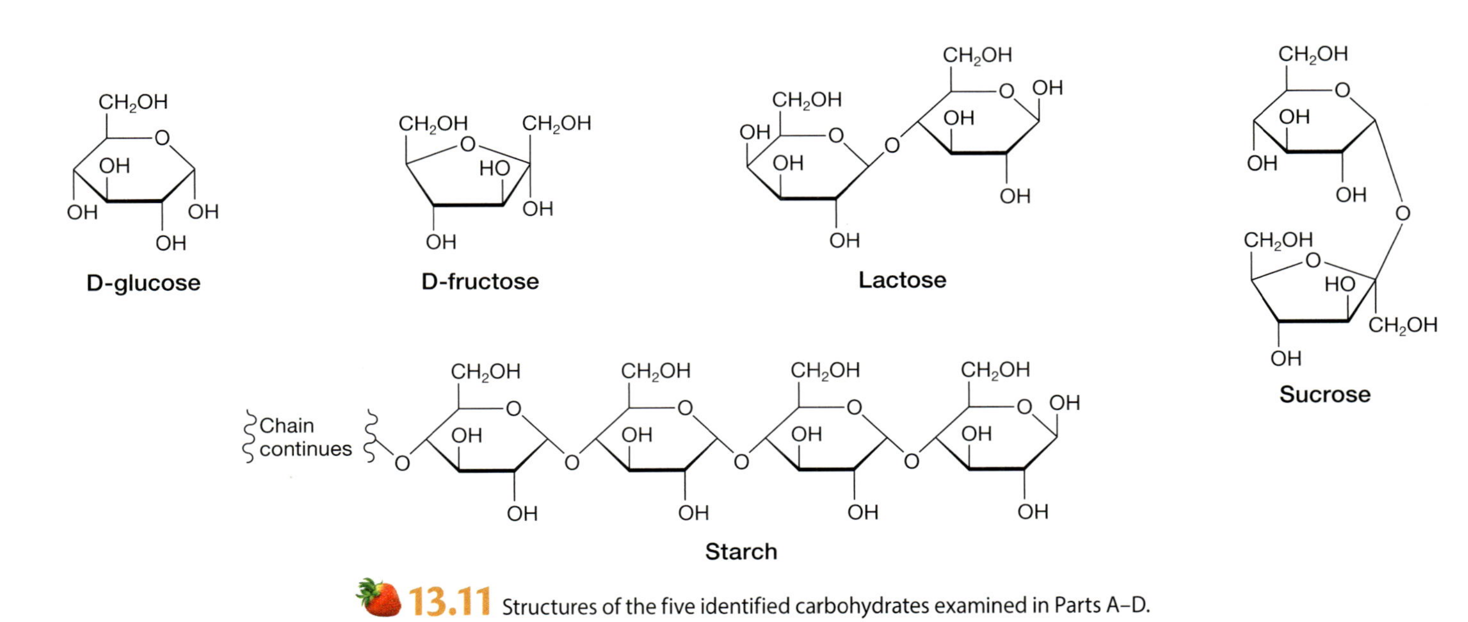

13.11 Structures of the five identified carbohydrates examined in Parts A–D.

Part A Benedict's Test for Reducing Sugars

1. Use a 250 mL beaker to prepare a boiling-water bath on a hot plate. Retain the boiling-water bath throughout the experiment.
2. Label seven test tubes 1–7.
3. To each test tube, add about 1 mL of the appropriate sample solution as indicated below:
   - Test Tube 1: Distilled water
   - Test Tube 2: Glucose
   - Test Tube 3: Fructose
   - Test Tube 4: Sucrose
   - Test Tube 5: Lactose
   - Test Tube 6: Starch
   - Test Tube 7: Mystery sample
4. Add approximately 2 mL of Benedict's reagent to each test tube and mix by gentle agitation (see Technique Tip 13.1 for safely shaking a test tube).
5. Place each test tube in a boiling-water bath for about 5 minutes.
6. While you are waiting, use Figure 13.11 to determine the expected outcome of the Benedict's test on each sample and record your predictions on the data sheet, page 201.

≡13

7. After the 5 minute heating period is over, record your observations on the data sheet and indicate whether each sample should be classified as a reducing sugar based on your data.
8. Dispose of the test tube contents as directed by your instructor.

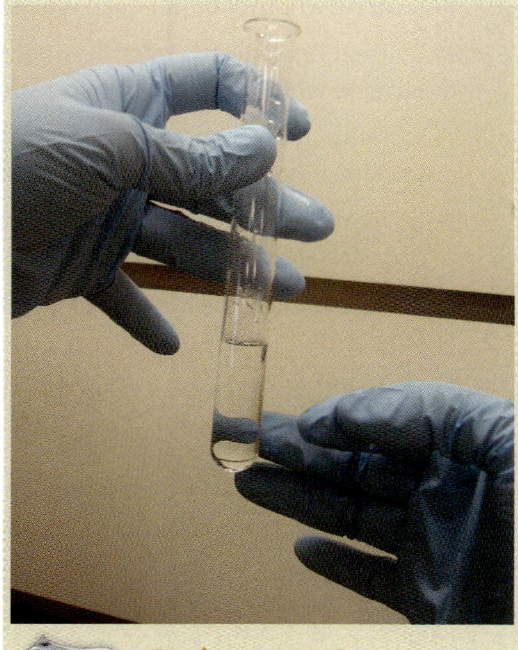

Technique Tip 13.1

To safely shake a test tube, hold it between your thumb and split first two fingers. Flick the test tube using a finger from the opposite hand.

Part B Barfoed's Test for Monosaccharides

1. Label seven test tubes 1–7.
2. To each test tube, add about 1 mL of the appropriate sample solution as indicated below:
   - Test Tube 1: Distilled water
   - Test Tube 2: Glucose
   - Test Tube 3: Fructose
   - Test Tube 4: Sucrose
   - Test Tube 5: Lactose
   - Test Tube 6: Starch
   - Test Tube 7: Mystery sample

3. Add approximately 2 mL of Barfoed's reagent to each test tube and mix by gentle shaking (see Technique Tip 13.1 for safely shaking a test tube).
4. Place each test tube in a boiling-water bath for about 5 minutes.
5. While you are waiting, use Figure 13.11 to determine the expected outcome of the Barfoed's test on each sample and record your predictions on the data sheet, page 202.
6. After the 5 minute heating period is over, record your observations on the data sheet and indicate whether each sample should be classified as a monosaccharide based on your data.
7. Dispose of the test tube contents as directed by your instructor.

Part C Seliwanoff's Test for Ketoses

1. Label seven test tubes 1–7.
2. To each test tube, add about 1 mL of the appropriate sample solution as indicated below:
   - Test Tube 1: Distilled water
   - Test Tube 2: Glucose
   - Test Tube 3: Fructose
   - Test Tube 4: Sucrose
   - Test Tube 5: Lactose
   - Test Tube 6: Starch
   - Test Tube 7: Mystery sample
3. Use Figure 13.11 to determine the expected outcome of the Seliwanoff's test on each sample and record your predictions on the data sheet, page 202.
4. Add approximately 2 mL of Seliwanoff's reagent to each test tube and mix by gentle agitation (see Technique Tip 13.1 for safely shaking a test tube).
5. Place each test tube in a boiling-water bath for up to 10 minutes. Carefully observe the sample and record the time it takes for any color change to occur. If no color change is noted for a sample after 10 minutes, write *none* on the data sheet. After recording your observations on the data sheet, indicate whether each sample should be classified as a ketose based on your data.
6. Dispose of the test tube contents as directed by your instructor.

Part D Iodine Test for Polysaccharides

1. Label seven test tubes 1–7.
2. To each test tube, add about 1 mL of the appropriate sample solution as indicated below:
   - Test Tube 1: Distilled water
   - Test Tube 2: Glucose
   - Test Tube 3: Fructose
   - Test Tube 4: Sucrose
   - Test Tube 5: Lactose
   - Test Tube 6: Starch
   - Test Tube 7: Mystery sample
3. Use Figure 13.11 to determine the expected outcome of the iodine test on each sample and record your predictions on the data sheet, page 203.
4. Add approximately 3 drops of an iodine solution to each test tube and mix by gentle agitation (see Technique Tip 13.1 for safely shaking a test tube).
5. Record your observations on the data sheet and indicate whether each sample should be classified as a polysaccharide based on your data.
6. Dispose of the test tube contents as directed by your instructor.

Part E Identification of the Mystery Sample

1. The mystery carbohydrate in your sample is one of those illustrated in Figure 13.12.
2. Using your results from Parts A–D, propose an identity for your sample, and record it on the data sheet, page 203. Check with your instructor to ensure that no tests need to be repeated.

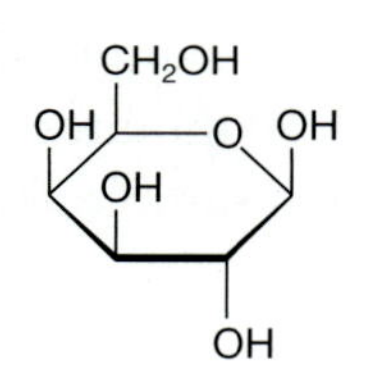

Galactose

Sorbose

Maltose

Trehalose

Glycogen

13.12 Possible identities for the unknown carbohydrate sample.

Part F Hydrolysis Reactions

1. Add approximately 3 mL of sucrose solution to one test tube and 3 mL of starch solution to another.
2. To each test tube, add approximately 0.5 mL of 3 M HCl.
3. Heat the test tubes in a boiling-water bath for 15 minutes.
4. While you are waiting for the hydrolysis reaction to finish, record your predictions about how the product mixtures you expect to obtain should behave in each of the tests indicated on the data sheet, pages 203–204.
5. When the 15-minute reaction period is complete, remove the test tubes from the boiling-water bath and add approximately 0.5 mL of 3 M NaOH to each in order to neutralize the acid.
6. To ensure that the pH of each solution is approximately 7, test the acidity of each using litmus paper: Add a drop of each solution to a piece of blue litmus paper. A neutral solution will not change the color of the paper. If the solution is still acidic, the paper will turn pink. If the paper turns pink, add 3 M NaOH 1 drop at a time to the test tube until the reaction solution no longer changes the color of the litmus paper.
7. Perform Benedict's test (see Part A) on the sucrose reaction mixture. Record your observations and interpret the results (+ or −) on the data sheet.
8. Pour half of the neutralized starch reaction mixture into a separate test tube. Perform Benedict's test (see Part A) on one half of the reaction mixture and the iodine test on the other half (see Part D). Record your observations and interpret the results (+ or −) on the data sheet.

Name ____________________

Lab Partner ____________________

Lab Section ____________________ Date ____________________

Lab 13 Pre-Laboratory Exercise

1 Provide a term that matches each description below.

a Monosaccharides that contain a ketone functional group. ____________________

b Carbohydrate composed of many monosaccharide subunits. ____________________

c Reaction that involves the breaking of large molecules into smaller units. ____________________

d Generic term for a carbohydrate that can be oxidized with Cu^{2+}. ____________________

e Carbohydrate composed of two monosaccharides linked through a glycosidic bond. ____________________

f Monosaccharides that contain an aldehyde functional group. ____________________

g Bond that joins two monosaccharides to each other. ____________________

2 Circle the anomeric carbon in each monosaccharide component of the carbohydrates shown below.

3 Which carbohydrate from Question 2 would match each set of test results?

Test	A	B	C
Benedict's	Red precipitate	Blue solution	Red precipitate
Barfoed's	Red precipitate in 2 minutes	Blue solution	Red precipitate forms very slowly
Seliwanoff's	No red color observed	Red solution in 3 minutes	Red solution in 2 minutes
Iodine	Yellow solution	Yellow solution	Yellow solution

Carbohydrate A: ____________________

Carbohydrate B: ____________________

Carbohydrate C: ____________________

4 Fill in Table 13.2 with the result (+ or –) that you would expect each carbohydrate below to exhibit in each of the four characterization tests.

TABLE 13.2 ■ Characterization Tests

	Carbohydrate	Benedict's Test (+ or –)	Barfoed's Test (+ or –)	Seliwanoff's Test (+ or –)	Iodine Test (+ or –)
a	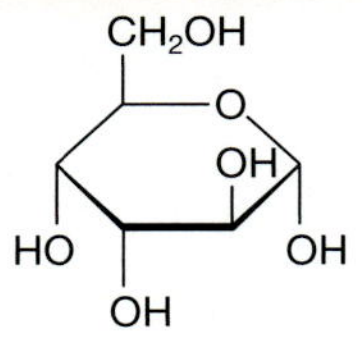 CH_2OH, O, OH, HO, OH, OH α-D-altrose				
b	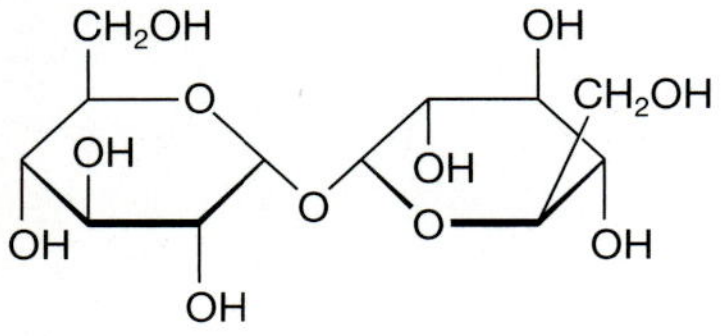 CH_2OH, O, OH, OH, OH, O, OH, CH_2OH, OH, O, OH Trehalose				

5 Draw the structures of the monosaccharide product(s) generated by the complete hydrolysis of the carbohydrates below.

CH_2OH O OH OH OH O CH_2OH O HO CH_2OH OH

Sucrose

$\xrightarrow[HCl]{H_2O}$ [ ] + [ ]

CH_2OH CH_2OH CH_2OH O OH O OH O OH O O O OH OH OH O CH_2OH CH_2OH CH_2 CH_2OH O OH O OH O OH O OH O O O O O OH OH OH OH

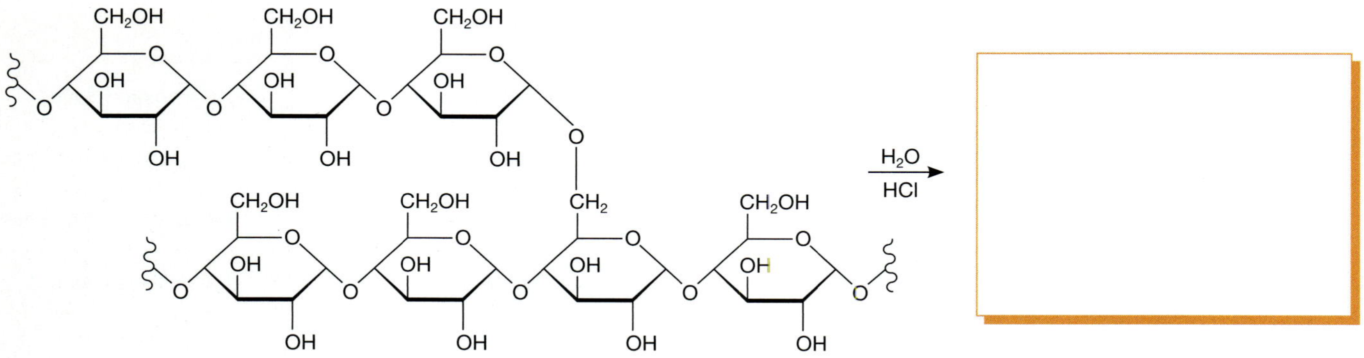

Glycogen

$\xrightarrow[HCl]{H_2O}$ [ ]

Name ______________________________

Lab Partner ______________________________

Lab Section ____________________ Date ____________

Lab 13 DATA SHEET

Data and Observations

Mystery carbohydrate sample identification code: ____________

Part A Benedict's Test for Reducing Sugars

Benedict's Test Results

Test Tube	Sample	Prediction	Observations	Reducing Sugar? (Y/N)
1	Distilled water			
2	Glucose			
3	Fructose			
4	Sucrose			
5	Lactose			
6	Starch			
7	Mystery sample	X		

Part B Barfoed's Test for Monosaccharides

Barfoed's Test Results

Test Tube	Sample	Prediction	Observations	Monosaccharide? (Y/N)
1	Distilled water			
2	Glucose			
3	Fructose			
4	Sucrose			
5	Lactose			
6	Starch			
7	Mystery sample			

Part C Seliwanoff's Test for Ketoses

Seliwanoff's Test Results

Test Tube	Sample	Prediction	Observations	Ketose? (Y/N)
1	Distilled water			
2	Glucose			
3	Fructose			
4	Sucrose			
5	Lactose			
6	Starch			
7	Mystery sample			

Name ______________________________

Lab Partner ______________________________

Lab Section ____________________ Date ______________

Lab 13
DATA SHEET
(continued)

Part D Iodine Test for Polysaccharides

Iodine Test Results

Test Tube	Sample	Prediction	Observations	Polysaccharide? (Y/N)
1	Distilled water			
2	Glucose			
3	Fructose			
4	Sucrose			
5	Lactose			
6	Starch			
7	Mystery sample	X		

Part E Identification of the Mystery Sample

Identity of your mystery carbohydrate sample: ______________________

Part F Hydrolysis Reactions

Hydrolysis Reactions—Sucrose

	Prediction	Observation	Result (+ or −)
Benedict's Test			

Hydrolysis Reactions—Starch

	Prediction	Observation	Result (+ or −)
Benedict's Test			
Iodine Test			

Reflective Exercises

1 In the table below, present a claim regarding the identity of your mystery carbohydrate. Describe your evidence and provide a rationale to justify your claim.

Claim	
Evidence	**Rationale**

2 In Part A, your results should have indicated that two of the carbohydrate samples were not reducing sugars. Identify the structural features of these two molecules that explain why each is not a reducing sugar.

Name

Lab Partner

Lab Section Date

Lab 13 DATA SHEET

(continued)

3 Why do you think the procedure directed you to perform each of the tests on a sample of distilled water in addition to the carbohydrate samples?

4 The structures of D-glucose and D-psicose are shown below. What test could be used to distinguish between solutions of these two carbohydrates? Explain your answer by predicting the results of the test for each sugar.

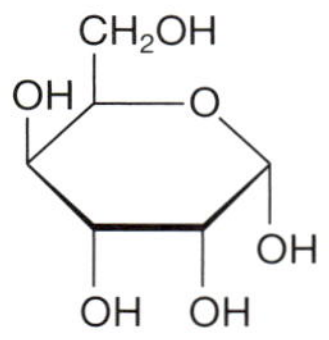

CH_2OH, OH, O, OH, OH OH

D-gulose

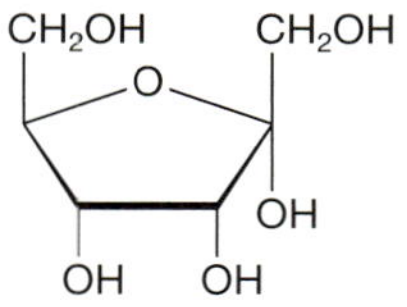

CH_2OH CH_2OH, O, OH, OH OH

D-psicose

5 In Part F, you performed hydrolysis reactions on sucrose and starch.

a Name the monosaccharide product(s) you should have obtained from the hydrolysis of sucrose.

b Explain how a Benedict's test can provide evidence that the hydrolysis of sucrose was successful.

c Explain why the procedure did not direct you to perform the Barfoed's, Seliwanoff's, or iodine tests to demonstrate the success of the sucrose hydrolysis reaction.

d Name the monosaccharide product(s) you should have obtained from the hydrolysis of starch.

e Explain how a Benedict's and iodine tests can provide evidence that the hydrolysis of starch was successful.

f Explain why the procedure did not direct you to perform the Barfoed's or Seliwanoff's tests to demonstrate the success of the starch hydrolysis reaction.

6 During digestion, disaccharides are hydrolyzed to produce monosaccharides, which are further metabolized to produce energy for the body. Honey is a mixture of glucose and fructose rather than a disaccharide like sucrose. Explain why eating a spoonful of honey would provide energy faster than eating spoonful of table sugar (sucrose).

7 Raffinose is a trisaccharide commonly found in peas, beans, and other vegetables. Consider the structure of raffinose and answer the following questions.

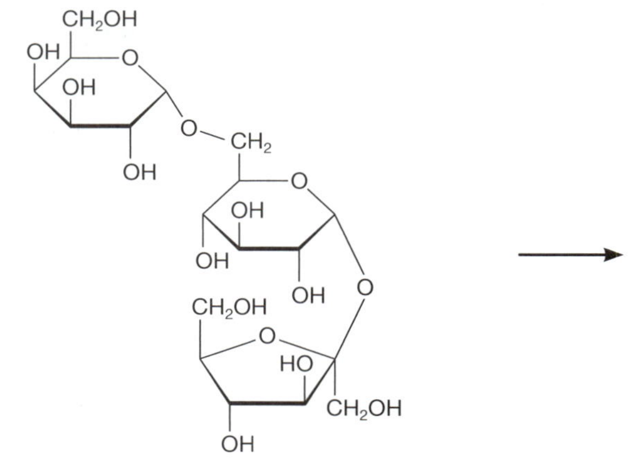

Raffinose

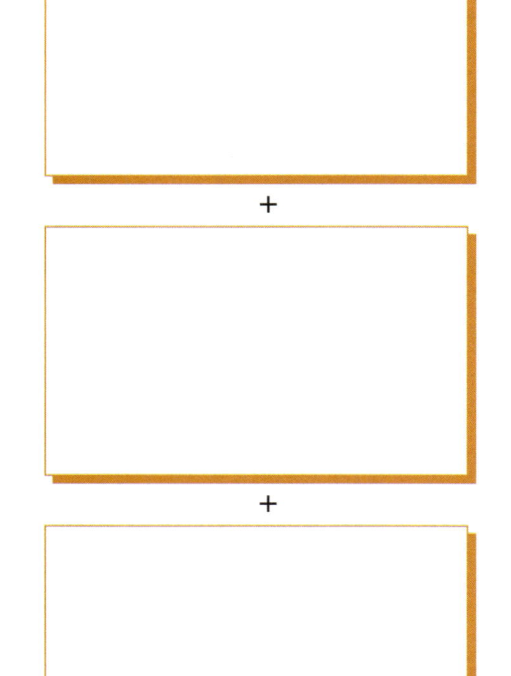

Name ______________________

Lab Partner ______________________

Lab Section ______________ Date ______________

Lab 13
DATA SHEET
(continued)

a In the boxes provided, draw the carbohydrates formed by the complete hydrolysis of raffinose.

b What observations would you expect when raffinose is treated with the following reagents?

Benedict's ______________________

Barfoed's ______________________

Seliwanoff's ______________________

Iodine ______________________

c If a sample of raffinose were completely hydrolyzed, what observations would you expect when the product solution is treated with the following reagents?

Benedict's ______________________

Barfoed's ______________________

Seliwanoff's ______________________

Iodine ______________________

d Which of the four reagents provides evidence that the raffinose sample was successfully hydrolyzed? Explain your answer.

▲ *Analgesic drugs are used to relieve pain.*

Lab 14

Take Your Medicine
Thin Layer Chromatography of Analgesic Drugs

Most substances we encounter on a daily basis are mixtures, and often it is useful to separate the components of these mixtures in order to obtain pure substances. Many of the medicines you are familiar with were first isolated by purifying cellular extracts taken from plants. One of the most common and powerful purification techniques is **chromatography,** which literally means "color writing." The name was chosen by Russian botanist Mikhail Tsvet, who discovered this technique while studying ways to separate plant pigments.

There are many different forms of chromatography, but they all operate on the same principle: compounds with similar polarities tend to attract each other. Chromatography takes advantage of this fact by supplying two different phases, a **mobile phase** and a **stationary phase**, with which a molecule might interact.

As the names imply, the mobile phase in a chromatographic system is allowed to flow over or through the stationary phase. The compounds in a mixture will have different affinities for the two phases, and they will pass through the system at a rate that depends upon the relative strengths of their interactions with each phase.

Consider this analogy: If a *Tyrannosaurus rex* and an orangutan fell into a deep river, the long-armed orangutan may be able to grab hold of some rocks or nearby tree limbs (Fig. 14.1). The *T. rex* would be unable to grab any stationary objects with its short arms. Which would be carried farther downriver by the current? The *T. rex*, of course!

In the same way, compounds that have stronger interactions with the stationary phase (i.e., the orangutan in our analogy) will spend less time moving with the mobile phase and therefore travel less distance. On the other hand, compounds that have stronger interactions with the mobile phase (i.e., the *T. rex* in our analogy) will spend more time moving and therefore travel farther faster. Figure 14.2 illustrates this principle for a chromatographic system.

In order to achieve good separation, the mobile and stationary phases in a chromatographic system must be complementary. That is, a polar stationary phase should be paired with a relatively nonpolar mobile phase.

Thin layer chromatography (TLC) is one of the simplest, fastest, and least expensive chromatographic techniques. The stationary phase in TLC is usually a

Objectives

After completing this lab you should be able to:

- Describe how compounds may be separated using thin layer chromatography (TLC).
- Properly set up and run a TLC experiment.
- Apply your knowledge of intermolecular forces to predict the relative order in which compounds will elute on a TLC plate.
- Calculate the R_f values for a TLC experiment.
- Identify the components of a mixture by comparing R_f values of the sample components to those of standard samples.
- Identify experimental factors that may influence the outcome of a TLC experiment.

very thin layer of polar adsorbent, such as silica (SiO_2), coated on a rectangular plate (see Fig. 14.2). The sample mixture is applied to the adsorbent as a small spot near one end of the TLC plate. After the sample has dried, the TLC plate is placed in a sealed chamber containing a small amount of liquid, which serves as the mobile phase (Fig. 14.3).

The mobile phase slowly rises up the TLC plate by capillary action. As it reaches the sample spot, the compounds in the mixture will begin traveling with the mobile phase at different rates. When the liquid nearly reaches the top of the plate, it is removed from the chamber. Ideally, the components of the mixture will have traveled different distances up the plate and now appear as separated spots.

Note that a paper wick (usually filter paper) is added to the TLC chamber prior to performing the separation in order to ensure that the atmosphere is saturated with vapors of the mobile phase. This prevents the mobile phase liquid from evaporating off of the TLC plate, which would inhibit efficient separation. It is also important that the sample spots start out above the level of the liquid in the chamber. If not, they may simply dissolve off the plate into the reservoir of the mobile phase.

The TLC plate in Figure 14.4 is an example of an effective separation of the constituent compounds in the hypothetical sample labeled 1. In a TLC experiment, it is important to use a pencil to mark the starting line, or **origin**, where the sample was first spotted. It is also necessary to mark the farthest distance traveled by the mobile phase immediately after removing the plate from the chamber before all the liquid evaporates. The line demarking the farthest edge of the mobile phase from the bottom of the plate is called the **solvent front**.

14.1 Analogy of chromatography. The two animals represent different compounds in a mobile phase (the river). The orangutan has the ability to grab a stationary object and will not travel as far downstream. Similarly, compounds with greater affinity for the stationary phase will not travel as far as those with greater affinity for the mobile phase.

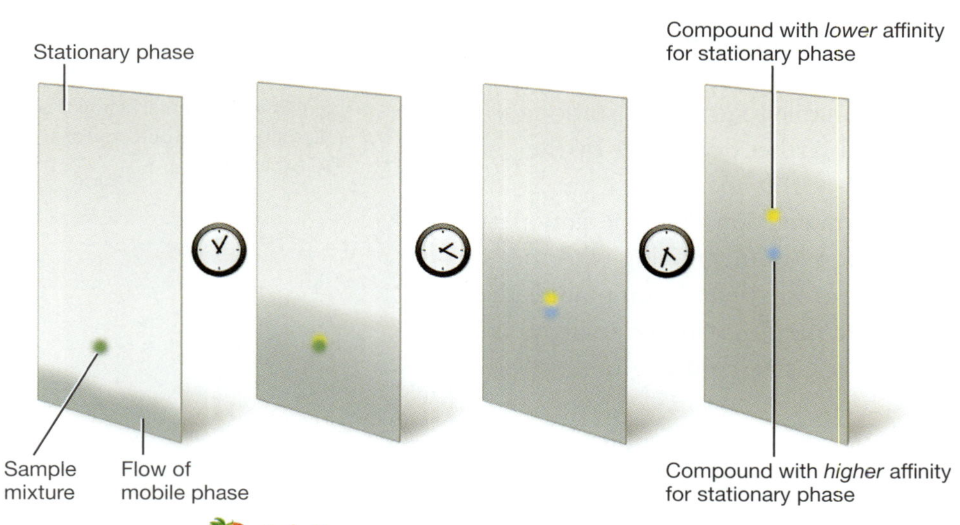

14.2 Overview of separation by chromatography.

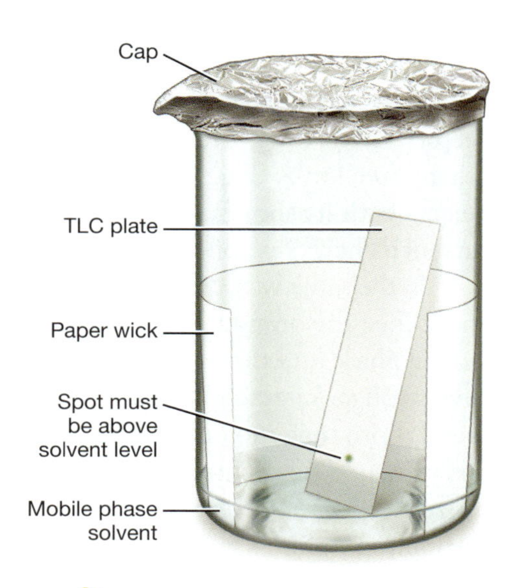

14.3 Overview of a thin-layer chromatography experiment.

Finally, any visible spots in the sample should be circled with a pencil as soon as the plate dries because spots may fade over time. Many compounds are not colored and will not be visible on the TLC plate to the naked eye. Most modern TLC plates are coated with a compound that fluoresces (glows) when exposed to ultraviolet (UV) light. Many sample compounds present on the plate mask the fluorescence, resulting in a dark spot where the sample is located. TLC plates may also be treated with a stain that colors any spots present.

To quantify the distances traveled by each compound on the TLC plate, the distance from the origin to the center of each spot is measured with a ruler. Similarly, the distance from the origin to the solvent front also is measured. The relative distance traveled by each compound is reported as the **retention factor** (**R_f**), which is simply the ratio of these two measurements (Eq. 1). For example, the R_f for spot B in Figure 14.4 would be 0.518.

$$R_f = \frac{\text{Distance traveled by sample}}{\text{Distance traveled by mobile phase}} \quad \textbf{[Eq. 1]}$$

The R_f value is a unitless physical constant for a given compound under the specific experimental conditions used (mobile phase, adsorbent type and thickness, quantity of spotted material, temperature, etc.). As such, it may help confirm the identity of a compound in a sample.

For example, if two different samples are run next to each other on the same TLC plate, spots that share the same R_f value may be the same compounds, although this is not definitive proof that they are identical. However, if two spots have different R_f values, then they must be different compounds.

Consider the TLC plate shown in Figure 14.5. Two different sample mixtures were run on this plate—one in the lane labeled 1 and the other in lane 2. Each sample separated into two individual spots. Spots A and C share the same R_f, indicating that these might be the same compound. Spots B and D have different R_f values and therefore must be different compounds. We may conclude from this TLC experiment that both samples are mixtures of at least two compounds. These two mixtures probably share one compound in common.

In this lab, you will carry out TLC experiments in order to identify a sample of medication by determining its active ingredients. Medicines used to alleviate pain are collectively referred to as **analgesics**. The chemical structures of common active ingredients found in over-the-counter analgesics are shown in Figure 14.6. Some analgesic brands are made of pure active ingredient while others are mixtures of two or more of these compounds. The analgesics found in five common brands of medication are given in Table 14.1.

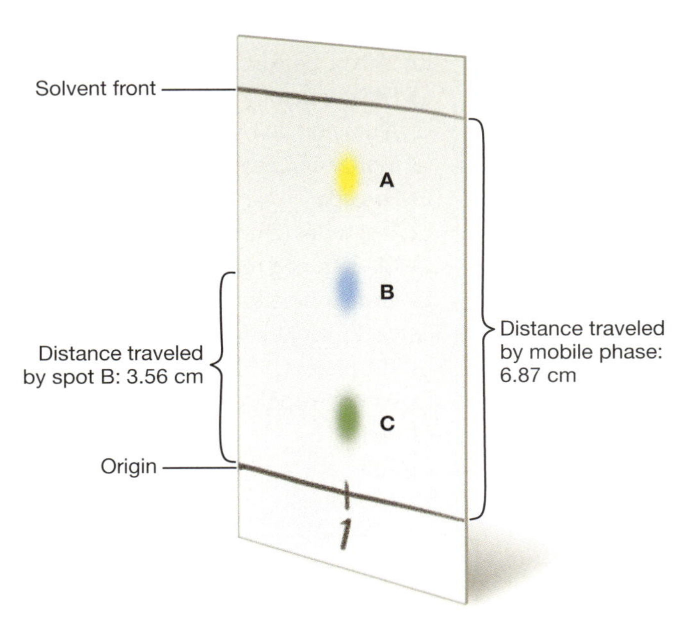

14.4 Important features of a TLC plate with effective separation of the compounds in hypothetical sample 1.

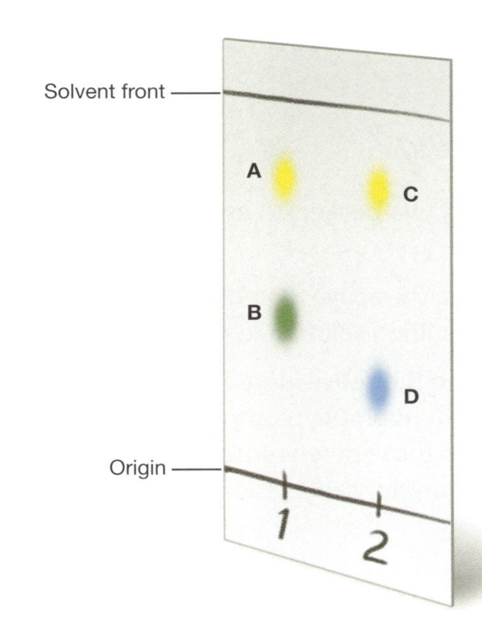

14.5 TLC results after separation of two different mixtures. One mixture was spotted in lane 1 and the other was spotted in lane 2.

Ibuprofen Caffeine Aspirin Acetaminophen

14.6 Chemical structures of common analgesic compounds.

To determine the identity of your assigned analgesic sample, you will extract the active ingredients from the drug and then perform a TLC analysis on your extract. On your TLC plate, you will spot your mystery sample along with authentic samples of the four compounds in Figure 14.6. The authentic samples should help you identify the active ingredients in your sample by a comparison of R_f values. You will perform the TLC analysis twice using two different mobile phases in order to examine the effect of the mobile phase composition on the efficiency of separation in a TLC experiment.

TABLE **14.1** ■ Active Ingredients in Some Common Over-the-Counter Medications

Brand Name	Active Ingredients
Anacin	aspirin, caffeine
Tylenol	acetaminophen
Excedrin	acetaminophen, aspirin, caffeine
Advil	ibuprofen
Bayer Aspirin	aspirin

Procedure

What Am I Doing?

Part A: Preparing two beakers to use as chromatography chambers in the experiment.

Part B: Extracting the active ingredient(s) in a mystery analgesic and filtering the resulting solution to remove any undissolved solids.

Part C: Separating the active ingredients in the mystery analgesic extract by thin layer chromatography; comparing the R_f values of the active ingredient(s) to those of reference samples in order to determine the identity of the mystery analgesic.

Materials

- ❑ 250 mL beakers (2)
- ❑ Filter paper
- ❑ Aluminum foil
- ❑ Mortar and pestle
- ❑ Forceps
- ❑ Test tubes (2)
- ❑ Glass stirring rod
- ❑ Pasteur pipette (2)
- ❑ Cotton or Kimwipe
- ❑ TLC plates (2)
- ❑ Wooden applicator stick
- ❑ Pencil
- ❑ Ruler
- ❑ Spotters (5)

Part A Preparing TLC Chambers

1. Obtain two 250 mL beakers to use as your TLC chambers. Cut two pieces of filter paper to provide one flat edge on each. Place one piece of filter paper in each TLC chamber with the flat edge on the bottom of the beaker.
2. Use Mobile Phase 1 (20% ethyl acetate in hexanes) in one TLC chamber and Mobile Phase 2 (ethyl acetate) in the other. In each chamber, add enough of the appropriate mobile phase to form a layer 0.5–1.0 cm deep.

Vapors from ethyl acetate and hexane should not be inhaled. Prepare the TLC chambers and perform the experiment in a fume hood or appropriately ventilated space!

3. Cover each chamber tightly with aluminum foil. Allow the chambers to equilibrate for at least 15 minutes while you prepare your mystery sample and your TLC plates (see Parts B and C). During this time, the liquid will travel up the filter paper by capillary action and saturate the air in the chamber with mobile phase vapors.

Part B Preparing the Mystery Analgesic Sample

1. Record the identification code for your mystery analgesic pill or powder on the data sheet, page 219.
2. If your mystery sample is a pill, place it in a mortar and grind it with a pestle until you have a relatively homogenous powder.
3. Transfer the powdered sample to a test tube and add about 2 mL of the extraction liquid (50% ethanol in dichloromethane).
4. Stir the mixture vigorously with a glass rod. Do not be concerned if some of the material does not dissolve. Insoluble binders are often used to hold together commercial analgesic tablets, and enough of the analgesic will dissolve in the extraction liquid for a TLC analysis. The solid material will be removed in Steps 5–6.
5. Prepare a microscale filter by plugging the tip of a Pasteur pipette with a small piece of cotton or Kimwipe. Use a wooden applicator stick to *gently* fit the plug material into the tip of a Pasteur pipette (Fig. 14.7A). A loose fit is sufficient and will allow the solution to flow easily. If the plug is pressed too tightly, the solution will flow very slowly.
6. Use another pipette to add about 1 mL of your analgesic solution through the top of the microscale filter. Collect the extract solution in a clean test tube as it filters through the plug (Fig. 14.7B). Save your mystery sample for TLC analysis in Part C.

Dichloromethane is a suspected carcinogen and its vapors should not be inhaled. Wear gloves and perform Part B in a fume hood or appropriately ventilated space!

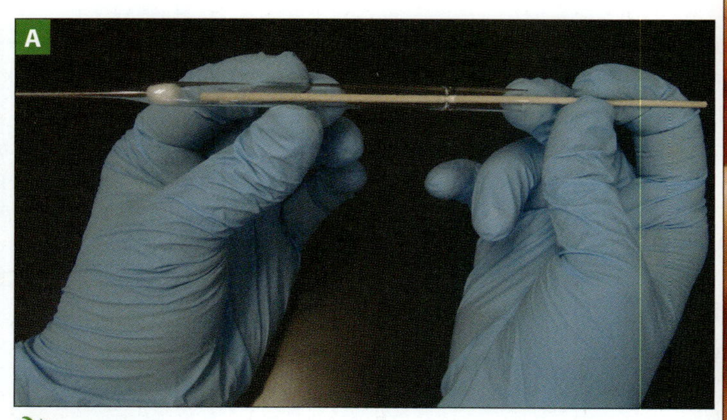

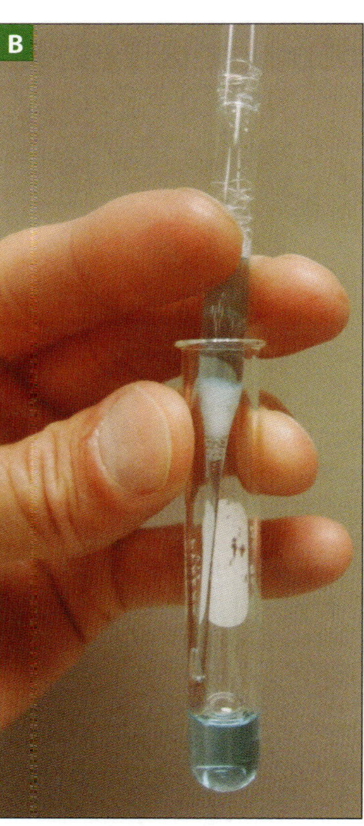

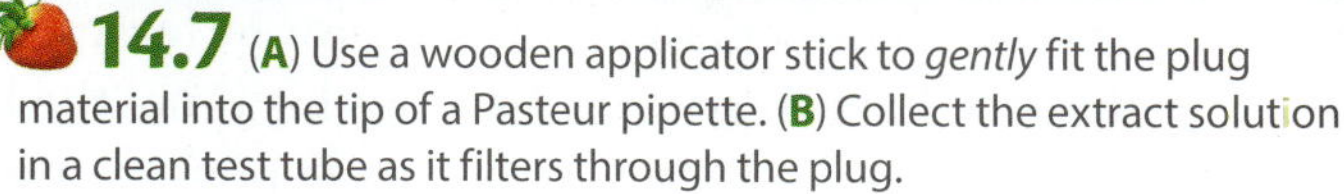

14.7 (**A**) Use a wooden applicator stick to *gently* fit the plug material into the tip of a Pasteur pipette. (**B**) Collect the extract solution in a clean test tube as it filters through the plug.

Part C TLC Analysis

1. Obtain two TLC plates. Handle these plates carefully and by the edges to prevent contamination or scraping the silica surface.
2. Orient your rectangular plate so the long side is vertical. Use a ruler and pencil to mark a light starting line (origin) approximately 1 cm from the bottom of the plate (Fig. 14.8). The origin must be above the liquid level in the TLC chamber when your TLC plate is standing in the mobile phase. Do not press too heavily! You do not want to gouge a trench into the silica surface.
3. On each TLC plate, use a pencil to label five lanes in which you will spot your five samples (four analgesics and one mystery sample) as illustrated in Figure 14.9. Sample lanes should not be too close to the edge of the TLC plate and should be separated from each other by approximately 0.5 cm.
4. Spot each of the five samples in the corresponding lane on the TLC plate by following the individual steps outlined below. Correct application of the sample spot is critical to the success of a TLC experiment. You should practice these steps on a paper towel several times before applying a sample to your TLC plate.
   - **a** Dip a spotter into the appropriate sample solution.
   - **b** Quickly and gently touch and remove the tip of the spotter to the surface of the TLC plate on the starting line at the appropriate position. Try to make as small a spot as possible, and be careful not to grind the micropipette into the surface.
   - **c** Wait until the spot dries and then reapply the sample to the same spot 2 to 3 more times. If you do not wait for the spot to dry, reapplication will enlarge the spot and decrease the efficiency of your separation.
   - **d** Hold your TLC plate under a UV lamp. You should see dark spots at the origin in each lane. Make sure that none of the sample spots are overlapping or too close (<0.5 cm) to the edge of the plate. If any spots are missing or considerably lighter than the others, reapply the sample 2 to 3 more times, and then recheck the plates to ensure that the spots are visible.

Never look directly into a UV light or shine one in anyone's eyes.

5. Once you have spotted each of your samples and all spots are dry, use forceps to pick up the TLC plate near the top edge (Fig. 14.10). Place one TLC plate in the chamber containing Mobile Phase 1 and the other in the chamber containing Mobile Phase 2. Make sure you position the plates with the origin toward the bottom and such that the side edges are not touching the filter paper wick.
6. Immediately cover the chambers with the aluminum foil and allow the mobile phases to rise up the plates.
7. When the mobile phase is approximately 0.5 cm from the top edge of the TLC plate, use forceps to remove it from the chamber. Be sure to handle the plate above the solvent front.
8. Quickly mark the position of the solvent front with a pencil before the liquid evaporates (Fig. 14.11).
9. Allow the TLC plate to dry, and then hold it under a UV light to visualize any analgesic spots that are not visible to the naked eye. Use a pencil to circle each compound on the TLC plate.
10. Illustrate your results on the TLC plate templates provided on your data sheet, page 219.
11. On each TLC plate, use a ruler to measure the distance from the origin to the solvent front and the distance from the origin to the center of each spot. Record the distances on the data sheet.
12. Calculate the R_f of each spot on the TLC plates and use your results to identify the active ingredients in your mystery analgesic. Use your results and Table 14.1 to determine the brand name of the sample.

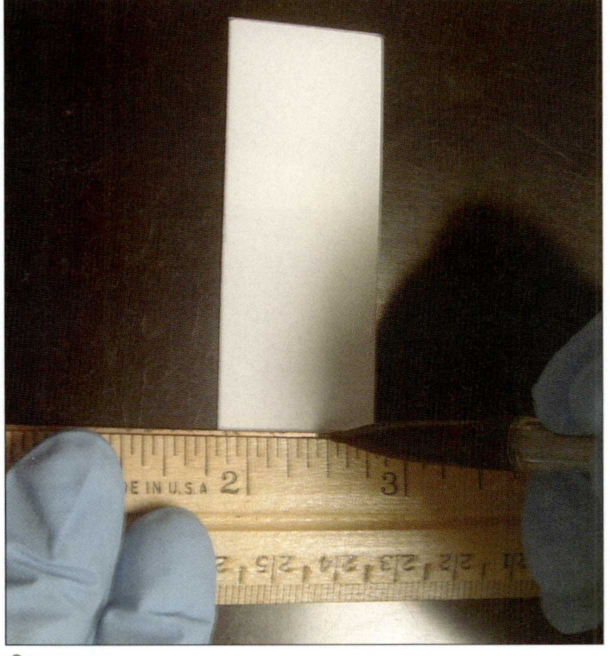

14.8 Use a ruler to guide you as you lightly draw the origin with a pencil.

14.9 Example TLC plate with five analgesic samples spotted in individual lanes.

14

14.10 Use forceps to insert the TLC plates into the appropriate chamber. Ensure that the origin is above the mobile phase and that the edges of the plate do not touch the filter paper wick.

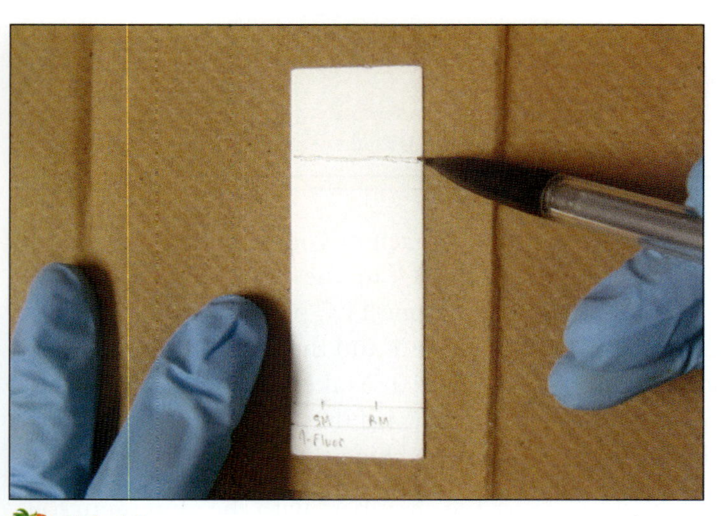

14.11 Mark the position of the solvent front on the TLC plate before it evaporates.

≡14

Name ______________________

Lab Partner ______________________

Lab Section ______________________ Date ______________________

Lab 14 Pre-Laboratory Exercise

1 Provide a term that matches each description below.

a Class of compounds that act as pain relievers. ______________________

b Phase in a chromatography system that remains fixed in position. ______________________

c Chromatography that uses a thin layer of adsorbent on a rectangular plate. ______________________

d Starting line on a TLC plate. ______________________

e Distance traveled by the mobile phase on a TLC plate. ______________________

2 Think about what happens when you put a drop of food coloring in a glass of water. Keeping that in mind, why do you think it is important for the liquid samples applied to a TLC plate to be as small as possible?

____

3 A student set up the TLC experiment illustrated here. Identify two errors in this setup, and describe how these errors might affect the results.

Error 1:

____

Error 2:

____

4 Consider the TLC plate illustrated here and calculate the R_f values for Spots A and B.

R_f of Spot A ______________

R_f of Spot B ______________

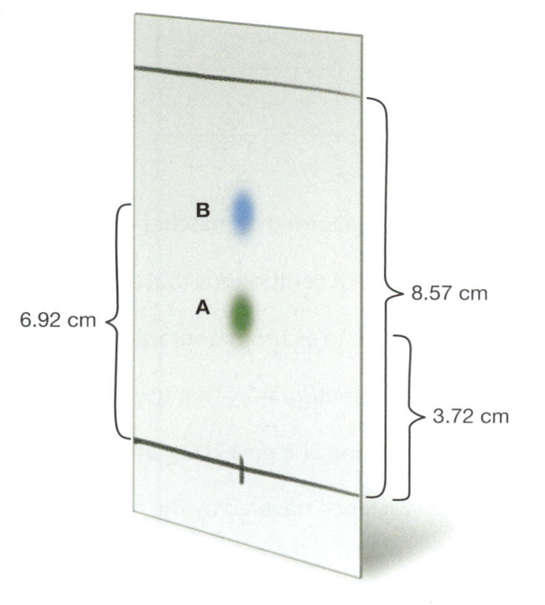

Name ______________________________

Lab Partner ______________________________

Lab Section ____________________ Date ____________________

Lab 14 DATA SHEET

Data and Observations

Part B Preparing the Mystery Analgesic Sample

Mystery Analgesic Sample ID: ____________________

Part C TLC Analysis

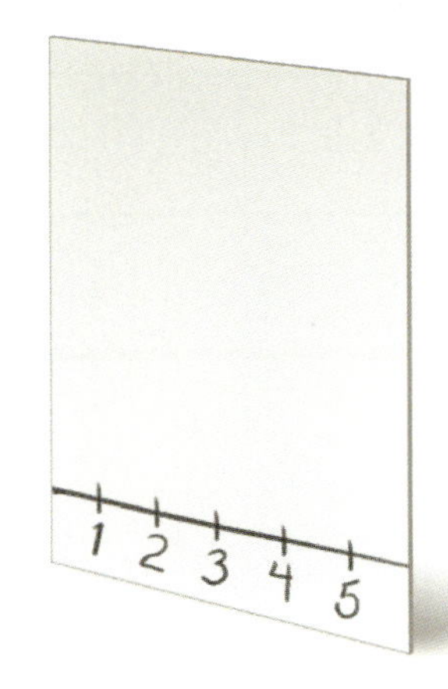

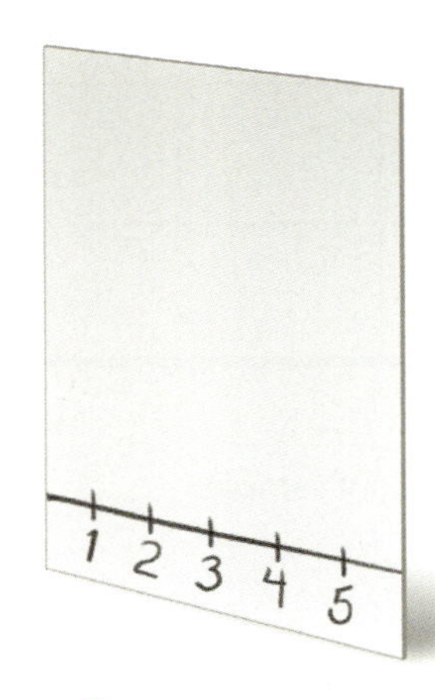

Mobile Phase 1

Distance Traveled by Mobile Phase: ____________________

Mobile Phase 2

Distance Traveled by Mobile Phase: ____________________

TLC Data: Mobile Phase 1

Lane Number	Distance Traveled	R_f	Identity
1			Ibuprofen
2			Caffeine
3			Aspirin
4			Acetaminophen
5			
5 (if additional spot present)			
5 (if additional spot present)			

TLC Data: Mobile Phase 2

Lane Number	Distance Traveled	R_f	Identity
1			Ibuprofen
2			Caffeine
3			Aspirin
4			Acetaminophen
5			
5 (if additional spot present)			
5 (if additional spot present)			

Active ingredient(s) present sample: ______________________________

Brand name of sample: ______________________________

Reflective Exercises

1 Were the two TLC plates equally useful in identifying the active ingredients in your analgesic sample? Explain your answer.

____

____

____

____

Name ______________________________

Lab Partner ______________________________

Lab Section ____________________ Date ______________

Lab 14
DATA SHEET
(continued)

2 In the table below, present a claim regarding the brand name of your mystery analgesic. Describe your evidence and provide a rationale to justify your claim.

Claim	
Evidence	**Rationale**

3 Ethyl acetate is more polar than hexane. Explain whether your results were consistent with this fact by comparing the TLC data obtained for Mobile Phases 1 and 2.

____

____

____

____

____

4 Determine each of the following from the TLC data you used to identify your analgesic sample:

a The analgesic ingredient that has a polarity most similar to the stationary phase. ______________

b The analgesic ingredient that has a polarity most similar to the mobile phase. ______________

5 Why is it not possible to have an R_f value greater than 1.00?

6 A student used a pen rather than a pencil to mark the origin on a TLC plate. Explain how this could be a problem for the student's TLC experiment.

7 Why do we allow the mobile phase to rise nearly to the top of the TLC plate rather than removing the plate when the mobile phase is only halfway up the plate?

8 After performing a TLC experiment, a student was surprised to find that the plate appeared blank under a UV light. The instructor suggested dipping the plate in a chemical reagent, and the samples suddenly appeared as bright pink spots. What might account for this result? What does this suggest about the reliability of using only one visualization method (such as exposure to a UV light) to gather data from a TLC experiment?

9 If a student performed a TLC experiment and failed to get a good separation of the compounds in a sample of medication, which of the following would be the best choice of experimental factors to change in order to achieve a better separation? Explain your answer.

a Mobile phase

b Size of TLC chamber

c Temperature

d Length of TLC plate

Explanation:

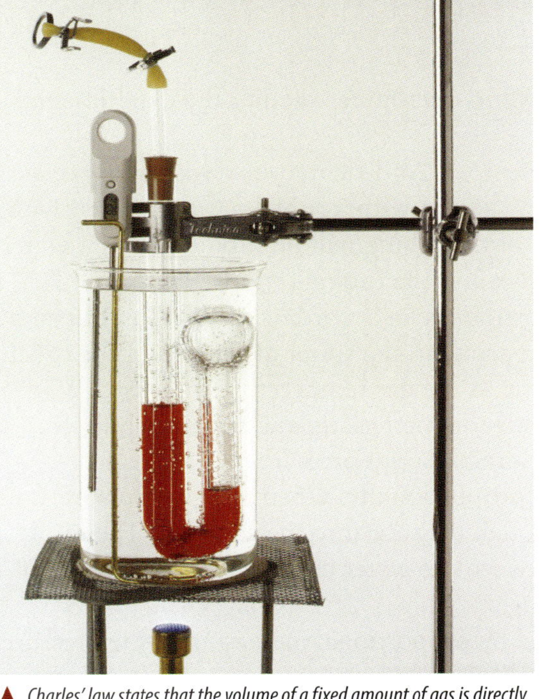

▲ *Charles' law states that the volume of a fixed amount of gas is directly proportional to the temperature.*

Lab 15

A Lot of Hot Air
Simple Gas Laws in Action

Objectives

After completing this lab you should be able to:

- Name and describe three simple gas laws.
- Predict and explain the behavior of gases using Charles' law, Boyle's law, and Gay-Lussac's law.
- Calculate gas properties using the three simple gas laws.
- Verify Charles' law from experimental data.

Gases, especially oxygen and carbon dioxide, are vital for most biological systems to function. Unlike solids and liquids, the attractive forces between gas particles are essentially nonexistent. As a result, the behavior of gases is significantly different than that of the other states of matter and does not depend on the identity of the gas.

Among the most conspicuous differences from other phases of matter is the fact that gases do not have a fixed volume but instead always expand to fill the volume of their container. In addition, the volume, temperature, and pressure of a fixed quantity of gas are interrelated properties; a change in any of these will alter the others. The simple gas laws summarized in Table 15.1 mathematically describe the relationships between individual pairs of these properties.

TABLE 15.1 ■ Simple Gas Laws: Descriptions of the Relationships between Volume, Temperature, and Pressure

Gas Law	Equation	Description
Boyle's law	$P_1V_1 = P_2V_2$	• P and V are inversely proportional (an increase in one causes a decrease in the other) • T must be constant
Charles' law	$\frac{V_1}{T_1} = \frac{V_2}{T_2}$	• V and T are directly proportional (an increase in one causes an increase in the other) • P must be constant
Gay-Lussac's law	$\frac{P_1}{T_1} = \frac{P_2}{T_2}$	• P and T are directly proportional (an increase in one causes an increase in the other) • V must be constant

Note: For an individual gas law to apply, the third property and the quantity of gas must remain constant. Variables that have a subscript of 1 correspond to properties under an initial set of conditions, and those with a subscript of 2 correspond to the values of those properties after a change in one of the variables. The Kelvin temperature must be used in any gas law calculation.

In this laboratory exercise, you will explore the behavior of air (a mixture of gases) and use the simple gas laws to explain and quantify your observations. Specifically, you will note the changes observed by heating and cooling a sample of air. You also will measure the volume of the air sample at two temperatures to determine whether the experimental data match **Charles' law**.

A schematic of the experimental design for this lab is presented in Figure 15.1. An Erlenmeyer flask will serve as the primary container for the air sample. The mouth of this "gas flask" will be sealed with a rubber stopper fitted with tubing that serves as a portal to the external environment. The opposite end of the tubing will be submerged in a water reservoir. The gas particles of the air sample will fill the volume of the gas flask and the tubing (Fig. 15.1A).

If the system is heated (Fig. 15.1B), the gas pressure will increase proportionally in accordance with **Gay-Lussac's law**. The increased pressure will move the only "flexible wall" of the gas compartment: the water at the end of the tube! As a result, some gas will escape as bubbles. This process will continue as long as the gas temperature increases. When the temperature stabilizes, the pressure of the gas remaining in the system will stop increasing, and no more bubbles will escape because the internal (gas) pressure exactly matches the external (atmospheric) pressure.

If the sample is then cooled (Fig. 15.1C), the gas pressure will decrease proportionally. When the gas pressure decreases below the external (atmospheric) pressure, water will be pushed through the tube into the gas flask. The result is a decrease in the volume of the gas, which is now confined to the space between the water collected in the bottom of the gas flask and the water-filled tube.

According to **Boyle's law**, the decreasing gas volume will be accompanied by proportional increasing of the pressure exerted by the gas. The water will therefore continue to flow into the system until the gas sample is reduced to a volume at which its pressure is great enough to "push back" and stop the flow of water. At that point, the gas pressure will be equal to the external pressure and its volume will be equal to the space between the water in the gas flask and the water-filled tubing.

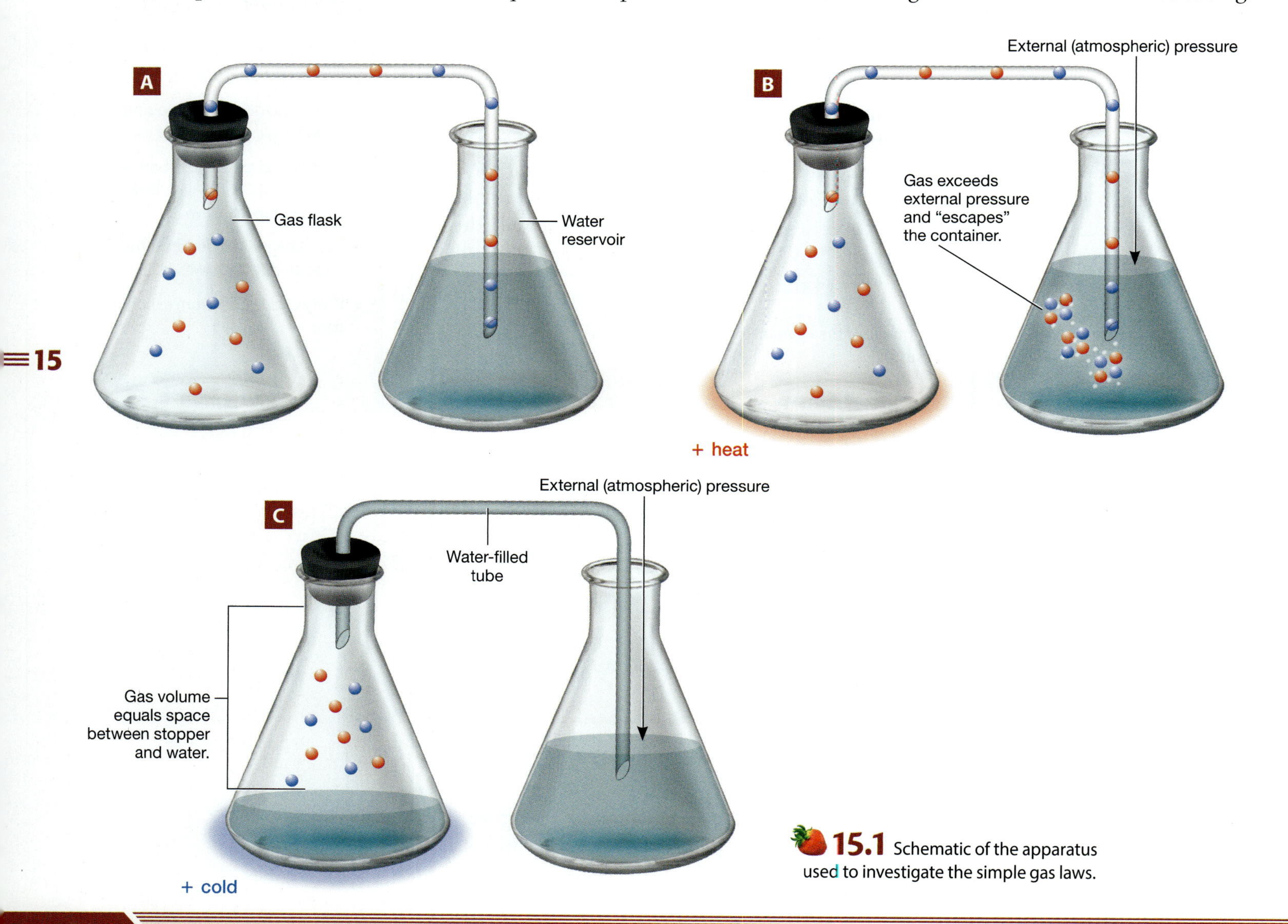

15.1 Schematic of the apparatus used to investigate the simple gas laws.

15

The change in gas volume associated with the decrease in temperature provides an opportunity to experimentally test the validity of **Charles' law**, which describes the relationship between the volume and temperature of a fixed quantity of gas at constant pressure. Specifically, the volume of a gas is directly proportional to the absolute temperature in Kelvin.

Mathematically, this relationship is usually stated as shown in Table 15.1. The expression may be rearranged to give Equation 1. Thus, a restatement of Charles' law indicates that when the temperature of a fixed amount of gas is changed (at constant pressure), the product of V_1 and T_2 divided by the product of V_2 and T_1 should equal 1.

$$\frac{V_1 T_2}{T_1 V_2} = 1 \quad \textbf{[Eq. 1]}$$

You will use Equation 1 and the experimental setup described in Figure 15.1 in an attempt to verify Charles' law by determining the volume of an air sample at two different temperatures (and constant pressure). You will heat the gas flask in a boiling-water bath and cool it in an ice-water bath. The gas temperatures under these two conditions are assumed to be equal to the bath temperatures.

The initial volume of hot gas (V_1) is equal to the full volume of the gas flask and the tubing. These are measured at the end of the experiment by filling the apparatus with water and using a graduated cylinder to measure the total water volume. To determine the final volume of chilled gas (V_2), you will subtract the volume of water drawn into the gas flask and tubing from the total volume of the system.

Inputting your values for V_1, T_2, V_2, and T_1 into Equation 1 should give you a value of 1. Of course, experimental error will prevent your result from matching this value exactly. To quantify how well your experimental data agree with Charles' law, you also will use Equation 2 to calculate the percent error in your data.

$$\text{Percent Error} = \left[\frac{1.00 - (\text{Value of Eq. 1})}{1.00}\right] \times 100 \quad \textbf{[Eq. 2]}$$

Procedure

Steps 1–7: Preparing the experimental apparatus for testing simple gas laws.

Step 8: Heating the gas contained in the apparatus and observing any resulting changes in the system.

Step 9: Chilling the gas sample in the apparatus and observing any resulting changes in the system.

Steps 10–12: Measuring the volume of water contained in the experimental apparatus after the heating/cooling processes.

Step 13: Measuring the volume of the gas flask by measuring the volume of water it can hold.

Steps 14–17: Calculating the values necessary to evaluate the consistency of your data with Charles' law and repeating the experiment two times.

Materials

- ❑ Ring stand
- ❑ Hot plate
- ❑ Clamp
- ❑ 600 mL and/or 1,000 mL beakers (2)
- ❑ 125 mL and 250 mL Erlenmeyer flasks
- ❑ 50 mL graduated cylinder
- ❑ Thermometer
- ❑ Wax pencil
- ❑ Tongs or hot mitt
- ❑ Rubber stopper/tubing assembly

Simple Gas Laws in Action

1. Use a 600 mL beaker to prepare a boiling-water bath on a hot plate. Use a separate 600 mL beaker to prepare an ice-water bath.
2. Obtain the rubber stopper and tubing assembly (Fig. 15.2). Fit the stopper into a clean, dry, 125 mL Erlenmeyer flask. We will refer to this flask as the *gas flask*.
3. On the outside of the gas flask, use a wax pencil to mark the position of the rubber stopper's bottom.
4. Add about 200 mL of tap water to a 250 mL Erlenmeyer flask. We will refer to this flask as the *water reservoir*.
5. Place the open end of the tube assembly from the gas flask into the water contained in the reservoir. Be sure the tube stays submerged in the water throughout the experiment.
6. When your boiling-water bath reaches a steady boil, record the temperatures of both water baths on the data sheet, page 231.
7. Clamp the gas flask securely on a ring stand above the boiling-water bath.
8. Lower the clamp until as much of the gas flask as possible is beneath the surface of the boiling water. At this point, your apparatus should resemble Figure 15.3, and you should immediately see bubbles flowing out of the tube into the water reservoir. Record your observations on the data sheet.
9. Allow the gas flask to remain beneath the surface of the boiling water for 3–4 minutes after bubbles cease to emerge from the tube, and then lift the gas flask out of the boiling-water bath. Remove it from the ring stand and quickly lower it into the ice-water bath so that as much of the gas flask as possible is submerged. During this process be careful to ensure that the end of the tube in the reservoir does not emerge from the water. Record any observed changes in the system on the data sheet, page 231.

SAFETY NOTE

The gas flask will be very hot! Handle it with tongs or a hot mitt and use caution while transferring it to the ice-water bath!

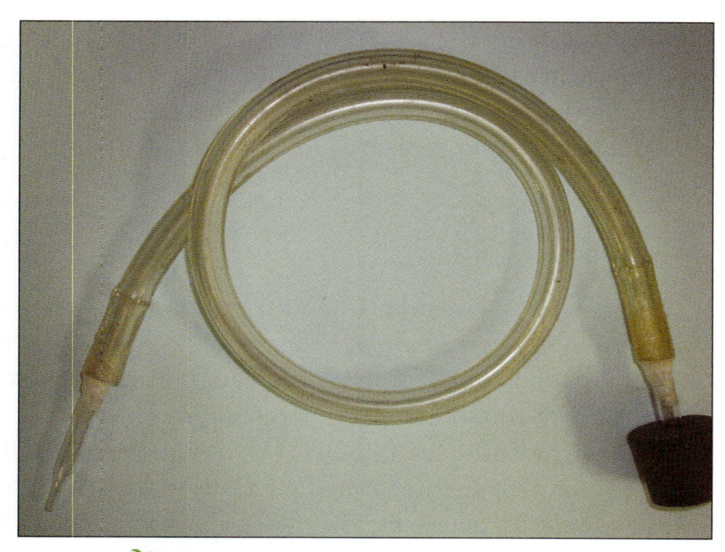

15.2 Rubber stopper and tubing assembly.

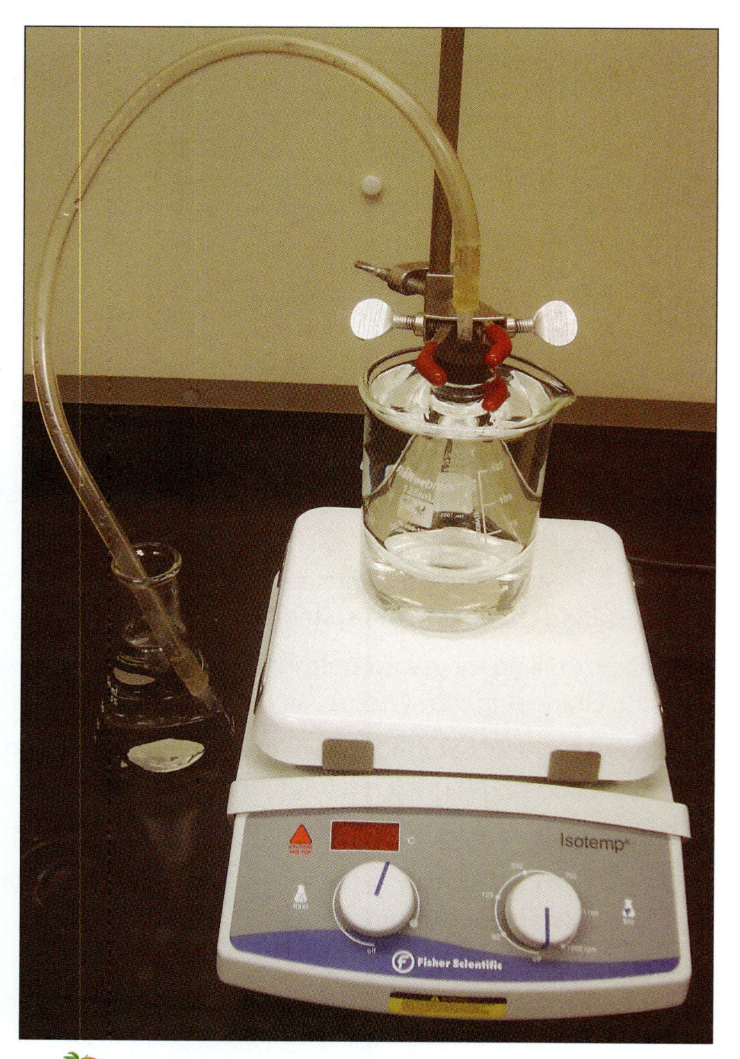

15.3 Fully assembled apparatus for gas laws experiment.

≡15

10. When water stops flowing into the gas flask, remove it from the ice-water bath and set it down on the bench top.
11. Use your finger to cover the end of the tube in the water reservoir. Move the end of the tube to a 50 mL graduated cylinder, being careful not to transfer water from your hand during the process. Remove your finger and allow the water in the tube to drain into the graduated cylinder. Lift the gas flask to ensure that all of the water is expelled from the tube. Record the tube volume (V_T) on the data sheet.
12. Remove the stopper and use the graduated cylinder to measure the volume of water drawn into the gas flask (V_W). Record this value on the data sheet.
13. Determine the total volume of the gas flask (V_{GF}) by filling it with water to the line you drew in Step 3 and using a graduated cylinder to measure the volume of water added. Record this value on the data sheet.
14. Calculate the values of T_1 and T_2 in Kelvin and record them on the data sheet.
    - The initial temperature of the air in the gas flask should have been equal to the temperature of the boiling-water bath.
    - The final temperature of the air should have been equal to the temperature of the ice-water bath.
15. Calculate the values of V_1 and V_2 and record them on the data sheet.
    - Determination of V_1: When the gas flask was in the boiling-water bath, the air completely filled the system, including the tubing. Thus, the initial volume of the air (V_1) was equal to the combined volume of the gas flask (V_{GF}) and the tube (V_T).
    - Determination of V_2: When the gas flask was placed in the ice-water bath, space was occupied by the water drawn into the system. The air volume was equal to the difference between the total system volume (V_1) and the combined volume of water in the gas flask and the tube ($V_W + V_T$).
16. Use your data to calculate your experimental value of Equation 1. Use Equation 2 to calculate the percent error between your data and the theoretical Charles' law value. Record your calculations on the data sheet, page 231.
17. Dry the gas flask and tubing assembly as much as possible. Repeat the experiment (Steps 3–16) two more times. Record your observations, data, and calculations on the data sheet, pages 231–232. Before beginning each trial, check that the water level in the boiling-water bath is still high enough to allow the gas flask to be submerged. If too much water has evaporated, add more and wait until a continuous boil is again observed.

Name ______________________________

Lab Partner ______________________________

Lab Section ____________________ Date ____________

Lab 15 Pre-Laboratory Exercise

1 Complete each sentence below.

a According to Gay-Lussac's law, the pressure and temperature of a fixed amount of gas are ____________________ proportional at constant pressure.

b When applying Charles' law, the gas temperature must be expressed in ____________________ units.

c Unlike solids and liquids, the ____________________ between gas particles are essentially nonexistent.

d According to Boyle's law, water flowing into the chilled gas flask shown in Figure 15.1 will eventually stop because decreasing the volume of gas will cause its pressure to ____________________.

2 Read the procedure for this experiment carefully and provide the following information:

a Name the gas or gas mixture that is employed in the experiment. ____________________

b Give the approximate temperature value (in °C) you should expect to measure when the gas is heated.

____

c Give the approximate temperature value (in °C) you should expect to measure when the gas is chilled.

____

d Name the piece of glassware that will serve as the gas flask. ____________________

e Give the approximate volume you should expect to measure for the gas when it is heated.

____

f Should the volume you calculate for the gas when it is chilled be greater or less than the volume of the gas flask?

____

3 If a balloon holds 5.4 L of gas at 29°C, what will be the volume of the balloon at 16°C? Assume that pressure is held constant. Show your work.

Name ____________________

Lab Partner ____________________

Lab Section ____________________ Date ____________________

Lab 15 DATA SHEET

Data and Observations

Simple Gas Laws in Action

Procedure Reference	Description	Trial 1	Trial 2	Trial 3
Step 6	Temperature: boiling water (°C)			
Step 6	Temperature: ice water (°C)			
Step 8	Observation: heating gas flask			
Step 9	Observation: cooling gas flask			
Step 11	Volume: tube (V_T)			
Step 12	Volume: water in gas flask (V_W)			
Step 13	Volume: gas flask (V_{GF})			

Calculations—Trial 1

Procedure Reference	Description	Show Calculations/Data Source	Value
Step 14	Temperature in Kelvin: hot air (T_1)	=	
Step 14	Temperature in Kelvin: cool air (T_2)	=	
Step 15	Volume: hot air (V_1)	=	
Step 15	Volume: cool air (V_2)	=	
Step 16	Experimental value of Eq. 1	=	
Step 16	Percent error (Eq. 2)	=	

Calculations—Trial 2

Procedure Reference	Description	Show Calculations/Data Source	Value
Step 14	Temperature in Kelvin: hot air (T_1)	=	
Step 14	Temperature in Kelvin: cool air (T_2)	=	
Step 15	Volume: hot air (V_1)	=	
Step 15	Volume: cool air (V_2)	=	
Step 16	Experimental value of Eq. 1	=	
Step 16	Percent error (Eq. 2)	=	

Calculations—Trial 3

Procedure Reference	Description	Show Calculations/Data Source	Value
Step 14	Temperature in Kelvin: hot air (T_1)	=	
Step 14	Temperature in Kelvin: cool air (T_2)	=	
Step 15	Volume: hot air (V_1)	=	
Step 15	Volume: cool air (V_2)	=	
Step 16	Experimental value of Eq. 1	=	
Step 16	Percent error (Eq. 2)	=	

Name ______________________________

Lab Partner ______________________________

Lab Section ____________________ Date ____________

Lab 15 DATA SHEET

Reflective Exercises

1 In the table below, present a claim regarding whether your data confirmed Charles' law. Describe your evidence and provide a rationale to justify your claim.

Claim	
Evidence	**Rationale**

2 What are some potential sources of error in the experimental procedure? Explain how these might have affected your results.

____

____

____

____

____

3 The gas you used in this experiment was a mixture rather than a pure substance. Explain why this should have no influence on your results.

____

____

____

4 For Charles' law to apply, only temperature and volume may vary under the initial and final conditions. The quantity of gas and pressure of the gas must be unchanged.

a When you heated the gas flask in the boiling-water bath, the bubbles emerging from the tube into the reservoir indicated that the quantity of gas in the system was changing. Explain why this change in gas quantity did not affect your Charles' law calculation.

b In the boiling-water bath, the gas pressure increased. In the ice-water bath, the gas pressure decreased. Explain why these pressure changes did not affect your Charles' law calculation.

5 Did your experiment provide evidence consistent with Gay-Lussac's and Boyle's laws? Explain your answer.

6 When canning jams and jellies, a small air space is left between the top of the jam and the jar lid, which is only loosely tightened. The jar is then submerged in boiling water for several minutes. During this time, bubbles are observed emerging from the jar. After the jar is removed from the hot water, the lid is soon heard "popping down" tightly on the jar to seal it.

a Explain why bubbles are observed to emerge from the jar when it is in the boiling water.

b Explain why the lid "pops down" on the jar top soon after removing the jar from the boiling water.

Name

Lab Partner

Lab Section Date

Lab 15 DATA SHEET

(continued)

7 A student who correctly performed the experiment in this lab obtained the data below. Use this student's data to verify Charles' law and calculate a percent error from the expected result. Show your work to justify your answer.

Temperature of boiling water = 97.8°C

Temperature of ice water = 2.9°C

Volume of water in gas flask (V_W) = 24.2 mL

Volume of gas flask (V_{GF}) = 143.9 mL

Volume of water in tube (V_T) = 12.6 mL

▲ *Properties of solutions, such as these chromium salt solutions, include concentration and solubility.*

Lab 16

It's No Problem
All About Solutions

Objectives

After completing this lab you should be able to:

- Describe and apply the "golden rule of solubility."
- Calculate the concentration of a solution in mass percent, volume percent, or molarity.
- Experimentally determine the concentration of an aqueous solution.
- Describe the processes of osmosis and dialysis.
- Define the terms *isotonic*, *hypertonic*, and *hypotonic*.
- Identify the direction in which molecules will travel across a semipermeable membrane through osmosis or dialysis.

A **solution** is a homogeneous mixture of one or more substances called **solutes** dissolved in a **solvent**, which comprises the majority of the mixture. The chemical reactions that govern biological systems occur primarily within **aqueous solutions**, for which the solvent is water. An understanding of such systems therefore depends on familiarity with the fundamental characteristics of aqueous solutions.

The ability of a solute to dissolve in a particular solvent is termed **solubility**. "Like dissolves like," the so-called golden rule of solubility, states that substances of similar polarity are more likely to be soluble in each other. For example, ionic salts and polar alcohols tend to be soluble in water, but oil and grease tend to be soluble in nonpolar solvents like alkanes. In addition to water, you may have several solvents of different polarity at home. The main ingredient in most fingernail polish removers is ethyl acetate, a relatively nonpolar substance. Isopropanol, commonly known as rubbing alcohol, has a polarity that is intermediate to water and ethyl acetate.

When working with a solution, it is often important to know the **concentration** (the *quantity* of solute relative to the amount of solution). Most medications, for example, are solutions of one or more active drugs in a solid or liquid mixture. To provide the correct dose of medicine, it is imperative to know the amount of each pharmaceutical (solute) present in a given quantity of the medication (solution). The concentration of any solution may be expressed in a variety of units. Three common concentration units are defined in Table 16.1.

TABLE 16.1 ■ Common Concentration Units and Their Definitions

Concentration Unit (Symbol)	Definition
Mass-mass percent (m/m %)	$\frac{\text{Mass of solute (g)}}{\text{Mass of solution (g)}} \times 100$
Mass-volume percent (m/v %)	$\frac{\text{Mass of solute (g)}}{\text{Volume of solution (mL)}} \times 100$
Molarity (M)	$\frac{\text{Moles of solute (mol)}}{\text{Liters of solution (L)}}$

The concentration of a solution may change if the amount of solvent is altered (by evaporation, for example) or if solute is added or removed. Living cells consist of an aqueous solution encapsulated in a bilayer of phospholipids, which forms the **cell membrane**. The membrane is **semipermeable**, which means that only certain substances may move freely across it while the movement of other substances across the membrane is inhibited.

If a semipermeable membrane separates two aqueous solutions of different concentrations, a process called **osmosis** causes the solvent molecules (water) to pass across the membrane in a direction that would equalize the concentrations on either side (Fig. 16.1). In effect, water migrates from an area of lower solute concentration to an area of higher solute concentration. When the concentrations of the two solutions separated by the membrane are equal, they are said to be **isotonic**. When they are unequal, the more concentrated solution is referred to as **hypertonic** and the less concentrated solution as **hypotonic**.

Dialysis, a process related to osmosis, involves the movement of *solutes* across a semipermeable membrane in addition to water. The solutes also will migrate across the membrane in a direction that equalizes the concentrations of the separated solutions. Thus, solutes will flow from a solution of higher solute concentration (hypertonic) to one with lower solute concentration (hypotonic). Because a membrane is *semi*permeable, not all solutes may participate in dialysis. Those that are able to migrate possess appropriate properties (e.g., size, polarity, and/or charge), as determined by the nature of the membrane.

An important application of this principle is kidney dialysis, which is a medical procedure used to remove waste products from the blood of patients who suffer from kidney malfunctions (Fig. 16.2). The patient's blood is pumped through a dialysis tube immersed in a **dialyzing solution**, which is isotonic to blood with respect to the concentrations of salt, glucose, and other vital compounds.

The dialysis tube has a semipermeable membrane that allows small molecules to pass through while large biomolecules, such as polysaccharides and proteins, are retained. Because the dialyzing solution is hypotonic with respect to waste compounds such as urea, these pass out of the blood through the membrane and into the dialyzing solution. The blood returned to the patient's body has therefore been effectively "cleaned" of waste products.

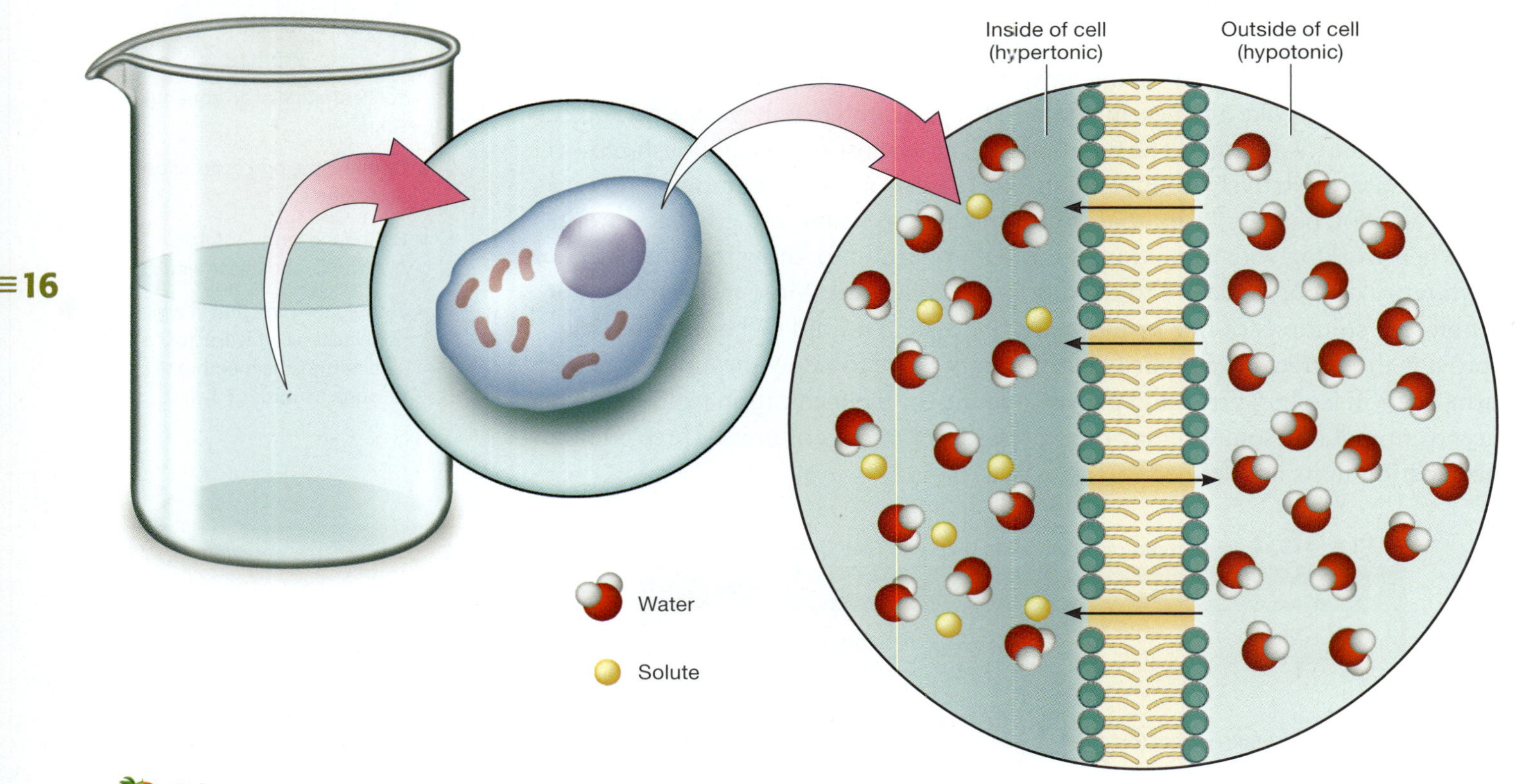

16.1 Osmosis involves the movement of water across a semipermeable membrane in a direction that equalizes the concentration of the two solutions. The inside of this cell has a higher solute concentration than the beaker of water in which it floats. Thus, there is a net movement of water across the membrane and into the cell by osmosis.

In this lab, you will examine the processes of osmosis and dialysis. You also will explore certain solution properties, such as solubility and concentration. In Part A, you will use a potato to study the process of osmosis across cell membranes. In Part B, you will perform dialysis on a mixture of salt, sugar, and starch using a dialysis tube that only allows the passage of small molecules or ions. You will analyze the results of the dialysis based on the presence of each solute in the dialysis tube and the external (dialyzing) solution. In Part C, you will test the concept of "like dissolves like" by examining the solubility of four common household substances in common solvents with different polarities (see Table 16.2). In Part D, you will experimentally determine the concentration of a saltwater solution.

TABLE **16.2** ■ Molecular Formulas and Household Sources of Substances Used in Part C

Substance	Molecular Formula	Household Source
Isopropanol	C_3H_8O	Rubbing alcohol
Ethyl acetate	$C_4H_8O_2$	Fingernail polish remover
Sodium chloride	NaCl	Table salt
Sucrose	$C_{12}H_{22}O_{11}$	Table sugar
Naphthalene	$C_{10}H_8$	Moth balls
Vegetable oil	Mixture of triacylglycerols	Vegetable oil

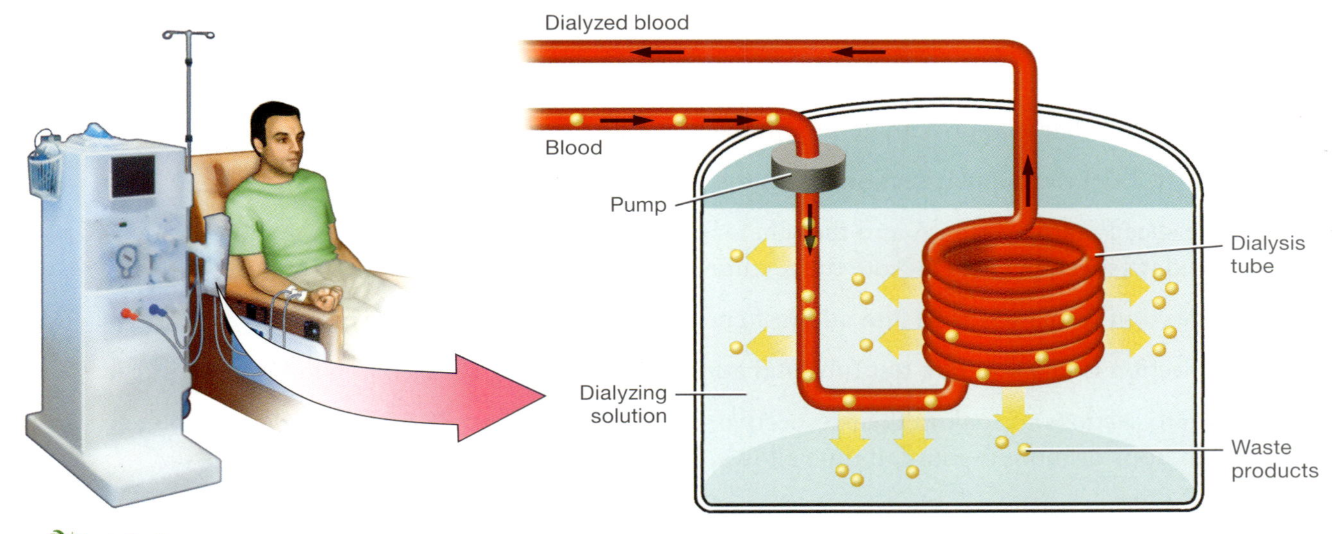

16.2 Kidney dialysis removes waste products by passing blood through a dialysis tube immersed in a dialyzing solution that is isotonic with blood. The waste products migrate out of the blood and into the dialyzing solution.

Procedure

What Am I Doing?

Part A: Examining mass and textural changes in potato pieces after soaking them in different solutions.

Part B: Soaking a dialysis tube filled with a mixture of salt, sugar, and starch in distilled water and then examining the internal and external solutions for the presence of each substance.

Part C: Observing the solubilities of four different household substances in three different liquids.

Part D: Evaporating water from a saltwater sample to determine the amount of salt it contains; calculating concentration.

Materials

- ❑ Small knife or cutting utensil
- ❑ 250 mL beaker
- ❑ 50 mL beaker
- ❑ 400 mL beaker
- ❑ Hot plate
- ❑ Test tubes (12)
- ❑ Funnel
- ❑ Microspatula
- ❑ Evaporating dish
- ❑ Glass stirring rod
- ❑ Dialysis tubing (~20 cm)
- ❑ Tongs or hot mitt
- ❑ 10 mL graduated cylinder

Parts A and B require waiting periods of 30 minutes or more. You may wish to set up the experiments in Parts A and B, and then perform Parts C and D during the wait period. Note that Part B requires a boiling-water bath after the waiting period. You may retain the water bath from Part D in order to complete Part B.

Part A Osmosis

1. Obtain a potato slice that is approximately 1 cm thick.
2. Cut two 4 cm × 1 cm rectangular sections from the potato slice.
3. Designate one slice Potato Section A and the other slice Potato Section B. Weigh each section and record their exact masses on the data sheet, page 245. Note the texture and appearance of the potato sections.
4. Place Potato Section A in a test tube and add just enough distilled water to cover it.
5. Place Potato Section B in a separate test tube and add just enough of a 10% NaCl solution to cover it.
6. Allow the potato sections to soak in their respective solutions for at least 30 minutes. It is even better to soak the sections for an hour or more if your lab time allows.
7. Following the soaking period, remove the potato sections from their respective solutions and quickly pat them dry using a paper towel.
8. Reweigh each potato section and record their exact masses on the data sheet. Also, record any changes in the texture or appearance of the potato sections.
9. Unless otherwise directed by your instructor, the salt solution may be poured down the drain and the potato sections may be thrown in the trash.

Part B Dialysis

1. In a small beaker, combine about 5 mL of each of the following solutions: 10% NaCl, 10% glucose, and 1% starch.
2. Obtain approximately 20 cm of dialysis tubing. Wet the tubing with distilled water to help open it, and tie a knot in one end of the tube.
3. Open the opposite end of the tube, insert a funnel, and pour in the mixture you prepared in Step 1. Do not use all of the solution if it would overfill the tube. Be sure to leave enough room at the end of the tube to tie another knot.

4. Twist the open end of the tube and close it by tying a knot.
5. Place the dialysis tube horizontally in a 400 mL beaker and add just enough distilled water to cover it entirely.
6. Allow the tube to sit in the water for 45 minutes. While waiting, label six test tubes 1–6.
7. After 45 minutes have elapsed, remove 10 mL of solution from the beaker. Divide the sample equally into the three test tubes labeled 1, 3, and 5.
8. Remove the dialysis tube from the beaker and untie or cut open one end. Divide the contents equally among the three test tubes labeled 2, 4, and 6.
9. Add 5 drops of 0.1 M $AgNO_3$ each to Test Tubes 1 and 2. Shake the tubes gently. Record your observations on the data sheet, page 245, and interpret the results to indicate the presence (+) or absence (−) of chloride in each sample. The formation of a white precipitate indicates the presence of chloride. Dispose of the test tube contents as directed by your instructor.
10. Add approximately 2 mL of Benedict's reagent each to Test Tubes 3 and 4. Place the test tubes in a boiling-water bath for about 5 minutes. Formation of a red to greenish-yellow precipitate indicates the presence of glucose. Record your observations on the data sheet and interpret the results to indicate the presence (+) or absence of glucose (−) in each sample. Dispose of the test tube contents as directed by your instructor.
11. Add 3 drops of iodine reagent each to Test Tubes 5 and 6. Mix the contents by swirling the tubes. Formation of a blue-black color indicates the presence of starch. Record your observations on the data sheet and interpret the results to indicate the presence (+) or absence (−) of starch in each sample. Dispose of the test tube contents as directed by your instructor.

Part C Like Dissolves Like

1. Label four dry test tubes 1–4.
2. To each test tube, add a rice grain-sized amount of the appropriate substance as indicated below:
   - Test Tube 1: Sodium chloride
   - Test Tube 2: Sucrose
   - Test Tube 3: Naphthalene
   - Test Tube 4: Vegetable oil

Avoid inhalation of vapors from isopropanol and ethyl acetate. Naphthalene is a suspected carcinogen. Wear gloves and perform this experiment in a fume hood or appropriately ventilated space!

3. Add approximately 3 mL of ethyl acetate to each test tube and mix the contents by gently agitating the tube.
4. Allow the test tubes to stand undisturbed for at least 1 minute and then record your observations on the data sheet, page 246. If the sample is soluble in the solvent, the mixture will appear uniform. If it is insoluble, then you will see solids, a cloudy mixture, or the formation of two separate layers. Use your observations to indicate whether the solute is soluble (S) or insoluble (I) in water on the data table.
5. Empty the test tubes into the appropriate waste container and rinse them with isopropanol.
6. Repeat Steps 2–4 using isopropanol in place of ethyl acetate.
7. Empty the test tubes into the appropriate waste container and rinse them with distilled water.

8. Repeat Steps 2–4 using distilled water in place of ethyl acetate.
9. Dispose of all the test tube contents in the appropriate waste container.

Part D Determining the Concentration of a Solution

1. Prepare a boiling-water bath using a hot plate and a 250 mL beaker half-filled with water.
2. Record the mass of a clean, dry evaporating dish on the data sheet, page 246.
3. Measure out approximately 10 mL of a saltwater solution and add it to the evaporating dish. Be sure to record the exact volume of your sample on the data sheet. Note that the solution is a mixture of sodium chloride dissolved in water.
4. Weigh the dish containing the saltwater sample and record the exact mass on the data sheet.
5. Place the evaporating dish on the boiling-water bath (Fig. 16.3). Allow the water to evaporate until the salt residue is completely dry. You will need to stir the mixture constantly with a glass rod after most of the water has evaporated to prevent the solution from spattering as the last few drops of water change into a gas. (Remember that you should retain your water bath for Part B analysis if you performed this step during the wait period.)
6. Remove the dish from the boiling-water bath and use a paper towel to remove any moisture from the outside surface.
7. Reweigh the dish containing the dry salt and record the exact mass on the data sheet.
8. Fill in the calculations table on the data sheet. Calculate the concentration of the saltwater solution by mass-mass percent, mass-volume percent, and molarity.
9. Dispose of the dry salt as directed by your instructor.

Steam from the boiling-water bath can cause severe burns. Exercise caution and use tongs or a hot mitt to remove the evaporating dish!

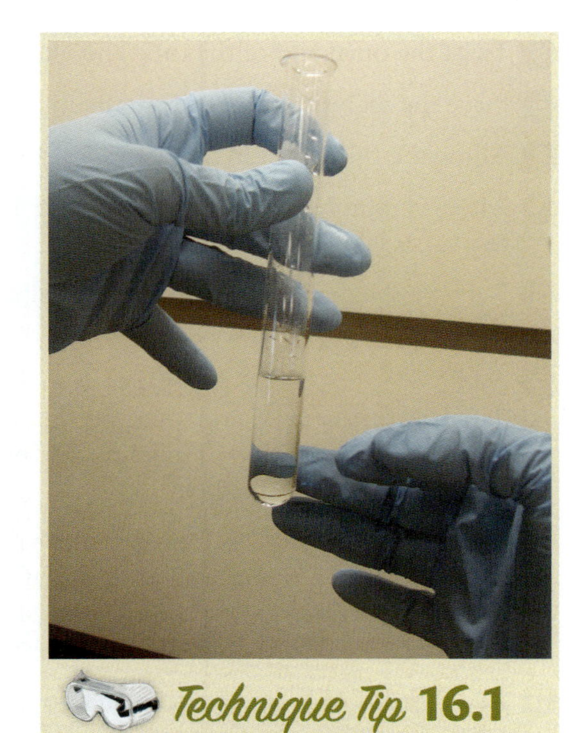

Technique Tip 16.1

To safely shake a test tube, hold it between your thumb and split first two fingers. Flick the test tube using a finger from the opposite hand.

16.3 Apparatus for evaporating water from the saltwater solution.

Name ______________________________

Lab Partner ______________________________

Lab Section ____________________ Date ____________

Lab 16
Pre-Laboratory Exercise

1 Provide a term that matches each description below.

a Involves the movement of water across a semipermeable membrane. ____________

b Majority component of a solution. ____________

c Describes the less concentrated solution within a pair. ____________

d Quantity of solute in a given amount of solution. ____________

e Involves the movement of solute across a semipermeable membrane. ____________

f Minority component of a solution. ____________

g Descriptor for two solutions of equal concentration. ____________

2 Wilted lettuce may be revived by placing the leaves in a bowl of water. Name the process that accounts for this and explain what is happening to cause the leaves to stiffen.

____

____

____

3 Consider two solutions separated by a semipermeable membrane, as shown in the illustration to the right. The membrane allows the passage of small molecules and ions, but not large molecules like polysaccharides or proteins. Solution A is composed of 10% glucose and 10% albumin protein dissolved in water. Solution B is a 5% solution of NaCl in water. Indicate whether each substance in the system would exhibit a net flow into Solution A, Solution B, or neither.

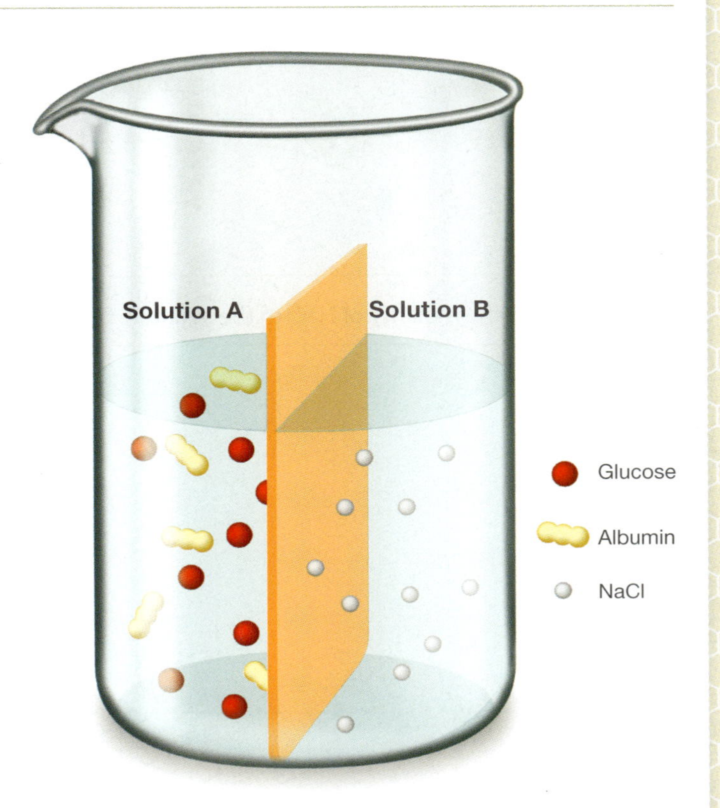

a Water ____________

b NaCl ____________

c Albumin ____________

d Glucose ____________

4 Why is it important to ensure that the test tubes used in Part C are dry before investigating the abilities of the substances to dissolve in ethyl acetate?

5 A 33.87 mL solution with an unknown salt concentration had a mass of 34.72 g. When the water was evaporated from the solution, 1.72 g of NaCl remained.

a Calculate the mass-mass percent NaCl concentration in the solution. Show your work.

b Calculate the mass-volume percent NaCl concentration in the solution. Show your work.

c Calculate the molarity of NaCl in the solution. Show your work.

Name ____________________

Lab Partner ____________________

Lab Section ____________________ Date ____________________

Lab 16 DATA SHEET

Data and Observations

Part A Osmosis

Testing Osmosis

Sample	Mass before Soaking (Step 3)	Mass after Soaking (Step 8)	Observations
Potato Section A			
Potato Section B			

Part B Dialysis

Testing Dialysis

Test Tube	Sample	Testing for	Observations	Result (+ or −)
1	Outside tube	Chloride		
2	Inside tube	Chloride		
3	Outside tube	Glucose		
4	Inside tube	Glucose		
5	Outside tube	Starch		
6	Inside tube	Starch		

Part C Like Dissolves Like

Testing Solubility

		Ethyl Acetate		Isopropanol		Distilled Water	
Test Tube	Sample	Observations	Soluble (S) or Insoluble (I)	Observations	Soluble (S) or Insoluble (I)	Observations	Soluble (S) or Insoluble (I)
1	Sodium chloride						
2	Sucrose						
3	Naphthalene						
4	Vegetable oil						

Part D Determining the Concentration of a Solution

Procedure Reference	Quantity	Value
Step 2	Mass: evaporating dish (g)	
Step 3	Volume: saltwater sample (mL)	
Step 4	Mass: evaporating dish + saltwater (g)	
Step 7	Mass: evaporating dish + dry salt (g)	

Saltwater Concentrations—Calculations Table

Quantity	Show Calculation	Result
Mass: saltwater sample (g)	=	
Mass: dry salt (g)	=	
Saltwater concentration: mass-mass percent	=	
Saltwater concentration: mass-volume percent	=	
Saltwater concentration: molarity (mol/L)	=	

Name ______________________

Lab Partner ______________________

Lab Section ______________________ Date ______________

Lab 16 DATA SHEET

(continued)

Reflective Exercises

1 Explain the cause of any changes in the masses of the potato sections you observed in Part A.

2 Did the texture of either of the potato sections in Part A change? If so, try to explain the cause of the difference at the molecular level.

3 Pickles are made by immersing cucumbers in a concentrated saltwater solution. Explain what happens to the cucumber in this process to cause it to shrink and taste salty.

4 Based on your results from Part B, which substances migrated through the walls of the dialysis tube? Which substances were not able to migrate through the tube walls? Use your data to justify your answer.

5 Among the substances that were able to migrate through the walls of the dialysis tube, did every molecule or ion of the substance in the tubes exit or did some stay in the tube? Use your data to support your answer.

6 Fill in the blanks to complete the following sentences.

a The distilled water was ______________________ relative to the cells of the potato in Part A.

b The solution in the Part B dialysis tube was initially ______________________ with respect to the distilled water bath.

c The driving force for both osmosis and dialysis is the generation of ______________________ solutions on either side of a membrane.

7 Explain why it is important that intravenous solutions used to deliver fluids or medications to patients are isotonic with respect to the patients' blood.

8 Consider your data from Part C, and answer the following questions:

a Propose an explanation for your observations regarding the solubility of sodium chloride in ethyl acetate, isopropanol, and water.

b Based on the data in Table 16.2 (p. 239), would you expect sucrose to be polar or nonpolar? What about naphthalene? Explain your answers.

c Were your results consistent with your answer to Question 6b? Explain.

d Table 16.2 described vegetable oil as a "mixture of triacylglycerols." What does your data suggest about the physical properties of triacylglycerols? What can you infer about the types of bonds in these molecules? Explain.

Name ______________________

Lab Partner ______________________

Lab Section ______________ Date ______________

Lab 16
DATA SHEET
(continued)

9 Vinegar is about 95% water mixed with acetic acid, another polar compound. Explain why salad dressings composed of olive oil and vinegar separate into two layers.

10 If you have ever peeled the label off of a glass jar, you may have noticed that the glue does not easily wash off with water. However, it can be easily removed with another common household solvent—fingernail polish remover (ethyl acetate). What does this tell you about the chemical composition of the glue? Explain your answer.

11 A solution is prepared by dissolving 10.98 g of $MgBr_2$ in 646 mL of water. The total mass of the solution is 656.98 g, and the total volume is 651 mL.

a What is the mass-mass percent concentration of $MgBr_2$ in the solution? Show your work.

b What is the mass-volume percent concentration of $MgBr_2$ in the solution? Show your work.

c What is the molarity of the $MgBr_2$ in the solution? Show your work.

12 You have a 2.92 M solution of NaCl in water. How many milliliters of the solution do you need if you plan to obtain 5.86 g of NaCl by evaporating the water? Show your work.

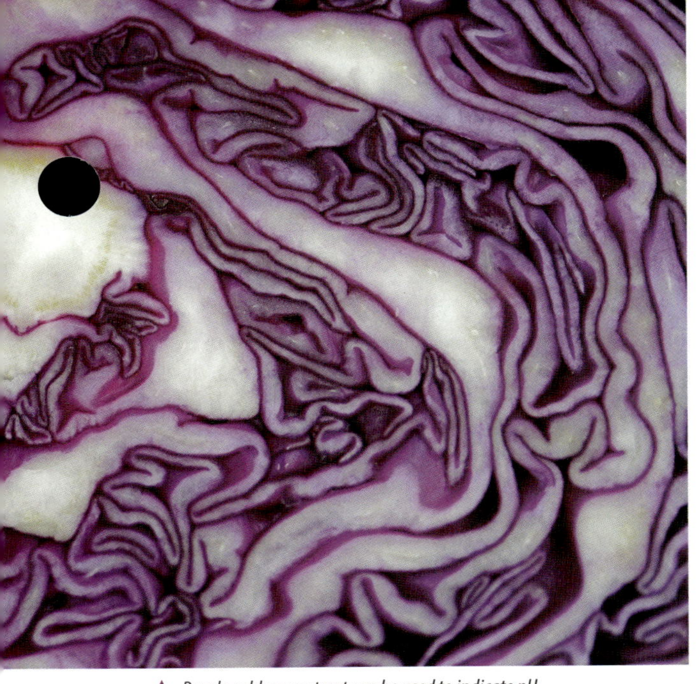

▲ *Purple cabbage extract can be used to indicate pH.*

Lab 17

True Colors

Acids, Bases, Buffers, and pH Indicators

Objectives

After completing this lab you should be able to:

- Describe how a natural organic dye may be used as an indicator of pH.
- Define the terms *acid*, *base*, and *pH*.
- Describe the pH scale.
- Identify conjugate acid-base pairs.
- Prepare a natural pH indicator from red cabbage and use it to measure the pH of various solutions.
- Analyze the relative effectiveness of various antacids.
- Write a balanced neutralization reaction between a strong acid and base.
- Describe the typical components and properties of a buffer solution.
- Describe the different behaviors of normal and buffer solution upon the addition of an acid or base.

The natural world offers an abundance of vibrant colors. Consider the rainbow of pigments you can see in the fruits and vegetables available at your local grocery store, in a picture of a coral reef, or during the seemingly magical change in foliage each autumn. In most cases it is the chemical structures of organic (carbon containing) compounds in these living organisms that give rise to the array of natural hues you have encountered. These structural features, in turn, may depend on the acid-base properties (i.e., pH) of their environment.

Acids and Bases

The classification of a compound as an acid or a base reflects its participation in a specific type of chemical reaction: the transfer of a hydrogen ion (H^+), usually referred to simply as a *proton*, from one species to another (Fig. 17.1). Such a reaction involves the asymmetrical breaking of a bond to hydrogen in one of the reactants (H–A), releasing a proton (hydrogen nucleus) that is unaccompanied by any electrons and therefore bears a positive charge. The other reactant (B:)

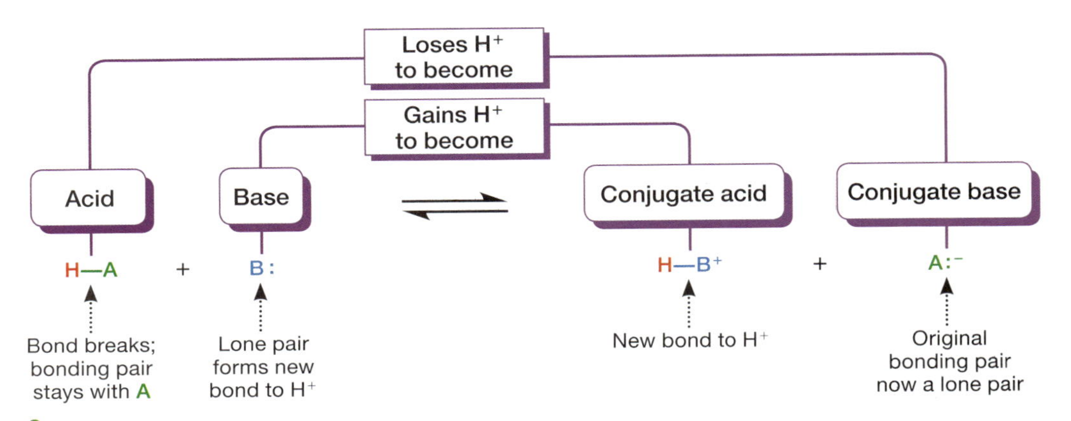

17.1 An acid-base reaction involves the transfer of H^+ from one species to another. In the generic reaction, H–A is the acid (proton donor) and B: is the base (proton acceptor). The reaction generates products that differ from the reactants only in the location of the proton. Loss of H^+ from the acid generates its conjugate base. Gain of H^+ by the base generates its conjugate acid.

uses a lone pair of electrons to form a new bond to the proton. The net result is the transfer of H^+ from one reactant to the other, generating the products (H–B^+ and A:$^-$). The reactant that donates the proton is called the **acid**, and the reactant that accepts the proton is the **base**.

Note the structural changes that accompany an acid-base reaction. First, the proton transfer is accompanied by a charge transfer. Upon donating the H^+, an acid will lose one unit of positive charge. Similarly, acceptance of H^+ by the base will cause it to gain one unit of positive charge. For example, neutral HCl loses H^+ to become negatively charged Cl^-, and neutral H_2O accepts H^+ to become positively charged H_3O^+ (Fig. 17.2).

Second, the reaction diagram in Figure 17.1 illustrates that the structures of the products differ from those of the reactants only in the location of the proton. Two chemical species that differ from each other by a single proton are said to constitute a **conjugate acid-base pair.** The substance formed when an acid loses a proton is called its **conjugate base.** The substance formed when a base gains a proton is called its **conjugate acid.** The Cl^- and H_3O^+ formed during the reaction of HCl with H_2O are the conjugate base and conjugate acid of the respective reactants (Fig. 17.2). Examples of other conjugate acid-base pairs are given in Figure 17.3.

$$\underset{\text{Acid}}{HCl} + \underset{\text{Base}}{H{-}\ddot{O}{-}H} \longrightarrow H_3O^+ + Cl^-$$

17.2 An example of an acid-base reaction. The acid (HCl) donates a proton to the base (H_2O), producing the conjugate base (Cl^-) and conjugate acid (H_3O^+) ions.

Conjugate Acids		Conjugate Bases
H—Cl:	lose H^+ ⇌ gain H^+	:Cl:$^-$
NH_4^+	lose H^+ ⇌ gain H^+	NH_3
H—O—H	lose H^+ ⇌ gain H^+	H—O:$^-$
H_3C—C(=O)—O—H	lose H^+ ⇌ gain H^+	H_3C—C(=O)—O:$^-$

17.3 Examples of conjugate acid-base pairs. When an acid loses a proton, its conjugate base is produced. When a base gains a proton, its conjugate acid is produced.

The pH Concept

Chemical reactions in biological systems take place in aqueous solutions, and the presence of acids or bases in these solutions can dramatically influence the nature of the chemical species present. For example, the ability of a protein molecule to function properly may be altered if one or more of its amino acid building blocks gains or loses a proton. As described in the opening paragraph, the structures and corresponding colors of many organic pigments depend on the acid-base properties of a solution.

The molar concentration of the hydronium ion (H_3O^+) was chosen to describe the acidity of an aqueous solution. Because this value can vary over a large range, chemists developed a corresponding quantity called **pH,** which is defined as the negative logarithm of the molar hydronium ion concentration (Eq. 1).

≡17

$$pH = -\log [H_3O^+] \quad \textbf{[Eq. 1]}$$

To understand the concept of solution pH, let us first consider the situation in pure water. Water has the ability to act as an acid (by donating a proton) or a base (by accepting a proton at one of the oxygen lone pairs). When two water molecules react with each other, a proton transfer may take place to produce one hydronium ion and one hydroxide ion (Fig. 17.4). This reversible reaction proceeds only to a small extent, and the concentrations of hydronium and hydroxide ions in pure water have been

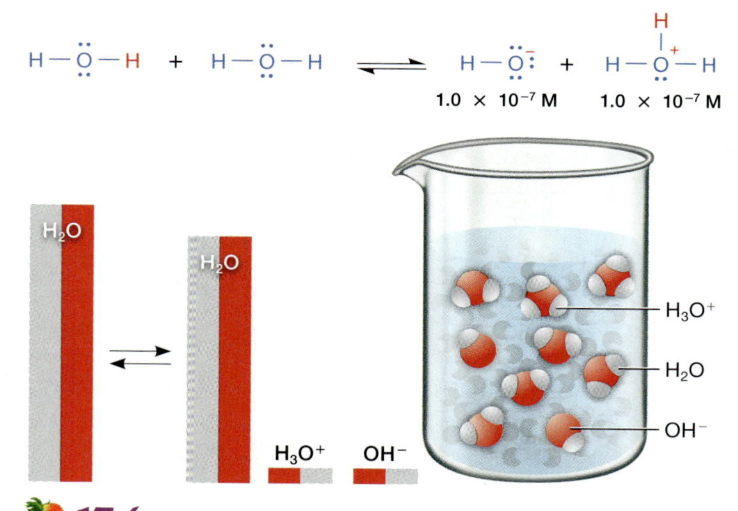

17.4 When two water molecules react with each other, one behaves as a base while the other behaves as an acid. This process occurs to a very small extent in water, producing small concentrations of hydronium and hydroxide ions. Each reaction produces one H_3O^+ and one OH^-, so the concentrations of the two species are equal in pure water.

determined to be 1.0×10^{-7} M. Water is said to be a **neutral solution** because the concentration of basic OH^- and acidic H_3O^+ ions are equal (Fig. 17.4).

When an acid is added to water, it will donate a proton to water molecules (and the small quantity of hydroxide ion present). As a result, the concentration of H_3O^+ will increase (and the concentration of OH^- will decrease). A solution in which the H_3O^+ concentration is *greater than* 1.0×10^{-7} is said to be **acidic** (Fig. 17.5). By contrast, the concentration of H_3O^+ decreases (and that of OH^- increases) when a base is added to water. Such a solution is **basic** and has a H_3O^+ concentration *less than* 1.0×10^{-7} (Fig. 17.5).

By applying Equation 1, we can calculate the pH of pure water to be 7. Any solution with a pH = 7 is considered neutral. A pH value less than 7 corresponds to a H_3O^+ greater than 1.0×10^{-7} M and means that the solution is acidic (contains more H_3O^+ ions than pure water). A pH greater than 7 corresponds to a H_3O^+ concentration less than 1.0×10^{-7} M and indicates that a solution is basic (contains fewer H_3O^+ ions than pure water). Most ordinary solutions have a pH between 0 and 14. Thus, the **pH scale**, illustrated in Figure 17.6, provides a simple gauge of a solution's acidity or basicity.

Acid-base indicators are compounds that vary in color depending upon the acidity of the solution and therefore may be used as a measure of pH. For many indicators, the compounds that make up a conjugate acid-base pair exhibit different colors (Fig. 17.7). As a result, the conjugate acid form (H-In) and its color is predominant at lower pH (high H_3O^+ concentrations) while the conjugate base form (In^-) and its color is predominant at higher pH values (low H_3O^+ concentrations). The specific H_3O^+ concentration required for the structure/color transition depends upon the acid-base properties of the indicator molecule. Certain indicators may undergo more complex structural changes that allow them to exhibit a variety of colors that correspond to various pH values.

The utility of an indicator as a measure of pH depends upon a number of factors, including knowledge of the compound's behavior at various pH values. Certain indicators provide only very general information. Litmus, for example, exhibits a red color in acidic solutions and a blue color in basic solutions. Others may provide more specific information since a color change is observed at a specific pH value.

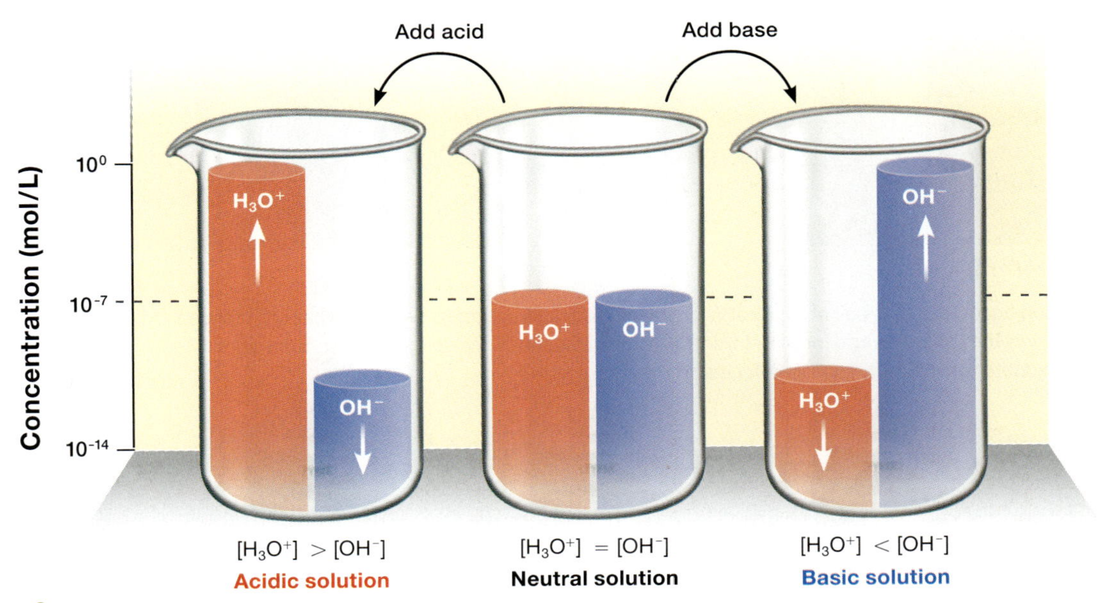

17.5 A neutral solution has an equal concentration (1.0×10^{-7} M) of H_3O^+ and OH^- ions. An acidic solution is one in which the concentration of H_3O^+ is greater than 1.0×10^{-7} M and the OH^- concentration. A basic solution is one in which the concentration of H_3O^+ is less than 1.0×10^{-7} M and the OH^- concentration.

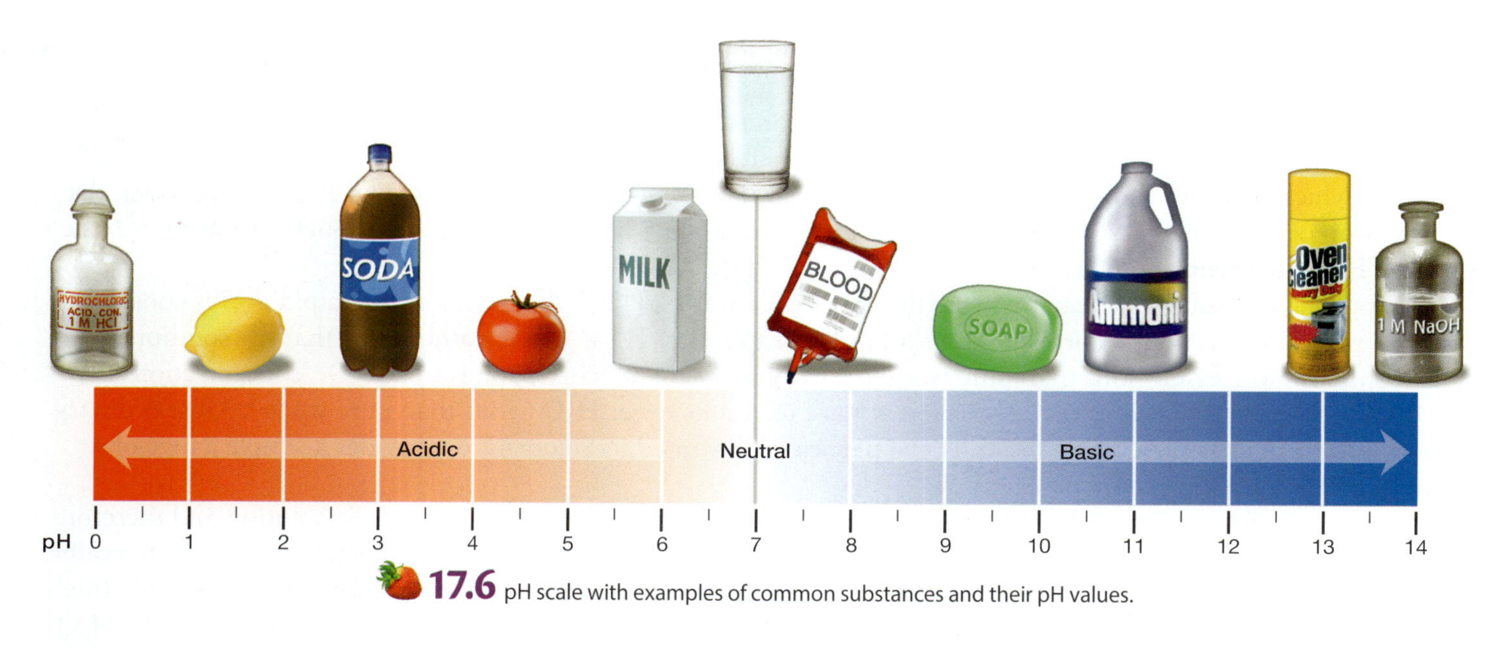

17.6 pH scale with examples of common substances and their pH values.

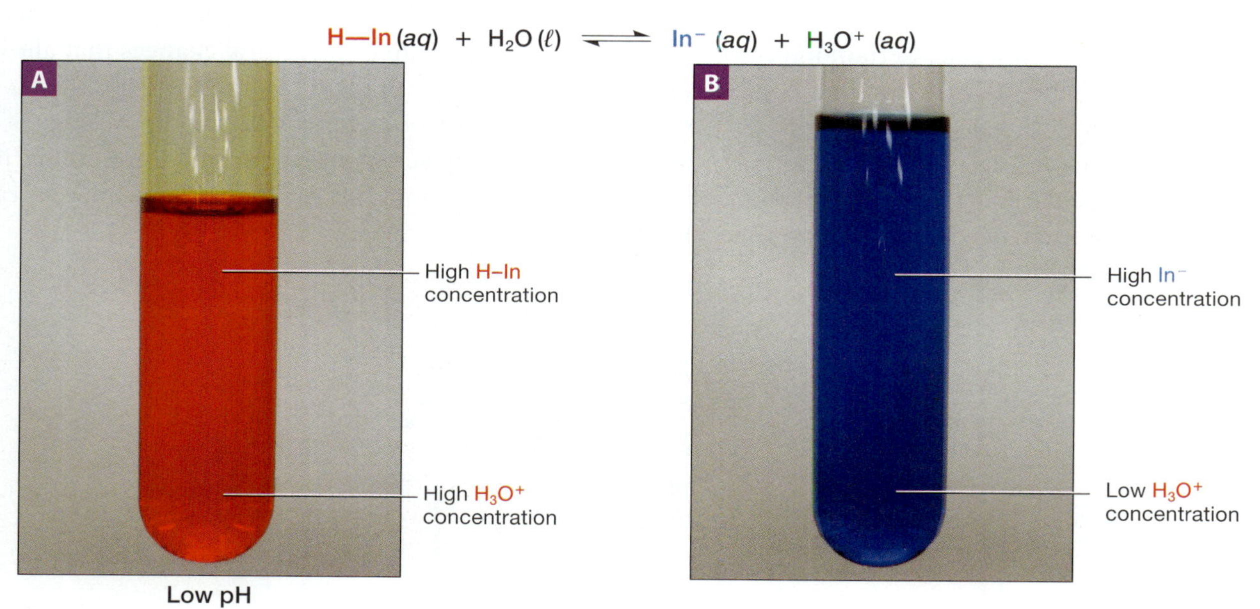

17.7 Conjugate acid-base pair of a typical indicator exhibits different colors. (**A**) When the H_3O^+ concentration is high, the equilibrium favors the conjugate acid form (H–In) and its color is evident. (**B**) When the H_3O^+ concentration is low, the equilibrium favors the conjugate base form (In^-) and its color is evident.

Neutralization Reactions and Buffers

The pH of your body and its various organs are very carefully controlled in order to maintain proper physiological function. A normal blood pH is about 7.4 and does not vary much. Your stomach, however, produces HCl to maintain the acidic environment necessary to digest foods. The stomach may sometimes overproduce acid, resulting in the condition we call "heartburn." Antacid medications used to combat heartburn are basic compounds that react with the excess HCl in a **neutralization reaction**, which involves the generation of neutral products from the combination of an acid and a base.

In most cases, the products of a neutralization reaction are a salt (ionic compound) and water. For example, the reaction of aqueous HCl with aqueous NaOH produces aqueous sodium chloride (NaCl) and water (Fig. 17.8). The reaction is essentially irreversible because the products are far less reactive and therefore strongly favored energetically. If HCl is treated with an exactly equal quantity of NaOH, then the product solution should have a neutral pH because it will not contain any excess acid or base.

Most solutions will exhibit significant pH changes when the level of acid or base in solution is altered even moderately. **Buffers** are solutions that resist large changes in pH upon addition of either acid or base (Fig. 17.9). Your blood, for example, is a buffered solution that maintains a physiologically vital pH near 7.4. Significant deviations from this value can cause serious health problems, including death.

Most buffers function by containing both an acidic and a basic component, typically a conjugate acid-base pair. The conjugate acid will neutralize any added base while the conjugate base will neutralize any added acid. An example of a buffer solution is a mixture of acetic acid (CH_3CO_2H) and acetate ($CH_3CO_2^-$) (Fig. 17.10). (Recall that charged species like acetate are always coupled with a counterion and are typically introduced into solution in the form of a salt such as sodium acetate.)

If an acid, such as HBr, is added to the buffer solution, the basic acetate ion will accept the proton and prevent an increase in the H_3O^+ concentration (Fig. 17.10A). On the other hand, an added base such as NaOH would be neutralized by reaction with the acetic acid, again preventing any change in H_3O^+ concentration (Fig. 17.10B).

In this lab, you will extract an organic pigment from red cabbage to use as a natural pH indicator for various solutions. In Part B, you will measure the pH of several household chemicals using a variety of methods, including the red cabbage indicator, litmus paper, a wide range pH paper, and a digital pH meter. This will allow you to compare the utility of the various measurement techniques. In Part C, you will use the red cabbage indicator to test the effectiveness of several commercial antacids at neutralizing an acid solution, and you will examine the special properties of a buffer solution in Part D.

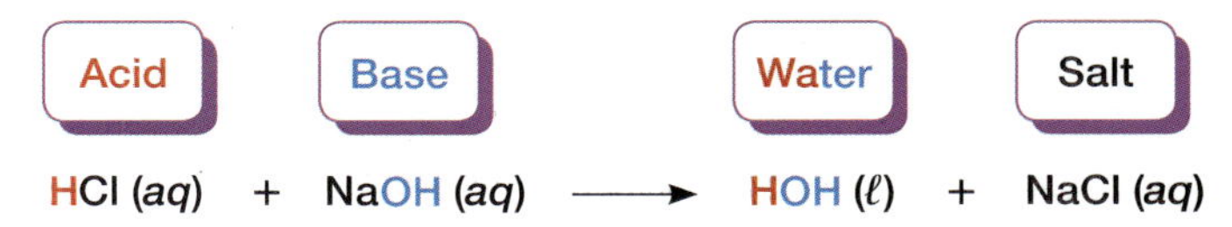

17.8 Neutralization reaction between HCl and NaOH produces water (H_2O) and a salt (NaCl). The reaction is essentially complete, which means that combining exactly equal quantities of the reactants will generate an aqueous NaCl solution with a neutral pH.

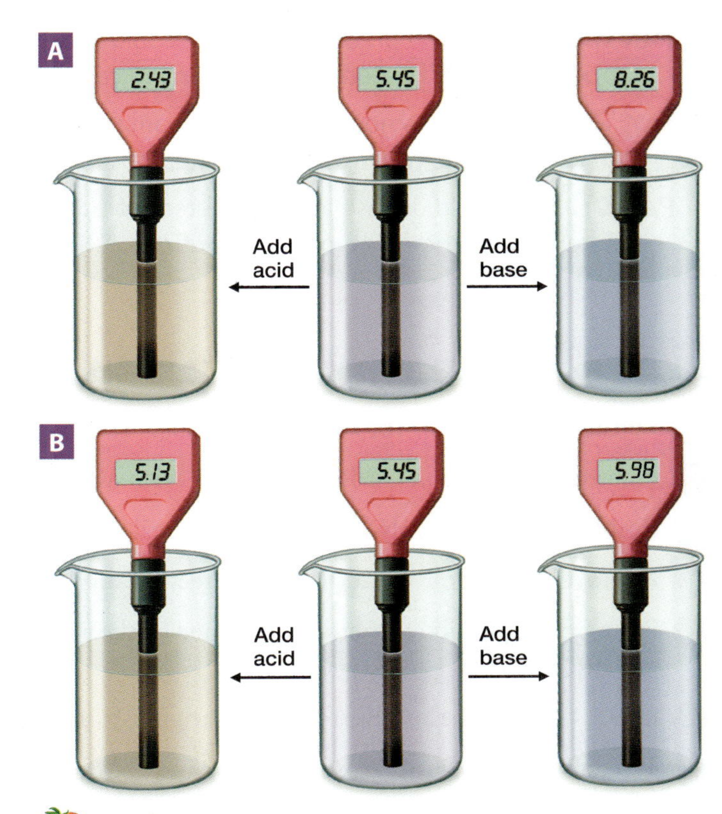

17.9 (**A**) Normal solutions experience large changes in pH when small amounts of acid or base are added. (**B**) A buffer is a solution that resists large changes in pH upon addition of acid or base.

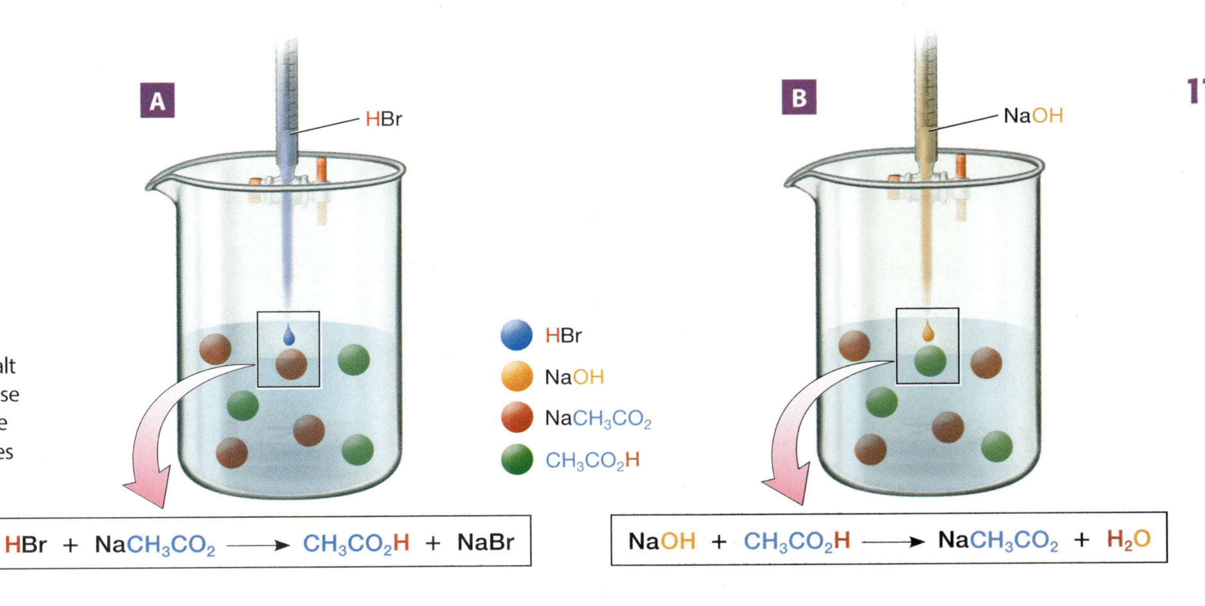

17.10 An acidic buffer usually contains an acid and a salt of its conjugate base. (**A**) The base neutralizes any added acid while (**B**) the conjugate acid neutralizes an added base.

Procedure

What Am I Doing?

Part A: Preparing a solution of indicator from red cabbage and a reference scale of the indicator colors at pH values between 1 and 14.

Part B: Measuring the pH of various household substances using a pH probe and variety of methods based on indicators.

Part C: Adding small amounts of different antacids to "stomach acid" (0.1 M HCl) to determine which sample can neutralize the acid most effectively.

Part D: Adding acid or base to four different solutions to observe the different pH changes in typical and buffer solutions.

Materials

- ❑ 250 mL beakers (2)
- ❑ Hot plate
- ❑ Test tubes (20)
- ❑ pH meter
- ❑ Litmus paper (red and blue)
- ❑ pH paper (wide range)
- ❑ Glass stirring rod
- ❑ Dropper
- ❑ Mortar and pestle
- ❑ 50 mL beakers (3)
- ❑ 10 mL graduated cylinders
- ❑ Microspatula

Part A Preparing the pH Indicator and a Reference Scale

1. Obtain a handful of shredded red cabbage leaves.
2. Place the cabbage in a 250 mL beaker with about 100 mL of distilled water.
3. Boil the mixture gently on a hot plate until the liquid becomes a dark color (about 10 minutes).
4. Remove the beaker from the hot plate and allow it to cool.
5. Pour off the liquid extract into another beaker, leaving behind the solid material. This liquid is your "red cabbage indicator."
6. Label 14 test tubes with pH values ranging from 1–14.
7. To each test tube, add 2 mL of a prepared buffer solution with the corresponding pH.
8. Add 2 mL of red cabbage indicator to each test tube.
9. Record the color of each solution on the data sheet, page 261. **Save this set of test tubes for reference throughout the experiment.**

Part B Comparing Methods of Measuring pH

1. Your instructor will supply various household chemicals. Add about 2 mL of each chemical to its own test tube or vial. You will test the pH of each substance using litmus paper, wide-range pH paper, red cabbage indicator, and a digital pH meter.
2. Test the pH of each substance using litmus paper by following the steps below:
   - **a** Wet a piece of red litmus paper with distilled water. Use a clean, dry glass stirring rod to transfer 1 drop of the substance to the paper. Record the resulting color of the paper and what it indicates about pH on the data sheet, page 262.
   - **b** Wet a piece of blue litmus paper with distilled water. Use a clean, dry glass stirring rod to transfer 1 drop of the substance to the paper. Record the resulting color of the paper and what it indicates about pH on the data sheet.
3. Test the pH of each substance using wide-range pH paper: Wet a piece of pH paper with distilled water. Use a clean, dry glass stirring rod to transfer 1 drop of the substance to the paper. On the data sheet, record the resulting color and pH by comparing it to the scale associated with the paper.

4. Test the pH of each substance using a digital pH meter: Rinse the probe with distilled water and then dip it directly into the sample. On the data sheet, record the pH value after the meter reading stabilizes. Remove the probe and rinse it with distilled water before testing the next substance.
5. Test the pH of each substance using the red cabbage indicator: To each sample, add approximately 2 mL of red cabbage indicator. Use the color scale of your red cabbage indicator reference set (created in part A) to determine the pH. Record the colors and associated pH values on the data sheet.
6. Discard the samples as directed by your instructor. **Do not discard your red cabbage indicator reference set.**

Part C Antacid Effectiveness

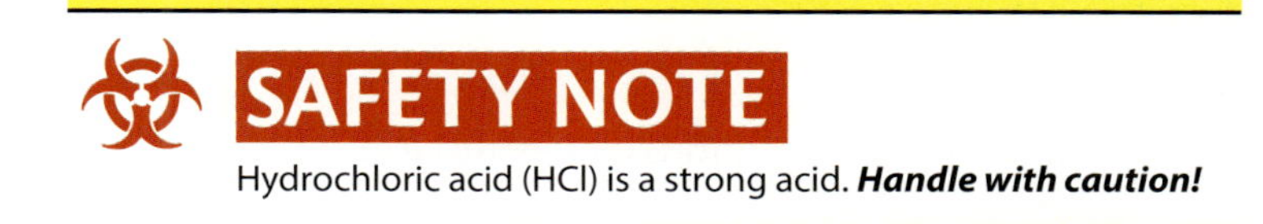

1. Obtain a single tablet of three different commercial antacids and designate them as Samples 1–3.
2. For each antacid sample, record on the data sheet, page 262, the name and mass(es) of the active ingredient(s) in one tablet according to the package label.
3. Weigh each tablet and record the mass on the data sheet.
4. Label three small beakers 1–3.
5. Use a mortar and pestle to grind antacid Sample 1 into a powder. Add the powdered sample to Beaker 1. Record the mass of the beaker with the antacid sample on the data sheet.
6. Clean and dry the mortar and pestle, then repeat Step 5 for Samples 2 and 3.
7. Label three test tubes 1–3.
8. Add 3 mL of 0.1 M HCl and 2 mL red cabbage indicator to each test tube.
9. To Test Tube 1, add rice-grain sized quantities of Antacid 1 until the color indicates that a neutral pH is achieved. With each addition, agitate the test tube well, ensure that all the antacid gets into the solution, and allow the antacid to react for about 2 minutes. Record your observations on the data sheet.
10. Reweigh Beaker 1 with the unused antacid and record this value on the data sheet.
11. Calculate the mass of Antacid 1 you used to neutralize the HCl solution.
12. Use Test Tubes 2 and 3 to repeat Steps 9–11 with the corresponding antacid samples.
13. Discard the samples as directed by your instructor. **Do not discard your red cabbage indicator reference set.**

Part D Buffer Solutions

1. Label four test tubes 1–4.
2. To each test tube, add the appropriate solutions as indicated below:
   - Test Tube 1: 3 mL of distilled water and 2 mL of red cabbage indicator
   - Test Tube 2: 3 mL of 0.1 M acetic acid and 2 mL of red cabbage indicator
   - Test Tube 3: 3 mL of 0.1 M sodium acetate and 2 mL of red cabbage indicator
   - Test Tube 4: 1.5 mL of 0.1 M acetic acid, 1.5 mL of 0.1 M sodium acetate, and 2 mL of red cabbage indicator
3. Use the color scale of your red cabbage indicator reference set to measure the initial pH of the solution in each test tube. Record these values on the data sheet, page 263.

4. To each test tube, add 1 drop of 0.1 M HCl. Note any color changes and record the pH of each solution on the data sheet.
5. To each test tube, add another 5 drops of 0.1 M HCl. Note any color changes and record the pH of each solution under the "6 drops" column on the data sheet.
6. To each test tube, add another 5 drops of 0.1 M HCl. Note any color changes and record the pH of each solution under the "11 drops" column on the data sheet.
7. To each test tube, add another 10 drops of 0.1 M HCl. Note any color changes and record the pH of each solution under the "21 drops" column on the data sheet.
8. To each test tube, add an additional 10 drops of 0.1 M HCl. Note any color changes and record the pH of each solution under the "31 drops" column on the data sheet.
9. Dispose of the test tube contents as directed by your instructor. Rinse the test tubes thoroughly.
10. Repeat Steps 1–9, using 0.1 M NaOH in place of 0.1 M HCl in Steps 4–8.

Name ______________________

Lab Partner ______________________

Lab Section ______________________ Date ______________________

Lab 17 Pre-Laboratory Exercise

1 Provide a term that matches each description below.

a Most common measure of solution acidity. ______________________

b Substance that readily donates a proton. ______________________

c Solution that resists a pH change when acid or base is added. ______________________

d Two substances related to each other by the presence of a single H^+. ______________________

e Substance that readily accepts a proton. ______________________

f Reaction between an acid and a base that produces neutral products. ______________________

2 Calculate the pH of each of the following solutions, and label each as acidic, basic, or neutral.

Solution	pH	Acidic/Basic/Neutral
$[H_3O^+] = 1.0 \times 10^{-4}$ M		
$[H_3O^+] = 0.01$ M		
$[H_3O^+] = 7.3 \times 10^{-8}$ M		

3 Circle all of the pairs of compounds below that represent a conjugate acid-base pair.

a H_2S/HS^-

b $Mg(OH)_2/MgCl_2$

c $H_2SO_4/KHSO_4$

d $CH_3CO_2H/NaCH_3CO_2$

e LiOH/NaOH

4 Write a balanced chemical equation for the complete neutralization reaction between each pair of reactants.

a CH_3CO_2H and NaOH

b HCl and $Mg(OH)_2$

c HCl and $CaCO_3$

5 After having a glass of red wine, a chemistry student rinsed her glass in the sink. When the tap water ran into the glass, the wine residue changed from a deep red to a light-blue color. How could this student explain what is causing this color change?

6 The acid H_2CO_3 decomposes to some extent in water to produce CO_2 gas. What should you expect to observe when the ingredients in an antacid react with HCl to produce H_2CO_3?

7 Consider the weak acid H_2A and its conjugate base HA^-. Which diagram below represents a buffer solution? Explain your answer.

= H_2A = HA^-

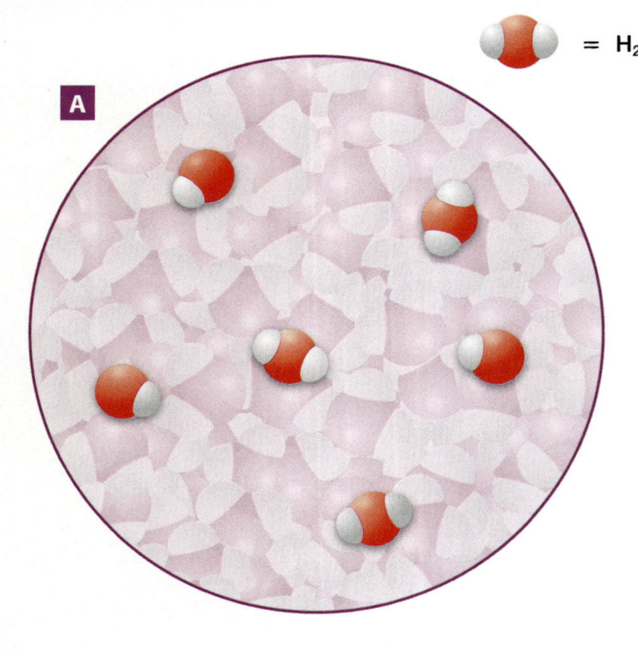

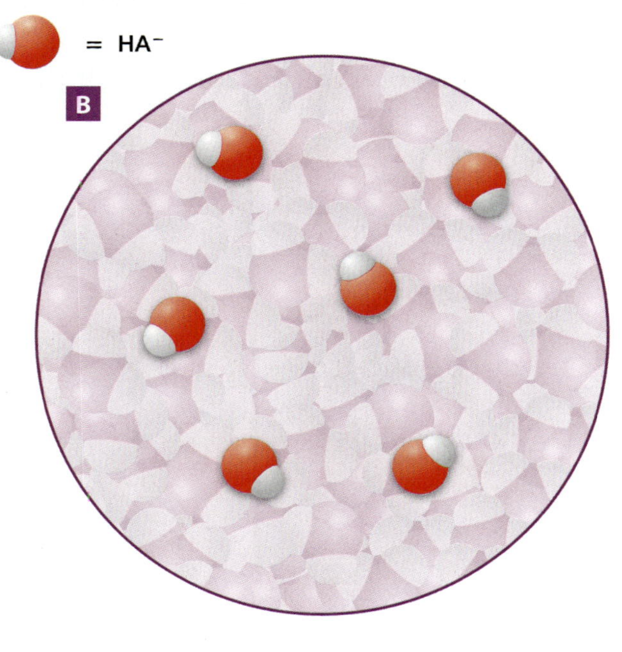

Name ____________________

Lab Partner ____________________

Lab Section ____________________ Date ____________________

Lab 17 DATA SHEET

Data and Observations

Part A Preparing the pH Indicator and a Reference Scale

Red Cabbage Reference Scale

Solution pH	Color with Red Cabbage Indicator
1	
2	
3	
4	
5	
6	
7	
8	
9	
10	
11	
12	
13	
14	

Part B Comparing Methods of Measuring pH

Household Chemicals pH Tests

	Measurement Method									
	Litmus Paper				pH Paper		Digital Meter	Red Cabbage	
	Red		Blue						
Household Chemical	Color	pH	Color	pH	Color	pH	pH	Color	pH

Part C Antacid Effectiveness

Testing Antacid Effectiveness

Procedure Reference	Description	Antacid 1	Antacid 2	Antacid 3
Step 2	Name/mass of active ingredient(s)			
Step 3	Mass: antacid tablet (g)			
Step 5	Mass: beaker + antacid (g)			
Step 9	Observations during addition			
Step 10	Mass: beaker + unused antacid (g)			
Step 11	Mass: antacid used to neutralize HCl (g)			

Name ______________________________

Lab Partner ______________________________

Lab Section ____________________ Date ____________

Lab 17
DATA SHEET
(continued)

Part D Buffer Solutions

Testing Buffer Solution pH: HCl Addition

Test Tube	Sample	pH after 0.1 M HCl addition					
		0 drops	1 drop	6 drops	11 drops	21 drops	31 drops
1	H_2O						
2	CH_3CO_2H						
3	$NaCH_3CO_2$						
4	$CH_3CO_2H + NaCH_3CO_2$						

Testing Buffer Solution pH: NaOH Addition

Test Tube	Sample	pH after 0.1 M NaOH addition					
		0 drops	1 drop	6 drops	11 drops	21 drops	31 drops
1	H_2O						
2	CH_3CO_2H						
3	$NaCH_3CO_2$						
4	$CH_3CO_2H + NaCH_3CO_2$						

Reflective Exercises

1 Red cabbage is not the only natural source of pigments that can be used as pH indicators. The flowers of hydrangea plants come in many different colors. Many people believe that the different color plants are separate species, but this is not true. In fact, some gardeners plant one color shrub and discover that it changes color over time. Suggest a reason this might be the case.

2 Based on your observations from Part B, identify one advantage and one disadvantage to each method of measuring pH.

Method	Advantage	Disadvantage
Litmus paper		
Wide-range pH paper		
Digital meter		
Red cabbage extract		

3 Would it be possible to use the color of a natural indicator to measure the pH of a dark household liquid, such as maple syrup? Explain your answer.

4 Some household chemicals, such as chlorine bleach, break down many organic molecules. Do you think it would be wise to try to measure the pH of these types of solutions using a natural pH indicator? Why or why not?

Name

Lab Partner

Lab Section Date

Lab 17 DATA SHEET

(continued)

5 The active ingredient in one dose of antacid A is 0.50 g $Al(OH)_3$. The active ingredient in one dose of antacid B is 0.50 g $CaCO_3$. Would one dose of antacid A or B neutralize more stomach acid? Explain your choice, and show a calculation to support your answer.

6 Use your data from Part C to answer the following questions.

a Use the tablet masses you measured to calculate the percent (by mass) of each active ingredient in the antacid tablets.

Antacid 1	Antacid 2	Antacid 3

b Using the values calculated in 6a, calculate the mass of each active ingredient you used to neutralize the acid sample.

Antacid 1	Antacid 2	Antacid 3

c Rank the antacids you tested by the mass of active ingredient(s) required to neutralize 3 mL of 0.1 M HCl.

d Are your rankings in 6c consistent with your data from 6a? If so, explain why. If not, explain how they are inconsistent and offer some possible explanations for the discrepancy.

7 Consider your data for Part D, and answer the following questions.

a Which solution was the most effective at resisting a change in pH when HCl was added? Offer an explanation for this difference in behavior compared to the other solutions.

b Which solution was the most effective at resisting a change in pH when NaOH was added? Offer an explanation for this difference in behavior compared to the other solutions.

c Did the buffer solution (Test Tube 4) stay at nearly its original pH for the entire experiment? If not, how much acid or base was added before a significant change was observed? Propose an explanation for why the solution might have stopped buffering against the acid or base addition at that point.

8 Aspirin is an acidic compound. Many aspirin products (e.g., Bufferin) contain antacids in addition to the drug itself. Why might this be a useful addition to the medication?

▲ *Vinegar contains acetic acid, which is used in this lab to conduct a titration.*

Lab 18

Pretty in Pink
Titration of Vinegar

Objectives

After completing this lab you should be able to:

- Explain the concept of an acid-base titration.
- Explain how the endpoint of an acid-base titration may be detected.
- Complete stoichiometric calculations required for a simple acid-base titration.
- Determine the concentration of an acidic solution using titration.

Vinegar is perhaps the most common household acid. It is commonly used in cooking as a flavoring agent, pickling liquid, or simply as a condiment. Many people also use vinegar to clean mineral deposits, which are effectively dissolved by the acid, from shower heads, bath tubs, coffeemakers, and other surfaces. Vinegar comes in many forms, depending on the ingredients and techniques used to produce it, but all vinegars are essentially dilute solutions of acetic acid (CH_3CO_2H) in water.

"White" vinegar generally contains 5–8% acetic acid. The exact concentration of acetic acid in a vinegar sample is readily determined by **titration** with a base, such as sodium hydroxide (Fig. 18.1). **Titration** is a method of quantitative chemical analysis in which increments of one reactant (the **titrant**) are added gradually to the sample until all of the **analyte** (substance being quantified) has been consumed. The **endpoint** of the titration is a sudden change in some physical property at the exact point that the reaction mixture contains a stoichiometric ratio of the reactants.

The reaction between CH_3CO_2H and NaOH involves a 1:1 stoichiometric ratio of the two reactants. Such a mixture should

$$CH_3CO_2H\ (aq) + NaOH\ (aq) \longrightarrow NaCH_3CO_2\ (aq) + H_2O\ (\ell)$$

Phenolphthalein color at the endpoint
pH ~ 7

18.1 Balanced equation for the reaction of acetic acid (CH_3CO_2H) with sodium hydroxide (NaOH). During a titration, a solution of NaOH is slowly added to vinegar (CH_3CO_2H solution) until the two reactants are present in equal quantities. The vinegar is initially acidic, and the phenolphthalein indicator added to the reaction mixture is colorless. At the endpoint of the titration, the pH abruptly changes from acidic to neutral and the phenolphthalein exhibits a faint pink color.

have a neutral pH because each acid and base molecule will have reacted to produce water. The endpoint of this acid-base titration may be detected by including a small amount of **indicator** (a compound that has a pH-dependent color) in the reaction mixture. Phenolphthalein is typically the indicator of choice because it changes from colorless in an acidic solution to faintly pink at neutral pH (Fig. 18.1).

In this lab, you will measure the quantity of acetic acid in a sample of commercial white vinegar and compare your result to the concentration reported on the label. You will first carry out a "test run" to familiarize yourself with the endpoint coloration as well as the quantity and rate at which the NaOH solution should be added. You will then complete the titration three times and report the average measured concentration of acetic acid. The multiple trials should enhance your confidence in the analysis and help you evaluate its reliability and sources of error.

To familiarize you with the necessary calculations, consider the example of a student whose titration of 6.8 mL vinegar required 12.67 mL of 0.50 M NaOH. The quantity of NaOH required is first converted into moles, and the reaction stoichiometry is then used to determine the moles of CH_3CO_2H consumed. Recall that rounding should be avoided until the end of a multistep calculation. To illustrate the process, the calculation below is broken into several steps. Note that the result in each step includes "extra" digits to avoid rounding error in the overall result.

Step 1: Determine the moles of NaOH added:

$$12.67 \cancel{\text{mL}} \times \frac{1 \cancel{\text{L}}}{1000 \cancel{\text{mL}}} \times \frac{0.50 \text{ mol NaOH}}{\cancel{\text{L}}} = 0.006335 \text{ mol NaOH} \quad \text{[Eq. 1]}$$

Step 2: Determine the moles of CH_3CO_2H consumed:

$$0.006335 \cancel{\text{mol NaOH}} \times \frac{1 \text{ mol } CH_3CO_2H}{\cancel{1 \text{ mol NaOH}}} = 0.006335 \text{ mol } CH_3CO_2H \quad \text{[Eq. 2]}$$

The mole quantity of CH_3CO_2H may then be used to calculate the solution concentration in various units. You should calculate the molarity of your solution as well as the mass-to-volume percent concentration to compare to the label value.

Step 3: Convert the moles of CH_3CO_2H to concentration:

Molarity:

$$\frac{0.006335 \text{ mol } CH_3CO_2H}{0.0068 \text{ L}} = 0.9316 \text{ M } CH_3CO_2H \rightarrow 0.93 \text{ M } CH_3CO_2H \quad \text{[Eq. 3]}$$

Mass-to-volume percent:

$$0.006335 \cancel{\text{mol } CH_3CO_2H} \times \frac{60.05 \text{ g } CH_3CO_2H}{1 \cancel{\text{mol } CH_3CO_2H}} = 0.3804 \text{ g } CH_3CO_2H \quad \text{[Eq. 4]}$$

18

$$\frac{0.3804 \text{ g } CH_3CO_2H}{6.8 \text{ mL}} \times 100 = 5.6\% \quad \text{[Eq. 5]}$$

The percent difference between your experimental mass-to-volume percent and the value on the vinegar label may be calculated using Equation 6:

$$\text{Percent difference} = \frac{(\text{Average mass-to-volume percent}) - (\text{Label percent})}{\text{Label percent}} \times 100 \quad \text{[Eq. 6]}$$

Procedure

What Am I Doing?

Steps 1–7: Performing a test run to become familiar with the endpoint of a vinegar titration with NaOH and phenolphthalein indicator.

Steps 7–13: Titrating the acetic acid in vinegar by adding NaOH solution from a burette to a solution of vinegar and phenolphthalein indicator until a faint pink color is produced; performing three separate titrations.

Materials

- ❑ Burette
- ❑ Ring stand
- ❑ Burette clamp
- ❑ 25 mL graduated cylinder
- ❑ 50 mL beaker
- ❑ 250 mL Erlenmeyer flask
- ❑ Funnel
- ❑ Dropper
- ❑ 5 mL volumetric or measuring pipette with bulb or pump
- ❑ Stirring plate with stir bar (optional)

Performing Vinegar Titrations

1. Obtain about 25 mL of vinegar in a 50 mL beaker. On the data sheet, page 273, record the percent concentration of acetic acid according to the label.
2. On the data sheet, record the concentration in molarity of the sodium hydroxide (NaOH) solution available to you. Use about 5 mL of this solution to rinse the inside of a burette. Discard the solution as directed by your instructor.

SAFETY NOTE

Sodium hydroxide is caustic. Wear goggles and ***handle it with care!***

3. Attach the burette to a ring stand using a burette clamp, ensuring that it is vertical (Fig. 18.2).
4. Using a funnel, fill the burette with the NaOH solution. Drain a little solution from the burette in order to expel any residual air from the tip. Ensure that the starting volume in the burette is at or below the 0.00 mL line.
5. Use a volumetric or measuring pipette to deliver 5.00 mL of vinegar into a 250 mL Erlenmeyer flask. Your instructor will demonstrate the proper measuring technique for the pipette available to you.
6. Add approximately 25 mL of distilled water and 2–3 drops of phenolphthalein to the flask.

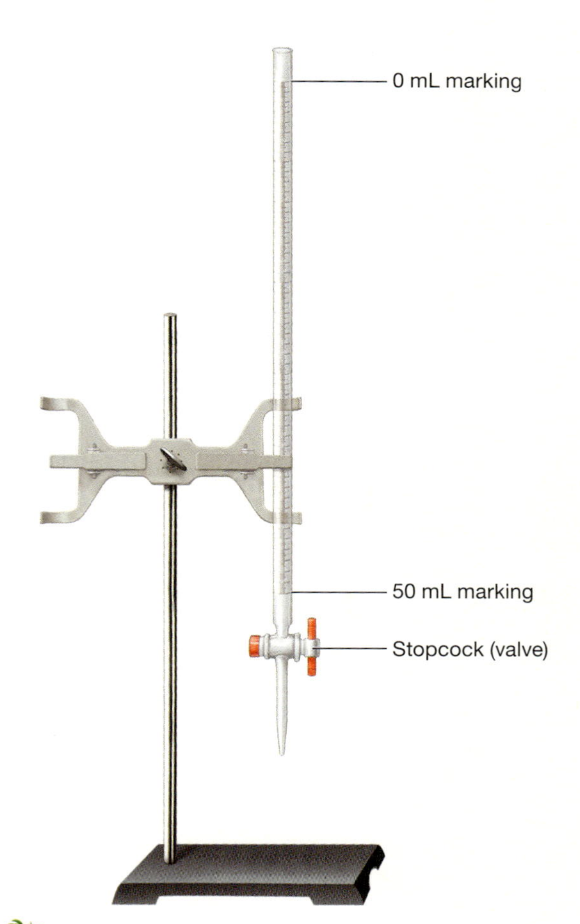

18.2 50 mL burette clamped to a ring stand. The volume of liquid dispensed through the stopcock is equal to the difference between the initial and final volumes measured on the burette scale.

7. Familiarize yourself with the appearance of the titration endpoint by carrying out a test run using the following steps:
   a. In order to better detect any change in solution color, place the flask beneath the burette on a white sheet of paper or on top of a white ceramic hot plate.
   b. If available to you, add a magnetic stir bar to the flask and turn on the stirring function of the stir plate. If this option is not available, you will need to control the burette stopcock with one hand while swirling the flask with the other.
   c. Open and close the burette stopcock repeatedly to add small portions of NaOH. At first, the pink color generated by the addition will dissipate quickly as the two solutions are mixed. As you near the endpoint, the pink color will dissipate more slowly. Begin adding the NaOH solution dropwise until the solution color changes permanently to a faint pink. This is the endpoint of the titration.
   d. On the data sheet, page 273, record your observations regarding the changes in appearance and the approximate quantity of NaOH you used.
   e. Add 1–3 drops of additional NaOH to the flask and note any changes in appearance on the data sheet.
8. Discard the vinegar solution as directed by your instructor. Using distilled water, thoroughly rinse the Erlenmeyer flask and stir bar.
9. Use the volumetric or measuring pipette to deliver 5.00 mL of vinegar into the Erlenmeyer flask. Record the exact volume of your sample on the data sheet.
10. Add approximately 25 mL of distilled water and 2–3 drops of phenolphthalein to the flask.
11. If necessary, add more NaOH solution to your burette. Ensure that you have enough to complete a titration.
12. Titrate the vinegar with the NaOH solution:
    a. Record the initial volume of the NaOH solution in the burette on the data sheet.
    b. Slowly add the NaOH solution to the vinegar, ensuring that the flask contents remain well mixed during the addition.
    c. As you near the endpoint, add the NaOH *dropwise* and stop the titration once a faint pink color is achieved.
    d. Record the final volume of the solution in the burette on the data sheet.
13. Repeat the titration (Steps 8–12) two times.
14. Complete the calculations table on the data sheet for each titration. Determine the average concentration of acetic acid in vinegar according to your data, and use Equation 6 to calculate the percent difference between your data and the value on the label.

Name ______________________________

Lab Partner ______________________________

Lab Section ____________________ Date ____________

Lab 18 Pre-Laboratory Exercise

1 Provide a term that matches each description below.

a Compound that has a pH-dependent color. ____________

b Compound that is the subject of analysis. ____________

c Method of determining the quantity of a substance in solution by adding increments of a reactant. ____________

d Sudden change in a physical property that indicates the completion of the experiment. ____________

e Reactant added to the system in small quantities during a titration. ____________

2 Identify the matching substance or glassware to be used in this lab.

a Analyte. ____________

b Glassware used to hold the sample of analyte during the titration. ____________

c Substance used to detect the endpoint. ____________

d Glassware used to deliver the titrant. ____________

e Titrant. ____________

3 In Step 5, the procedure directs you to add *approximately* 25 mL of distilled water to the vinegar sample before the titration. Explain why the exact amount of water added does not affect the titration results.

4 Why does the procedure direct you to use a volumetric (or measuring) pipette to obtain the 5.00 mL vinegar sample rather than a graduated cylinder?

5 A sample of vinegar contains 6.2% (mass-to-volume) acetic acid.

a How many grams of acetic acid are in 8.3 mL of the vinegar?

b How many moles of acetic acid are in 8.3 mL of the vinegar?

c How many moles of NaOH would be required to react completely with the acetic acid in the 8.3 mL vinegar sample?

d How many milliliters of a 0.25 M NaOH solution would be required to react completely with the acetic acid in the 8.3 mL vinegar sample?

Name ______________________________

Lab Partner ______________________________

Lab Section ____________________ Date ______________

Lab 18 DATA SHEET

Data and Observations

Test Run Observations

Procedure Reference	Description	Observations
Step 1	Label concentration of vinegar (mass-to-volume %)	
Step 2	NaOH concentration (M)	
Step 7d	Observations/approximate volume added at endpoint	
Step 7e	Observations upon further addition of NaOH	

Titration Data

Procedure Reference	Quantity	Trial 1	Trial 2	Trial 3
Step 9	Vinegar (mL)			
Step 12a	NaOH initial volume (mL)			
Step 12d	NaOH final volume (mL)			

Calculations—Trial 1

Quantity	Calculation	Value
NaOH added (mL)	=	
NaOH added (moles)	=	
CH_3CO_2H present (moles)	=	
CH_3CO_2H concentration (M)	=	
CH_3CO_2H concentration (mass-to-volume %)	=	

Calculations—Trial 2

Quantity	Calculation	Value
NaOH added (mL)	=	
NaOH added (moles)	=	
CH_3CO_2H present (moles)	=	
CH_3CO_2H concentration (M)	=	
CH_3CO_2H concentration (mass-to-volume %)	=	

Calculations—Trial 3

Quantity	Calculation	Value
NaOH added (mL)	=	
NaOH added (moles)	=	
CH_3CO_2H present (moles)	=	
CH_3CO_2H concentration (M)	=	
CH_3CO_2H concentration (mass-to-volume %)	=	

Quantity	Calculation	Value
Average concentration CH_3CO_2H (molarity)	=	
Average concentration CH_3CO_2H (mass-to-volume %)	=	
Percent difference experiment vs. label (Eq. 6)	=	

Name ____________________

Lab Partner ____________________

Lab Section ____________________ Date ____________________

Lab 18 DATA SHEET

(continued)

Reflective Exercises

1 In the table below, make a claim regarding the concentration of your vinegar sample. Describe your evidence and provide a rationale for your claim.

Claim	
Evidence	**Rationale**

2 Describe the accuracy of your titration data assuming the label value is the correct concentration of acetic acid. What are some sources of error that might account for the difference between your measurement and the label value?

3 Describe the precision (agreement between measurements) of the data you obtained during your three trials. Identify any sources of error that might account for any variation.

4 A lab mate performing the titration of vinegar tells you that he has added twice the expected volume of NaOH to his vinegar sample without any color change.

a What would you suggest is the most likely mistake that would account for your lab mate's problem?

b How would you suggest your lab mate quickly determine whether your suspicion is correct? What result would indicate that you were correct?

5 Phenolphthalein is just one of many indicators that change color based on the pH of the solution. Which of the indicators provided in the table would be a good alternative to indicate the endpoint of an acid-base titration like the one you performed? Explain your answer.

Indicator	pH Range for Color Transition
Congo red	3.0–5.0
Methyl red	4.4–6.2
Phenol red	6.8–8.4
Thymophthalein	9.3–10.5

6 A chef says that he uses rice vinegar because it "contains less acid than white vinegar." You decide to test this claim by titrating a 5.6 mL sample of rice vinegar. You find that it takes 16.04 mL of 0.25 M NaOH to reach the endpoint. Is the chef's claim correct? Show your calculations to justify your answer.

▲ *Soaps are prepared from naturally occurring fats or oils.*

Lab 19

Clean It Up

Formation and Evaluation of Soap

You are no doubt familiar with many of the properties of soaps. As with all other chemicals, the nature of a soap is a consequence of its structure. In general, a **soap** is an ionic compound formed by the interaction of a sodium or potassium cation with the conjugate base (carboxylate anion) of a long-chain fatty acid.

Examples of these types of molecules are illustrated in Figure 19.1. Notice that the soap is amphipathic, which means that it has a dual nature. The long hydrocarbon "tail" of the carboxylate is nonpolar and hydrophobic (water-fearing), but the small "head" portion is ionic and hydrophilic (water-loving).

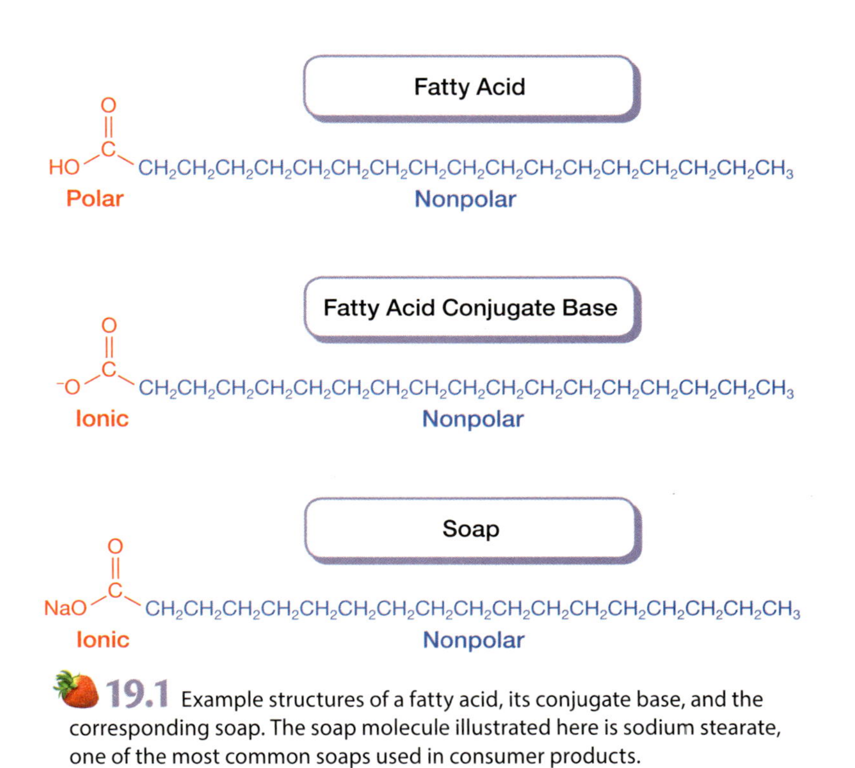

19.1 Example structures of a fatty acid, its conjugate base, and the corresponding soap. The soap molecule illustrated here is sodium stearate, one of the most common soaps used in consumer products.

Objectives

After completing this lab you should be able to:

- Describe the structure of a soap molecule and relate the structure to the function of the soap.
- Explain how hard water is related to the production of soap scum.
- Describe and explain the different solubility properties of soaps and detergents.
- Write a saponification reaction to describe the synthesis of a soap from a triacylglycerol molecule.
- Prepare a sample of soap from a common fat or oil source.

Recall that the principle of "like dissolves like" states that substances of similar polarity are more likely to be soluble in each other. Most dirt, oil, and grease cannot be washed away with water alone because the substances are nonpolar and therefore insoluble in water.

The amphipathic nature of soap allows it to act as an **emulsifier**, which is a substance that causes polar and nonpolar compounds to be suspended in the same mixture. The hydrophobic tails of the soap surround the nonpolar grease molecules, leaving the hydrophilic heads pointed outward, where they experience a favorable ion-dipole interaction with the surrounding water molecules (Fig. 19.2). When the water is washed away, it carries the soap and the encapsulated grease molecules with it.

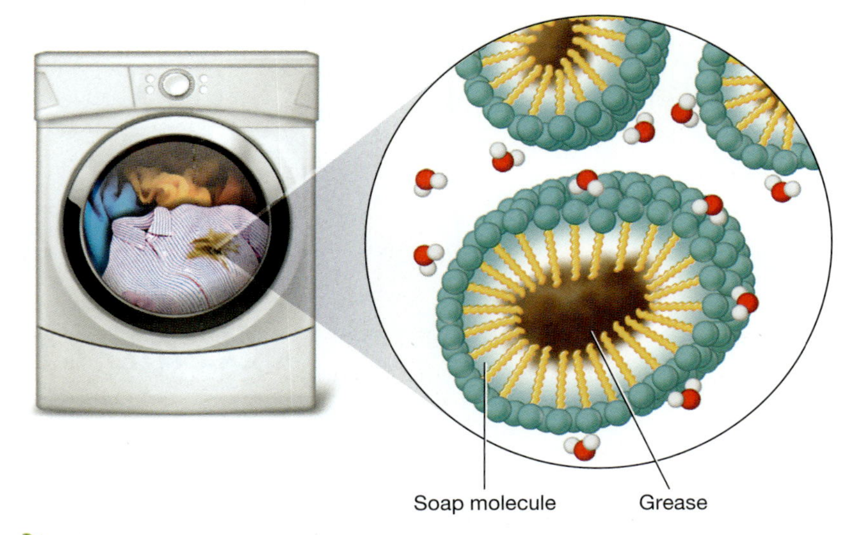

19.2 Soap is capable of cleaning away dirt and grease by acting as an emulsifier. The nonpolar tails surround the nonpolar grease In three dimensions, and the hydrophilic heads interact with the surrounding water.

Note that soap must have an appreciable solubility in water in order to be an effective cleaning agent. "Hard water" contains relatively high concentrations of mineral cations such as Ca^{2+}, Mg^{2+}, and Fe^{3+}. These ions are capable of displacing the Na^+ or K^+ ions in a soap, forming the insoluble solids you know as "soap scum" (Fig. 19.3).

$$3\ NaO_2C(CH_2)_{16}CH_3\ (aq) + FeCl_3\ (aq) \longrightarrow Fe[O_2C(CH_2)_{16}CH_3]_3\ (s) + 3\ NaCl\ (aq)$$

19.3 Chemical equation for the formation of soap scum by the reaction of a soap with $FeCl_3$.

For millennia, humans have produced soaps by treating animal fats or plant oils with lye (NaOH) or potash (KOH) through a process called **saponification**. The fats and oils belong to the **triacylglycerol** class of lipids, which is characterized by a glycerol molecule bound to three long-chain fatty acids through ester bonds (Fig. 19.4).

The saponification reaction involves treating the triacylglycerol with an aqueous base (e.g., NaOH), which causes the ester groups to hydrolyze, generating glycerol and the conjugate bases of the three fatty acids. Ion pairing between these carboxylate anions and the aqueous cation (e.g, Na^+) yields the soap.

In this lab, you will use a source of triacylglycerols in the form of animal fat or plant oil to perform a saponification reaction. You will isolate the product soap and then evaluate some of its properties in comparison to a commercial detergent and two soaps made from other sources by your lab mates.

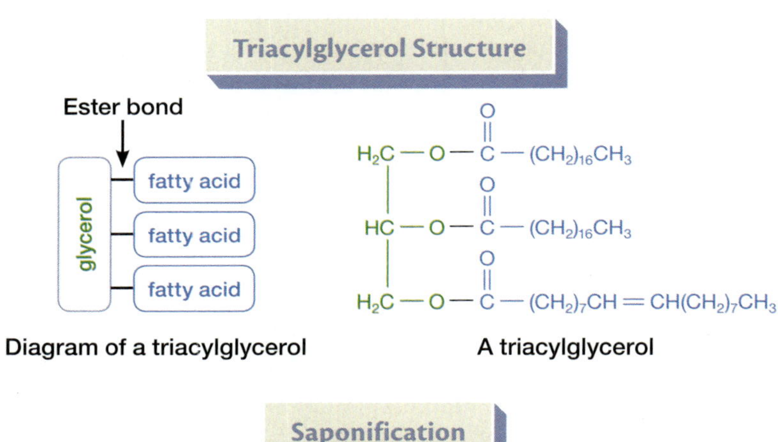

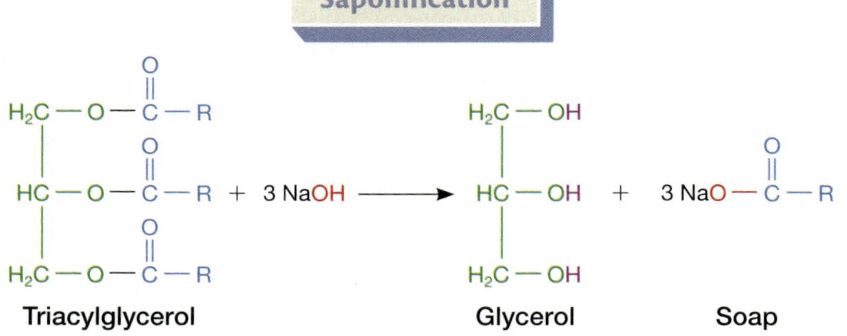

19.4 Overview of triacylglycerol structure and a generic saponification reaction. R = a long-chain hydrocarbon. Note that the fatty acid tails in a triacylglycerol molecule are not necessarily identical.

Procedure

What Am I Doing?

Part A: Reacting a fat or oil with aqueous NaOH in ethanol to produce a soap; isolating the soap by vacuum filtration.

Part B: Testing your soap against a commercial detergent and two soaps prepared by your lab mates for pH, cleaning effectiveness, and tendency to form soap scum with $CaCl_2$.

Materials

- ❑ 600 mL (or large) beaker for ice/hot baths (3)
- ❑ Ring stand
- ❑ Hot plate
- ❑ Clamp
- ❑ 250 mL Erlenmeyer flasks (2)
- ❑ 25 mL graduated cylinder
- ❑ 125 mL Erlenmeyer flask
- ❑ Test tubes (4)
- ❑ Wide-range pH paper
- ❑ Watch glass
- ❑ Glass stirring rod
- ❑ Filter paper
- ❑ Büchner funnel with rubber adapter
- ❑ Filter flask with tubing
- ❑ Scoopula
- ❑ 50 mL graduated cylinder

Part A Synthesis of Soap

1. Assemble the apparatus illustrated in Figure 19.5.
   a. Fill a 600 mL beaker or other large container approximately two-thirds full of water and then place it on a hot plate.
   b. Heat the water bath to a gentle boil. You will add the Erlenmeyer flask after filling it with the reactants.
2. Record your choice of triacylglycerol source on the data sheet, page 285. Measure approximately 10 g of the fat into a 250 mL Erlenmeyer flask.
3. Add approximately 20 mL of 95% ethanol to the Erlenmeyer flask.
4. Add approximately 20 mL of 6 M NaOH to the Erlenmeyer flask.

SAFETY NOTE

6 M sodium hydroxide is caustic. Wear goggles and ***handle it with care!***

5. Clamp the flask to the ring stand and lower it into the hot water. Use a glass stirring rod to stir the mixture vigorously for several minutes. Heat the reaction mixture for at least 30 minutes. Leave the stirring rod in the flask, and use it to occasionally stir the solution to ensure thorough mixing and to prevent any material from escaping the flask if the solution begins to froth.

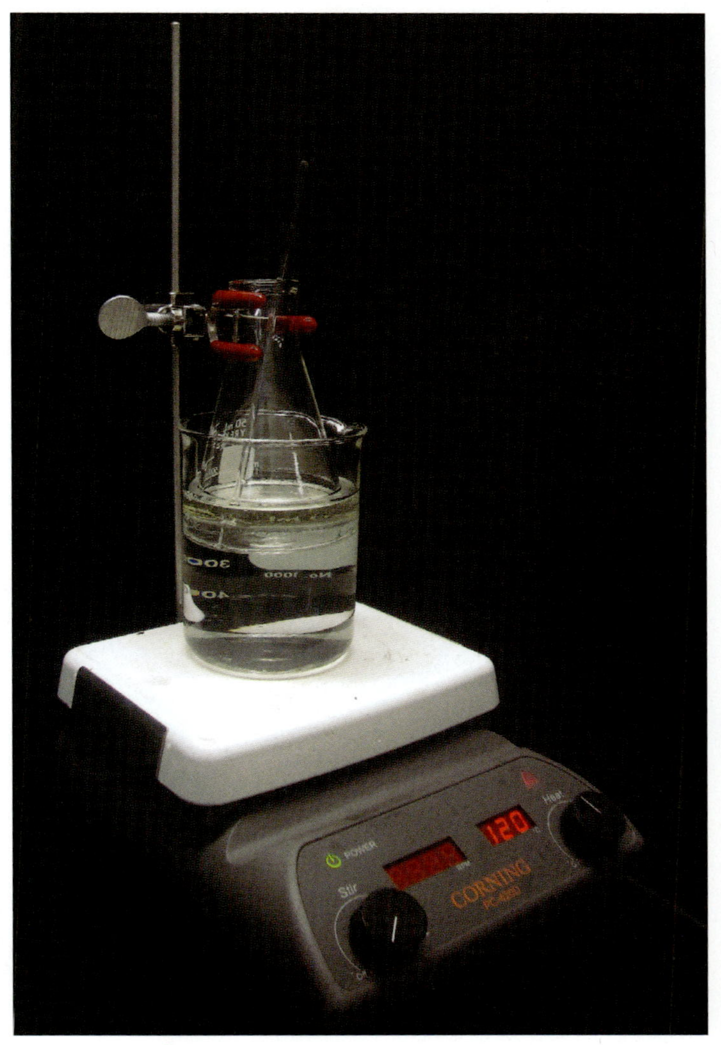

19.5 Apparatus for saponification.

6. During the 30-minute heating period, use an ice-water bath to chill about 100 mL of a saturated NaCl solution and 40 mL of distilled water in separate Erlenmeyer flasks. You may also use this time to set up the vacuum filtration apparatus (Fig. 19.6):
   a. Clamp a filter flask to a ring stand.
   b. Fit the filter flask with a rubber adapter and a Büchner funnel.
   c. Attach a vacuum hose from the vacuum source to the port on the vacuum flask.
   d. Place an appropriately sized piece of filter paper flat in the funnel. Ensure that all the perforations are covered and that the paper does not curl up the sides.
7. At the end of the 30-minute heating period, remove the Erlenmeyer flask from the hot-water bath and allow it to cool on the bench top. The solution should now be a pasty, semisolid mixture.
8. When the product reaches room temperature, pour the 100 mL of chilled NaCl solution into the flask. Use the glass rod to stir the mixture vigorously. You should see a solid precipitate form and float on top of the liquid.
9. Use some of the chilled distilled water to wet the filter paper in the funnel and then turn on the vacuum source. This should seal the filter paper to the bottom of the funnel.
10. Pour the soap mixture into the funnel, and wash the solid with the remaining chilled distilled water.
11. Allow the solids to dry in the funnel with the vacuum on for at least 10 minutes, and then break the vacuum by removing the vacuum tube from the port on the filter flask. Turn the vacuum source off.
12. On the data sheet, page 285, record your observations about the appearance and texture of your soap once it is dry.

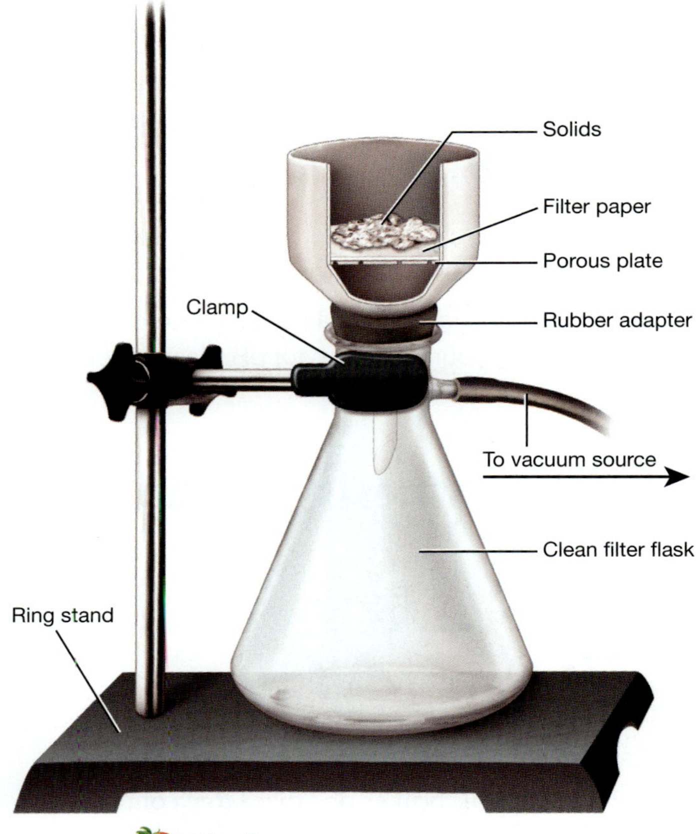

19.6 Apparatus for vacuum filtration.

SAFETY NOTE

Handle your soap with gloves because it may contain unreacted NaOH.

19

Part B Evaluation of Soap and Detergent Samples

1. Find two of your lab mates who used different triacylglycerol sources to prepare their soaps. Obtain a scoopful of each of their soap samples. Also obtain a similar amount of a commercial detergent. On the data sheet, page 285, record the appearance and texture of the soap and detergent samples.
2. Label four test tubes 1–4.
3. Add a pea-sized amount of the appropriate soap to each of the test tubes as indicated below:
    - Test Tube 1: Your soap
    - Test Tube 2: Soap from lab mate 1
    - Test Tube 3: Soap from lab mate 2
    - Test Tube 4: Commercial detergent supplied by your instructor
4. Add ~5 mL of distilled water to each test tube. To dissolve the soap, stir each solution with a glass rod and heat it in a warm-water bath if necessary.
5. Test the pH of the samples with a wide-range pH paper by dipping a clean glass stirring rod into the solution and then touching it to the piece of pH paper. Record the results on the data sheet.
6. Add 3 drops of grease supplied by your instructor to a watch glass and spread it around to form a thin layer. Add 5 drops of the soap solution from Test Tube 1 and use it to wipe the grease from the watch glass. On the data sheet, record your observations regarding the cleaning effectiveness of the soap.
7. Dry the watch glass with a paper towel and repeat Step 6 three times using the other three samples (Test Tubes 2–4). Record your observations on the data sheet.
8. Add 2 drops of 0.5 M $CaCl_2$ to each test tube and record your observations on the data sheet.
9. Dispose of all your soap solutions and any remaining soap samples as directed by your instructor.

Name

Lab Partner

Lab Section Date

Lab 19 Pre-Laboratory Exercise

1 Provide a term that matches each description below.

a Describes a molecule that has both polar and nonpolar regions.

b Describes a molecule or portion of a molecule that does not mix well with water.

c Hydroxide-promoted hydrolysis reaction used to generate soap.

d Substance that facilitates a nonpolar compound dispersing in a polar solvent.

e Triple ester composed of glycerol and three fatty acids.

f Sodium or potassium salt of a long-chain carboxylate.

g Describes a molecule or portion of a molecule that is attracted to water.

2 Soap compounds often are depicted in the cartoon fashion shown below. Label the portions of the cartoon that correspond to the polar and nonpolar portions of a soap.

3 When a soap is added to water, the soap molecules form a structure called a micelle, illustrated below. Micelles are spheres in which the interior is nonpolar and the exterior surface is extremely polar. Explain the intermolecular interactions that cause soap micelles to form in water.

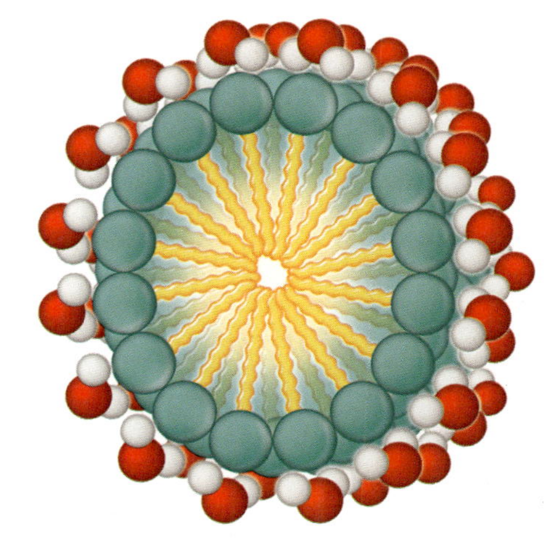

4 Draw the products of the following saponification reaction in the boxes provided.

$$\begin{array}{l} H_2C-O-\overset{O}{\overset{\|}{C}}-C_{15}H_{31} \\ \;| \\ HC-O-\overset{O}{\overset{\|}{C}}-C_{15}H_{31} \\ \;| \\ H_2C-O-\overset{O}{\overset{\|}{C}}-C_{15}H_{31} \end{array} \xrightarrow[CH_3CH_2OH]{3\ NaOH} \boxed{} + \boxed{}$$

Name ______________________________

Lab Partner ______________________________

Lab Section ____________________ Date ____________________

Lab 19 DATA SHEET

Data and Observations

Part A Synthesis of Soap

Triacylglycerol source	
Appearance of soap sample	

Part B Evaluation of Soap and Detergent Samples

Soap Sample Evaluations

Test Tube	Sample	Appearance	pH	Cleaning Effectiveness	Addition of $CaCl_2$
1	Your soap				
2	Soap from lab mate 1				
3	Soap from lab mate 2				
4	Commercial detergent				

Reflective Exercises

1 Ethanol, which has an intermediate polarity, is not directly involved in the saponification reaction you performed. Explain why it was necessary to add ethanol to your reaction mixture.

2 The triacylglycerols in liquid plant oils contain mostly unsaturated fatty acid chains while semisolid animal fats contain mostly saturated fatty acids chains. Consider the various soaps you examined in Part B. Were there any obvious differences in the appearance or texture of soaps made from different triacylglycerol sources that might be explained by the difference in fatty acid structures?

3 Which of the samples you tested was the most effective cleaner in your experiment? What can you infer about the relationship between the structure of this cleaning agent and the grease?

4 Was the soap you prepared acidic, basic, or neutral? Explain this result.

5 In acidic solution, soaps will undergo the reaction shown below. Explain how this would affect the ability of soap to act as a cleaning agent in acidic water.

$$\mathrm{NaO{-}\overset{\displaystyle O}{\overset{\|}{C}}{-}R}\,(aq) + \mathrm{H_3O^+}\,(aq) \longrightarrow \mathrm{HO{-}\overset{\displaystyle O}{\overset{\|}{C}}{-}R} + \mathrm{Na^+}\,(aq) + \mathrm{H_2O}\,(\ell)$$

Name

Lab Partner

Lab Section   Date

Lab 19
DATA SHEET
(continued)

6 A friend lives in an area where the tap water is very hard due to high levels of calcium ion. She notices that a bath soap containing sodium palmitate leaves a heavy film of soap scum on her shower. The chemical reaction below is responsible for this observation. Complete and balance the chemical equation, and explain the source of the soap scum.

$NaO_2C(CH_2)_{14}CH_3\ (aq) + CaCl_2\ (aq) \longrightarrow$

7 Synthetic detergents, such as sodium dodecylbenzenesulfonate, were developed in order to overcome the production of soap scum in hard water. These cleaners are structurally similar to ordinary soaps, but the polar region often is a sulfonate group rather than a carboxylate.

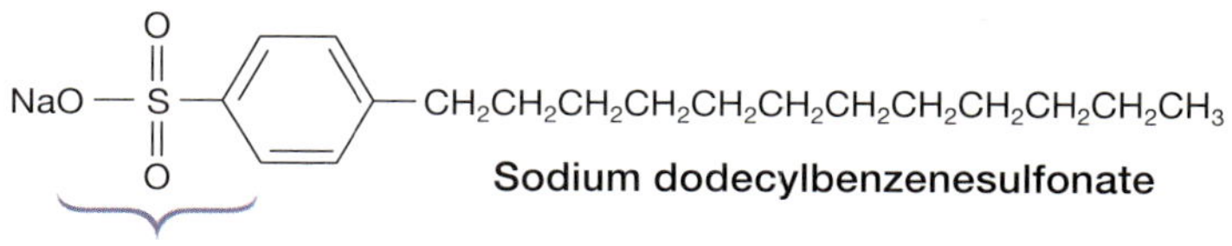

NaO—S(=O)(=O)—C₆H₄—$CH_2CH_2CH_2CH_2CH_2CH_2CH_2CH_2CH_2CH_2CH_2CH_3$

Sodium dodecylbenzenesulfonate

Sulfonate group

a According to your results from Part B, did the detergent behave differently than the natural soap samples when treated with $CaCl_2$?

b Would you expect a detergent such as sodium dodecylbenzenesulfonate to experience similar chemical attraction to hard water cations (e.g., Ca^{2+}) as an ordinary soap molecule? Explain.

c Given your answer to 7b, does the contrast in solubilities between soaps and detergents reflect a difference in the chemical properties or the physical properties of these substances? Explain your answer.

▲ *Milk contains all the biomolecules needed for life: carbohydrates, lipids, and proteins.*

Lab 20

Why It Does a Body Good
The Components of Milk

Objectives

After completing this lab you should be able to:

- Describe and test the chemical composition of milk.
- Explain the chemical basis for separating milk curds from whey.
- Justify the characterization of milk as a "complete food."

Milk is widely considered a "complete" food because it contains water and remarkably high concentrations of all the major classes of nutrients, including simple carbohydrates, proteins, lipids, vitamins, and minerals. The exact chemical composition of milk varies between animal species, breeds, individuals within a breed, and even seasonally within a single individual! The typical composition of cows' milk is summarized in Table 20.1.

TABLE 20.1 ■ Average Chemical Composition of Cows' Milk

Component	Average
Water	87%
Lipids	3.9%
Proteins	3.4%
Simple carbohydrates	4.8%
Other (vitamins and minerals)	0.9%

Although it may appear to be a homogenous solution, milk is more precisely described as an **emulsion**—a mixture of two normally immiscible liquids. Emulsions are a type of **colloid**, a term that refers to two-phase systems of any state in which one component is permanently suspended in the other. The milk emulsion consists of nonpolar lipids (i.e., triacylglycerols) and proteins dispersed as small droplets within an aqueous solution.

If left untreated, the lower density lipids in fresh milk will rise and form a separate layer of cream on the surface of the aqueous milk fraction. **Homogenization** is a mechanical treatment process that prevents this separation by breaking the lipids into smaller globules that remain suspended in the aqueous layer for extended periods of time.

About 80% of the proteins in milk are classified as **caseins**. The caseins carry a net negative charge at the normal pH of milk (about 6.6) and tend to repel each other. When acid is added to milk, the anionic functional groups in the proteins may become protonated. At a certain pH, called the **isoelectric point,** the net charge of the proteins will reach zero (Fig. 20.1). At this point, the dipolar proteins are attracted to each other and will precipitate from the liquid. This is part of the process you observe when milk goes sour. Bacteria in the milk consume the sugars and produce lactic acid, which decreases the pH and results in "curdling" of the milk. The isoelectric point for the casein proteins in milk is approximately 4.7. At lower pH values, the excess acid will cause the proteins to acquire a net positive charge, repel each other, and remain suspended in solution.

During the first step of the cheese-making process, acid (or an enzyme) is added to milk to precipitate the caseins and lipids (called "curds"), which are then separated from the aqueous solution (called "whey"). The non-casein proteins in milk are often called **whey proteins** because they do not precipitate at their isoelectric points and remain suspended in the aqueous fraction when the curds are removed. Heating the whey, however, will denature the proteins and cause them to precipitate.

The aqueous fraction of milk also contains a variety of dissolved mineral ions and polar vitamins. The disaccharide lactose, by far the most abundant carbohydrate in milk, also is water-soluble and is fully dissolved in the aqueous portion of milk. Lactose is responsible for the sweet taste of milk.

In this lab, you will isolate the lipids, casein, and whey proteins from a sample of whole cows' milk. You also will test the separated aqueous fraction of your milk sample for the presence of simple sugars and two important mineral ions (calcium and phosphate). In addition, you will determine the total water content of your milk sample.

The separation process you will use is outlined in Figure 20.2. Treating warm milk with acetic acid will cause the curds (lipids + casein protein) to separate from the whey (aqueous mixture containing soluble proteins, sugars, and ions). The curds can then be treated with a nonpolar solvent (hexane) to separate the nonpolar lipids from the casein. Heating the whey causes precipitation of the whey proteins, and centrifugation then separates these denser solids them from the aqueous solution that contains primarily simple carbohydrates and soluble ions.

Each fraction will be examined by qualitative tests to indicate the presence of the biomolecules of interest. Triacylglycerols in the lipid fraction will be detected by a combination of the grease spot test and the hydroxamic acid test. The grease spot test involves adding a few drops of dissolved sample to a piece of paper and allowing it to dry. Unlike water, which completely evaporates from the paper, the lipids will leave a visible "grease spot" after solvent evaporation. The hydroxamic acid test involves reaction of the ester groups in the triacylglycerol molecules to generate a magenta or burgundy color.

20.1 Relationship between pH and charge of a casein protein. As the pH decreases, the anionic protein reacts with H^+ ions until the isoelectric point is reached. The neutral protein then precipitates. If the pH continues to decrease, the excess protons will react with the protein to give it a net positive charge, the molecules repel each other and the protein is soluble again.

Proteins are detected by the Biuret test, in which the reagent generates a purple color upon reaction with proteins. The Benedict's test is used to detect simple sugars (e.g., lactose) which react with the reagent to produce a red to greenish-yellow precipitate. An iodine solution indicates the presence of polysaccharides by changing to a dark-blue or black color. The tests for calcium and phosphate ions cause the precipitation of white and yellow solids, respectively.

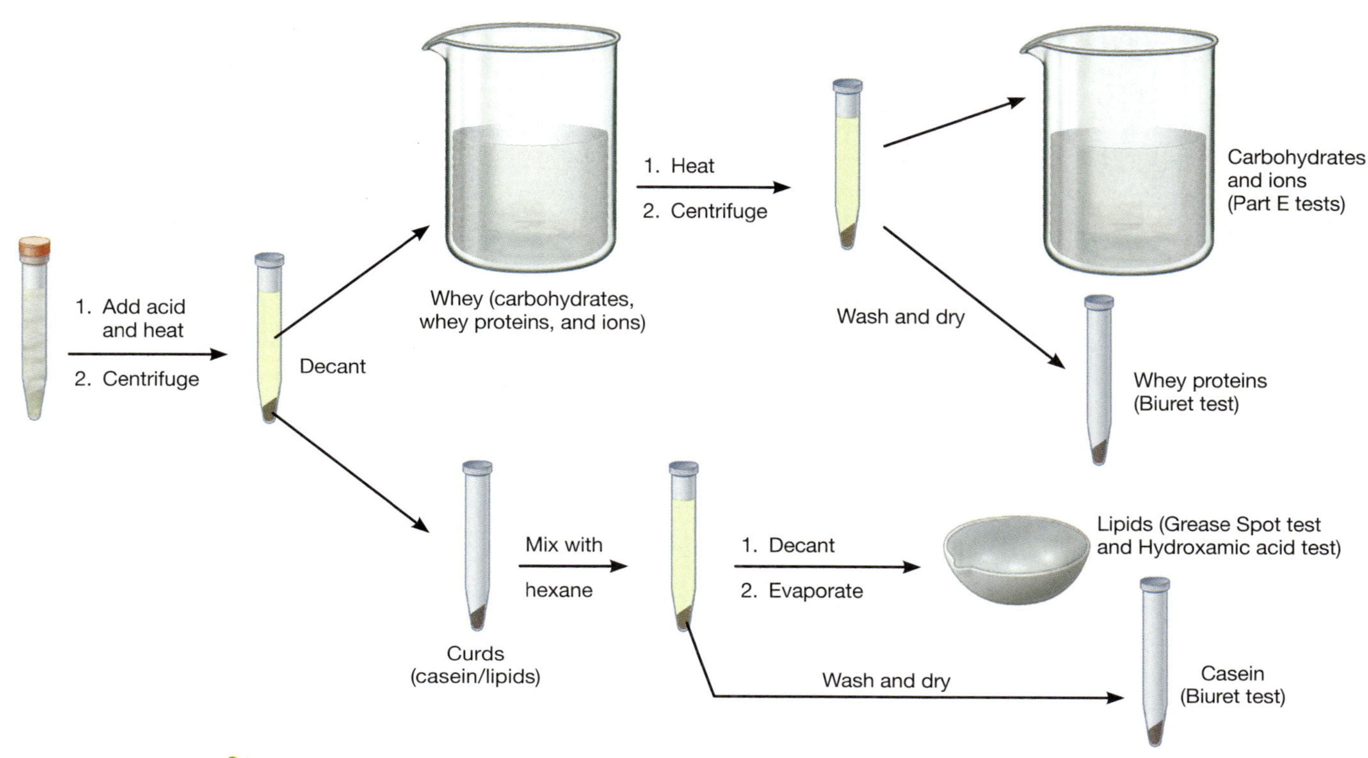

20.2 Outline of the procedure used to separate the primary chemical components of milk.

Procedure

What Am I Doing?

Part A: Separating solid casein/lipid mixture (curds) from the aqueous portion of milk (whey).

Part B: Extracting the lipids from the casein/lipid solid mixture into hexane; removing the hexane to determine the quantity of lipids obtained; testing the isolated material for evidence that it is a lipid.

Part C: Purifying the solid casein and testing the isolated material for evidence that it is a protein.

Part D: Precipitating proteins from the whey portion of milk using heat; isolating the proteins by centrifugation; testing the isolated material for evidence that it is protein.

Part E: Performing qualitative tests to determine if the aqueous portion of milk (whey) contains carbohydrates, proteins, and/or ions.

Part F: Evaporating the water from milk to determine the quantity of water in the sample.

Materials

- ❑ Hot plate
- ❑ 250 mL beaker
- ❑ 10 mL graduated cylinder
- ❑ 125 mL Erlenmeyer flask
- ❑ 50 mL beakers (2)
- ❑ Thermometer
- ❑ Centrifuge tubes (2)
- ❑ Centrifuge
- ❑ Test tubes (3)
- ❑ Microspatula
- ❑ Glass stirring rod
- ❑ Boiling stick
- ❑ Evaporating dish (2)
- ❑ Dropper
- ❑ Tongs or hot mitt
- ❑ Test-tube holder
- ❑ Filter paper
- ❑ Spot plate

Note: The separation and analysis of milk proceeds in several phases, not all of which must be carried out sequentially. To maximize your efficiency, you may want to skip ahead to subsequent steps during any downtime (e.g., waiting for samples to heat or evaporate). Part D can be started at any point after Part A. Part F can be performed at any time during the rest of the experiment.

Part A Precipitation of Casein Protein and Lipids from Whole Milk

1. Use a 250 mL beaker on a hot plate to prepare a warm-water bath with a temperature of approximately 55°C. It is preferable to set up the water bath in a fume hood, if possible. **Retain the warm-water bath throughout the experiment because you will need hot water in several different steps.**
2. Record the mass of a centrifuge tube on the data sheet, page 301.
3. Add approximately 5 mL of whole milk to the tube. Record the mass of the centrifuge tube containing the milk sample on the data sheet. (You will need an Erlenmeyer flask or beaker to hold the filled centrifuge tube while it is on the balance.) Calculate the mass of your milk sample, and record it on the data sheet.
4. Place the centrifuge tube in the warm-water bath. After the milk sample has warmed up for several minutes, add 10 drops of 10% acetic acid to the tube. Use a glass stirring rod to ensure that the sample is well mixed. You should observe the formation of solid chunks of casein/lipid as well as a relatively clear aqueous solution.
5. Add 10 drops of 1 M sodium acetate to the sample and return it to the warm-water bath for another 3–4 minutes. On the data sheet, page 301, record your observations about the changes in appearance to the original milk sample.
6. Remove the tube from the warm-water bath and centrifuge the sample for about 5 minutes. Be sure your tube is counterbalanced in the centrifuge by a lab mate's sample or a blank tube filled with water. While you are waiting, increase the heat on your warm-water bath so the water begins to boil.
7. At the end of the centrifuge period, you should observe the solid casein/lipid mixture deposited at the bottom of the tube with the relatively clear aqueous solution on top of it. Decant the aqueous solution into a 50 mL beaker. Take care not to allow any pieces of the casein/lipid pellet to fall into the beaker. You may need to use a stirring rod to hold back small chunks of solid. **DO NOT discard the aqueous layer in the beaker. You will use it in Part D.**
8. Add about 4 mL of methanol to the centrifuge tube. Use a glass stirring rod to break up the solid pellet and mix it with the methanol. The solvent will help remove residual water from the pellet.
9. Centrifuge the tube again for about 5 minutes.
10. At the end of the centrifuge period, you should observe the casein/lipid pellet again deposited at the bottom of the tube with a methanol layer on top of it. Decant the methanol and dispose of it as directed by your instructor. **The solid will be used in Parts B and C to isolate the lipids and the casein protein.**

20

Part B Isolation of Milk Lipids

SAFETY NOTE

Perform this part of the experiment in the fume hood or appropriately ventilated space. Hexane is a flammable liquid!

1. Add about 4 mL of hexane to the solid casein/lipid pellet in the centrifuge tube. Use a dry glass stirring rod to break up the solid pellet and mix it with the hexanes. The solvent will extract the lipids from the pellet.
2. Centrifuge the tube again for about 5 minutes. While the tube is in the centrifuge, record the mass of a clean, dry evaporating dish on the data sheet, page 301.
3. At the end of the centrifuge period, you should observe a pellet of casein deposited at the bottom of the tube with a hexane layer containing the extracted lipids on top of it. Carefully decant the hexane layer into the pre-weighed evaporating dish. **DO NOT discard the solid. You will use it in Part C.**
4. In a fume hood, place the evaporating dish on top of the boiling-water bath and heat it until the solution no longer boils and the volume stops changing (Fig. 20.3).

SAFETY NOTE

Steam from the boiling-water bath can cause severe burns. Use tongs or a hot mitt to remove the evaporating dish and exercise caution!

20.3 Apparatus for evaporating a solvent above a boiling-water bath.

5. Use tongs or a hot mitt to remove the dish from the water bath and allow it to cool. On the data sheet, page 301, record the mass of the dish with the isolated lipid.
6. Calculate the mass of lipid you obtained and the percent of lipid in your milk sample. Record your results on the data sheet.
7. Dissolve the lipid in a few milliliters of acetone.
8. Perform the grease spot test using the following steps:
   - **a** Add 3 drops of distilled water to a piece of filter paper.
   - **b** In a separate location, add 3 drops of the lipid sample to the filter paper.
   - **c** Wait 3–5 minutes and observe the locations where you spotted the samples.
   - **d** Record your observations on the data sheet.

9. Perform the hydroxamic acid test using the following steps:
   - **a** Add 2 drops of distilled water to one well on a spot plate and 2 drops of the lipid sample to another.
   - **b** Add 2 drops of 6 M NaOH followed by 2 drops of hydroxylamine hydrochloride to each well.
   - **c** Wait 3–4 minutes, and then add 3 drops of 6 M HCl to each well.
   - **d** Add 2 drops of 0.1 M $FeCl_3$ to each well.
   - **e** The development of a burgundy or dark-purple color indicates the presence of an ester. Record your observations on the data sheet, page 301.

Hydroxylamine hydrochloride is toxic! Wear gloves when you handle this reagent and avoid skin contact. NaOH is caustic and HCl is a strong acid. *Wear goggles and handle them with care!*

10. Dispose of the lipid as directed by your instructor.

Part C Isolation of Casein Protein

1. The solid remaining in the centrifuge tube is casein protein. Use a glass stirring rod to break up the solid pellet. Ensure that there are no pockets of solvent remaining underneath the pellet.

Solvent trapped beneath the solid pellet in the centrifuge tube may cause the solids to pop when the liquid evaporates. Break up the solids so there is no solvent left trapped beneath the pellet.

2. Using a test-tube holder, place the centrifuge tube in the boiling-water bath. The heat will evaporate residual solvent and dry the casein. Allow the tube to remain in the boiling water until no vapors are observed leaving the tube and the appearance of the solids stops changing.
3. Remove the tube from the boiling-water bath and allow it to cool.
4. Ensure that the outside of the tube is dry, and then record the mass of the tube with the dry casein on the data sheet, page 302.
5. Calculate the mass of the casein you obtained, and record your results on the data sheet.

6 M sodium hydroxide is caustic. *Wear goggles and handle it with care!*

6. Verify that the solid is a protein by performing a Biuret test using the following steps:
   - **a** In a spot plate well, dissolve a few flakes of the solid in about 5 drops of 6 M NaOH.
   - **b** Add 2–3 drops of Biuret reagent.
   - **c** The appearance of a purple color is a positive test for proteins. Record your observations on the data sheet.
7. Dispose of the solid as directed by your instructor.

20

Part D Precipitation of Whey Proteins from the Aqueous Milk Fraction

1. Add a boiling stick to the beaker containing the aqueous milk fraction from Part A (Step 7).
2. Use a hot plate to gently boil the mixture on mild heat for a few minutes. You should see the formation of precipitate in the solution. These particles are the denatured and coagulated whey proteins.
3. When you can see the precipitated whey proteins, immediately remove the beaker from the heat and allow it to cool to room temperature.

Remove the beaker from the hot plate as soon as precipitation is evident! If too much water evaporates from the beaker, popping or splattering can occur as the residual water evaporates from beneath the precipitate solids.

4. Record the mass of a clean, dry centrifuge tube on the data sheet, page 302.
5. Pour the aqueous mixture into the pre-weighed centrifuge tube. If necessary, use an additional 1–2 mL of distilled water to help wash the precipitate into the tube.
6. Centrifuge the tube for about 5 minutes.
7. At the end of the centrifuge period, you should observe a pellet of whey proteins deposited at the bottom of the tube with the aqueous layer on top of it. **Decant the aqueous layer into a 50 mL beaker and save it for qualitative testing in Part E.**
8. Add about 4 mL of methanol to the centrifuge tube. Use a glass stirring rod to break up the solid pellet and mix it with the methanol. The solvent will help remove residual water from the pellet.
9. Centrifuge the tube again for about 5 minutes.
10. At the end of the centrifuge period, you should observe the whey protein pellet again deposited at the bottom of the tube with a methanol layer on top of it. Decant the methanol and dispose of it as directed by your instructor.
11. Use a glass stirring rod to break up the solid pellet. Ensure that there are no pockets of solvent remaining underneath the pellet.
12. Using a test-tube holder, place the centrifuge tube in the boiling-water bath. The heat will evaporate residual solvent and dry the proteins. Allow the tube to remain in the boiling water until no vapors are observed leaving the tube and the appearance of the solid stops changing.

Solvent trapped beneath the solid pellet in the centrifuge tube may cause the solids to pop when it evaporates. Break up the solids so there is no solvent left trapped beneath the pellet.

13. Remove the tube from the boiling-water bath and allow it to cool.
14. Ensure that the outside of the tube is dry and then record the mass of the tube with the dry whey protein on the data sheet, page 302.
15. Calculate the mass of the whey protein you isolated, and record your results on the data sheet.

16. Calculate the combined mass of casein and whey protein in your sample, and then calculate the mass percent of protein in your milk sample.
17. Verify that the solid is a protein by performing a Biuret test using the following steps:
    - **a** Dissolve the residual solid in about 1 mL of distilled water.
    - **b** Add 5 drops of 6 M NaOH and mix the tube by swirling it gently.
    - **c** Add about 1 mL of Biuret reagent.
    - **d** The appearance of a purple color is a positive test for proteins. Record your observations on the data sheet.
18. Dispose of the sample as directed by your instructor.

Part E Tests for Proteins, Carbohydrates, and Ions in the Aqueous Milk Fraction

Test the aqueous milk fraction you saved in Part D for proteins, carbohydrates, and mineral ions by following each of the procedures below.

1. Perform the Biuret test using the following steps:
    - **a** In a spot plate well, add 5 drops of the sample, 3 drops of 6 M NaOH, and 2–3 drops of Biuret reagent.
    - **b** The appearance of a purple color is a positive test for proteins. Record your observations and conclusion on the data sheet, page 303.
2. Perform the iodine test using the following steps:
    - **a** In a spot plate well, add 5 drops of the sample and 1 drop of iodine reagent.
    - **b** The development of a blue-black color is a positive test for polysaccharides. Record your observations and conclusion on the data sheet.
3. Perform the Benedict's test using the following steps:
    - **a** Add about 1 mL of sample and 2 mL of Benedict's solution to a test tube.
    - **b** Place the tube in a boiling-water bath for about 5 minutes.
    - **c** The appearance of a red or yellow to green solid is a positive test for simple carbohydrates such as lactose. Record your observations and conclusion on the data sheet.

Nitric acid is a strong acid. *Handle with caution!*

4. Perform the calcium test using the following steps:
    - **a** Add about 1 mL sample and 5 drops of ammonium oxalate [$(NH_4)_2C_2O_4$] solution to a test tube.
    - **b** The formation of a white precipitate is a positive test for the presence of calcium ion. Record your observations and conclusion on the data sheet, page 303.

=20

5. Perform the phosphate test using the following steps:
   - **a** Add about 1 mL of sample and 10 drops of 6 M HNO_3 to a test tube.
   - **b** Place the test tube in a boiling-water bath for about 5 minutes.
   - **c** Remove the test tube from the boiling-water bath and allow it to cool to room temperature.
   - **d** Add about 1 mL of the ammonium molybdate solution to the test tube.
   - **e** The formation of a yellow precipitate is a positive test for the presence of phosphate ion. Record your observations and conclusion on the data sheet, page 303.
6. Dispose of all test tube and spot plate contents as directed by your instructor.

Part F Determining the Percent of Water in Milk

1. Weigh a clean, dry evaporating dish and record the mass on the data sheet, page 303.
2. Add 5 mL of fresh whole milk to the evaporating dish.
3. Reweigh the dish with the milk and record the mass on the data sheet. Calculate the mass of the milk sample and record your result on the data sheet.
4. Set the evaporating dish with milk on top of a boiling-water bath (see Fig. 20.3). Stir the milk continuously with a glass stirring rod to prevent burning.
5. Stop heating when the water is gone from the dish, which you will recognize because steam will no longer escape from the dish. The dried milk may look pasty, but its appearance should stop changing at this point.
6. Scrape any dried milk from the stirring rod onto the edge of the evaporating dish.
7. Use tongs or a hot mitt to remove the evaporating dish from the boiling-water bath and dry the dish with a paper towel.
8. Weigh the dish with the dried milk and record the mass on the data sheet, page 303. Calculate the mass of the dry milk and record your result on the data sheet.
9. Calculate the mass of water in the milk sample, and record your results on the data sheet.
10. Calculate the mass percent of water in the milk sample, and record your results on the data sheet.
11. Dispose of the solids as directed by your instructor.

Steam from the boiling-water bath can cause severe burns. Use tongs or a hot mitt to remove the evaporating dish and exercise caution!

Name ______________________________

Lab Partner ______________________________

Lab Section ____________________ Date ____________

Lab 20 Pre-Laboratory Exercise

1 Provide a term that matches each description below.

a Uniform mixture of immiscible liquids. ____________________

b Reagent used to test for the presence of polysaccharides. ____________________

c pH at which the net charge of a protein is zero. ____________________

d Chemical test for the presence of simple carbohydrates such as lactose. ____________________

e Major class of proteins found in milk. ____________________

f Chemical test for the presence of proteins. ____________________

g Solid mixture of protein and lipid formed by adding acid to milk. ____________________

h Aqueous solution formed by removing proteins and lipid from milk. ____________________

2 Complete each sentence below.

a The first step in the separation scheme for milk involves removing the curds by treating the whole milk with ____________________ and ____________________.

b The curds are removed from the aqueous portion (whey) by the technique called ____________________.

c ____________________ are removed from the curds by mixing the solids with hexane and pouring off the liquid, leaving behind a relatively pure solid protein called ____________________.

d Water-soluble ____________________ are precipitated by heating the whey, leaving an aqueous mixture primarily composed of ____________________ and ____________________.

3 Explain why hexane, a nonpolar solvent, can be used to separate the nonpolar lipids from the casein protein.

____

____

____

____

4 When using a centrifuge, what should you always check prior to starting the spin? Why?

____

____

5 In several steps of this lab, the procedure calls for drying solids by heating them in a centrifuge tube. Why is it important to make sure that the solids are well broken up prior to heating them?

Name ______________________

Lab Partner ______________________

Lab Section ______________ Date ______________

Lab 20 DATA SHEET

Data and Observations

Part A Precipitation of Casein Protein and Lipids from Whole Milk

Procedure Reference	Description	Calculation	Observation/Measurement
Step 2	Mass: centrifuge tube (g)		
Step 3	Mass: centrifuge tube + milk (g)		
Step 3	Calculated mass: milk (g)	=	
Step 5	Observation: changes in milk sample		

Part B Isolation of Milk Lipids

Procedure Reference	Description	Calculation	Observation/Measurement
Step 2	Mass: evaporating dish (g)		
Step 5	Mass: evaporating dish + lipids (g)		
Step 6	Calculated mass: lipids (g)	=	
Step 6	Mass percent: lipid in milk	=	
Step 8	Grease spot test		
Step 9	Hydroxamic acid test		

Part C Isolation of Casein Protein

Procedure Reference	Description	Calculation	Observation/Measurement
Step 4	Mass: centrifuge tube + casein (g)		
Step 5	Calculated mass: casein (g)	=	
Step 6	Biuret test		

Part D Precipitation of Whey Proteins from the Aqueous Milk Fraction

Procedure Reference	Description	Calculation	Observation/Measurement
Step 4	Mass: centrifuge tube (g)		
Step 14	Mass: centrifuge tube + whey protein (g)		
Step 15	Calculated mass: whey protein (g)	=	
Step 16	Combined mass: whey protein + casein (g)	=	
Step 16	Mass percent: protein in milk	=	
Step 17	Biuret test		

Name ______________________

Lab Partner ______________________

Lab Section ______________ Date ______________

Lab 20 DATA SHEET

(continued)

Part E Tests for Proteins, Carbohydrates, and Ions in the Aqueous Milk Fraction

Aqueous Milk Fraction Tests

Procedure Reference	Test	Observations	Conclusions
Step 1	Biuret		
Step 2	Iodine		
Step 3	Benedict's		
Step 4	Calcium		
Step 5	Phosphate		

Part F Determining the Percent of Water in Milk

Procedure Reference	Description	Calculation	Observation/Measurement
Step 1	Mass: evaporating dish (g)		
Step 3	Mass: evaporating dish + milk (g)		
Step 3	Calculated mass: milk (g)	=	
Step 8	Mass: evaporating dish + dry milk (g)		
Step 8	Calculated mass: dry milk (g)	=	
Step 9	Calculated mass: water (g)	=	
Step 10	Mass percent: water in milk	=	

Reflective Exercises

1 In the table below, make a claim regarding your ability to confirm the assertion that milk is a "complete food." Describe your evidence and provide a rationale for your claim.

Claim	
Evidence	**Rationale**

2 Identify potential sources of error from your experiment that could explain any differences between your experimental values for the nutrient and water content of milk and the average values reported in Table 20.1.

3 Explain the chemical cause of the change you observed when you added acetic acid to the milk sample.

Name ______________________

Lab Partner ______________________

Lab Section ______________________ Date ______________________

Lab 20 DATA SHEET

(continued)

4 Some people like to add lemon juice to their hot tea while others prefer milk. What might happen when lemon juice and milk are mixed together?

5 Describe how powdered milk might be prepared from whole milk. Would you expect the nutritional value of the powdered milk to be any different than that of the whole milk? Justify your answer.

6 Do you think you were able to remove every protein molecule from the milk in Parts A through D? Did your tests indicate the presence of any proteins in the aqueous milk solution? Explain your answer.

7 Consider the qualitative tests you performed on the fractions of milk you separated.

a Were you able to detect anything specific about the structures of the lipids in milk? Explain.

b Were you able to detect anything specific about the structures of carbohydrates in milk? Explain.

c Were you able to detect anything specific about the structures of the proteins in milk? Explain.

8 Both calcium and phosphate ions are important nutrients for the development of healthy bones and teeth. According to your data, would drinking milk promote the healthy development of bones and teeth? Justify your answer.

▲ Spirulina *tablets have the protein phycocyanin.*

Lab 21

The Blue-Green Protein

Denaturation of Phycocyanin

Proteins are large, biologically active compounds composed of many **amino acid** subunits linked through amide bonds. Amino acids share a common structural core in which a central carbon atom, called the **alpha** (α) carbon, is bound to both an amine and a carboxylic acid functional group (Fig. 21.1). At physiological pH, the carboxylic acid group is typically deprotonated (i.e., a carboxylate) and the amine is protonated. The differences between the structures and properties of the amino acids arise from a **side chain** (R-group), which is also bound to the α carbon.

Amines and carboxylic acids may undergo a condensation reaction to produce an amide functional group (Fig. 21.2). The product formed by the condensation of two amino acids is called a **dipeptide**, and the amide linkage between them is known as a **peptide bond.**

For example, the reaction of glycine and tyrosine might generate the dipeptide shown in Figure 21.2B. There are 20 different amino acids commonly incorporated into natural proteins, which exhibit remarkable diversity since the protein chain may vary in length, combination, and sequence of amino acid subunits.

α-carbon

$H_3\overset{+}{N}-C(H)(R)-C(=O)-O^-$

A generic amino acid

$H_3\overset{+}{N}-C(H)(CH_3)-C(=O)-O^-$

Alanine

$H_3\overset{+}{N}-C(H)((CH_2)_3-NH-C(=NH_2^+)-NH_2)-C(=O)-O^-$

Arginine

21.1 Shown is a generic amino acid structure with two examples. Each has a carboxylic acid, amine, and variable side chain bound to a central carbon atom. At physiological pH, the carboxylic acid is usually deprotonated and the amine group is protonated.

Objectives

After completing this lab you should be able to:

- Describe the structure of an amino acid.
- Draw the hydrolysis or condensation products involving two amino acids.
- Determine whether a protein is present in a sample using the Bradford test.
- Describe the four levels of protein structure.
- Explain how proteins may be denatured and the levels of protein structure affected by denaturation.
- Extract the protein phycocyanin from *Spirulina* and examine its denaturation.

A Carboxylic acid + Amine ⇌ (Condensation / Hydrolysis) Amide + H_2O (Amide bond)

B Glycine + Tyrosine ⇌ (Condensation / Hydrolysis) Dipeptide + H_2O (Peptide bond)

21.2 (**A**) Condensation of a carboxylic acid with an amine produces an amide. (**B**) Condensation of two amino acids produces a dipeptide. The amide bond that joins two amino acids is called a peptide bond.

In order to function properly, most proteins must adopt a specific three-dimensional shape. A protein's shape is described by considering up to four levels of structural organization that correspond to progressive increases in scale (Fig. 21.3). The smallest scale, called the **primary structure**, is the specific sequence of amino acids joined by peptide bonds in the protein polymer (Fig. 21.3A).

The next two levels of structural organization correspond to the concept that an individual polypeptide chain does not usually remain linear in solution. Instead, non-covalent attractive forces (such as hydrogen bonding) within the chain cause it to fold into a three-dimensional structure (e.g., Fig. 21.3C).

Secondary structure (Fig. 21.3B) describes localized folding into a regular pattern along a small section of the polypeptide, such as coiling into a helix. Structural arrangements of this kind arise from hydrogen bonding among the N–H and C=O portions of different amino acid monomers in the chain.

Tertiary structure (Fig. 21.3C) describes the folding and packing of an entire polypeptide chain into a compact, well-defined, three-dimensional shape stabilized by non-covalent attractions involving the amino acid side chains. Only certain proteins exhibit the highest level of organization, **quaternary structure**, which is the association of two or more individual protein chains into a complex that exhibits biological activity (Fig. 21.3D).

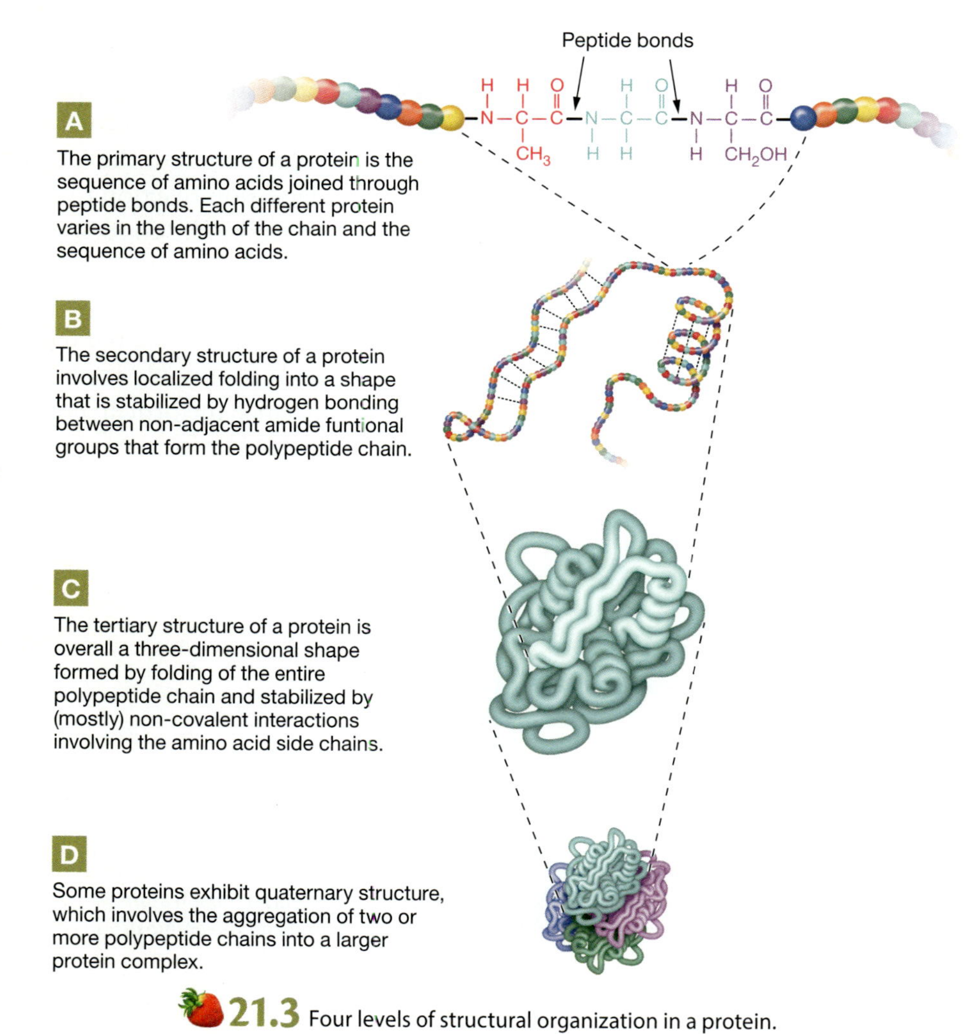

21.3 Four levels of structural organization in a protein.

Figure 21.4 illustrates some examples of the non-covalent interactions that stabilize the secondary and tertiary structures of a protein. The folding of a polypeptide chain generally maximizes the stability of the protein's shape in the aqueous cellular environment. As a result, polar or charged (i.e., hydrophilic) amino acid side chains are often exposed on the outer surface of a protein, where they interact with water molecules through dipole-dipole and hydrogen bonding forces.

Nonpolar (i.e., hydrophobic) side chains are often clustered on the protein's interior surface where they interact via dispersion forces. In other regions, polar or charged side chains may also interact with each other through hydrogen bonding, dipole-dipole, or electrostatic attractions.

In addition, there is one type of covalent bond that may help stabilize tertiary structure. When the side chains of two cysteine amino acid residues are oxidized, a **disulfide bond** may form between their two sulfur atoms (Fig. 21.5). Because they are true covalent bonds, these interactions are among the strongest of stabilizing features within a protein structure.

The peptide bonds that constitute the primary structure of a protein are relatively strong. They may be hydrolyzed by prolonged heating in an aqueous acid or base and by the action of various enzyme catalysts. The non-covalent interactions responsible for the secondary, tertiary, and quaternary structures are comparatively weak and may be disrupted by less vigorous means. This process, called **denaturation**, alters the overall three-dimensional shape of a protein and generally causes loss of the protein's biological activity (Fig. 21.6).

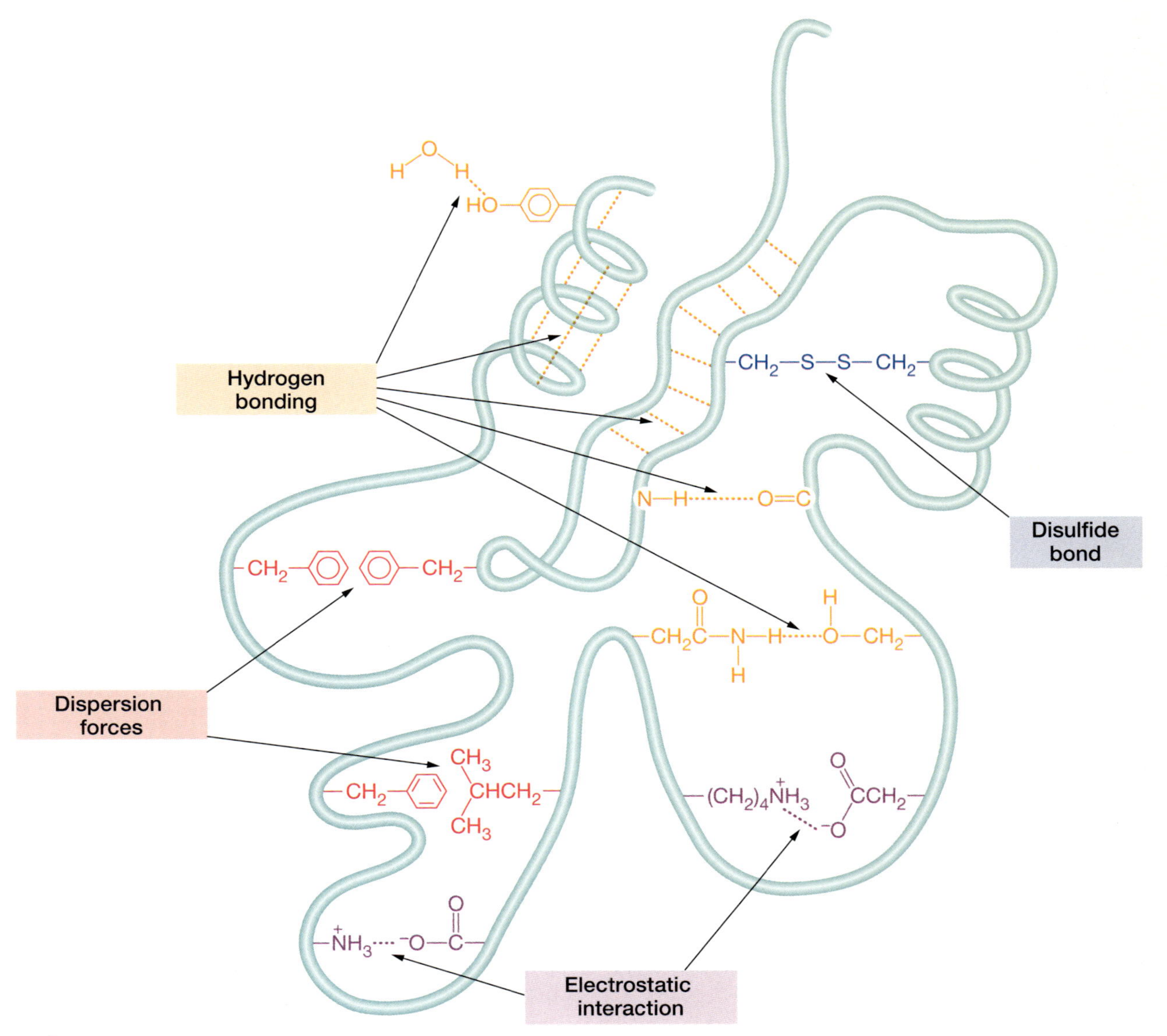

21.4 Examples of interactions that stabilize the folding of a polypeptide chain into secondary and tertiary structures.

The primary structure is unchanged upon denaturation. Common denaturing agents are heat, acid, and base. The heat will add enough energy to the system to overcome the various non-covalent attractive forces, while the acid or base may interfere with hydrogen bonding or electrostatic interactions. Heavy metals and other reagents similarly disrupt the attractions of polar side chains and may reduce disulfide bonds. Organic compounds may obstruct stabilizing interactions between water and the side chains on the protein surface.

In this lab, you will examine the protein phycocyanin, which is easily extracted from cells of *Spirulina* cyanobacteria, formerly known as "blue-green algae" (Fig. 21.7). These species perform photosynthesis using a variety of pigments in addition to the chlorophyll you are familiar with from green plants. The phycocyanin protein serves as a scaffold that holds one of those pigments, phycocyanobilin, in the shape required for light harvesting during photosynthesis.

In the appropriately folded protein, phycocyanobilin is a blue color and also exhibits **fluorescence**, which means that it emits light. The phycocyanobilin fluorescence is visible to the naked eye as a reddish glow. When phycocyanin is denatured, the scaffolding required to hold phycocyanobilin in the necessary shape is lost and the pigment loses its characteristic blue color and fluorescence. Thus, the color of a phycocyanin solution is a reliable indicator of the protein's structural integrity.

In Part A of this lab, you will extract phycocyanin from a capsule of *Spirulina* cells, which is readily available as a nutritional supplement. In Part B, you will test the extract and a series of other samples for the presence of protein. Although various qualitative tests for proteins are available, this lab uses the Bradford reagent because it is very sensitive (detects small amounts of protein) and robust (operates under a range of conditions). The Bradford reagent contains Coomassie Blue dye, which changes from reddish brown to royal blue in the presence of a protein. In Part C, you will expose your phycocyanin extract to potential denaturing agents, and use the Bradford test to determine whether the protein's primary structure remains intact upon denaturation.

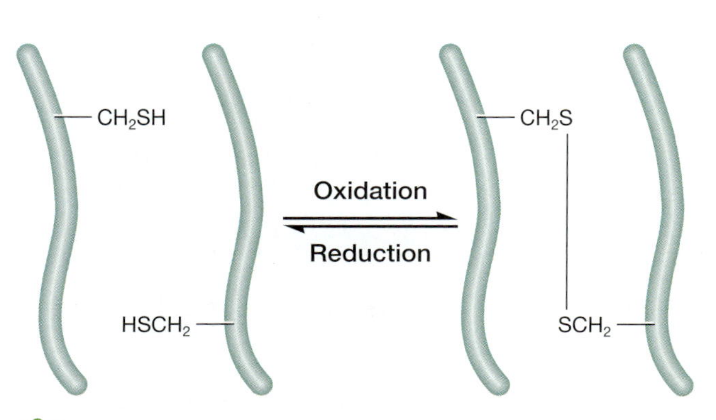

21.5 Oxidation may generate a disulfide bond between cysteine side chains.

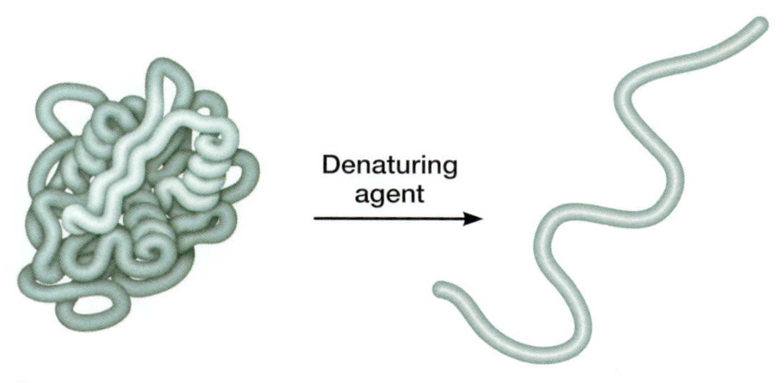

21.6 Denaturation of a protein disrupts the secondary, tertiary, and quaternary structures.

21.7 *Spirulina* are available in capsules sold as nutritional supplements. A solution of phycocyanin exhibits a blue color and visible red glow.

Procedure

Part A: Extracting the phycocyanin protein from a sample of *Spirulina*.

Part B: Examining several solutions with the Bradford test in order to determine the presence of protein.

Part C: Treating samples of phycocyanin with heat or various additives to determine whether they cause denaturation of the protein as indicated by a color change; testing denatured protein samples with the Bradford reagent.

Materials

- ❑ 150 mL and 250 mL beakers
- ❑ Hot plate
- ❑ Test tubes (10)
- ❑ Mortar and pestle
- ❑ Glass stirring rod
- ❑ Funnel
- ❑ Filter paper
- ❑ Ring stand
- ❑ Iron ring
- ❑ 50 mL Erlenmeyer flask
- ❑ Dropper
- ❑ Spot plate
- ❑ 25 mL graduated cylinder

Part A Phycocyanin Isolation

1. Open a capsule of *Spirulina* and pour the contents into a mortar.
2. Obtain a sample of sand approximately equal in volume to your *Spirulina* sample and add it to the mortar.
3. Use the pestle to grind the *Spirulina* and sand together for several minutes. Use the walls and bottom of the mortar to efficiently grind the material and break open the cell walls.
4. Transfer the mixture to a 150 mL beaker and add 25 mL of pH 7 sodium phosphate (Na_3PO_4) buffer.
5. Vigorously mix the contents of the beaker with a glass stirring rod, and then allow the mixture to stand for about 5 minutes.
6. Using a fluted piece of filter paper (see Technique Tip 21.1) in a funnel, gravity filter the mixture and collect the liquid in a 50 mL Erlenmeyer flask. The funnel should be supported by an iron ring attached to a ring stand. It may take 10–15 minutes for all the liquid to run through the filter. You may proceed to Part B during the filtration.
7. After the filtration is complete, you should obtain a clear, dark-blue solution that emits a reddish glow under strong light. The filter paper and solids may be discarded as directed by your instructor.

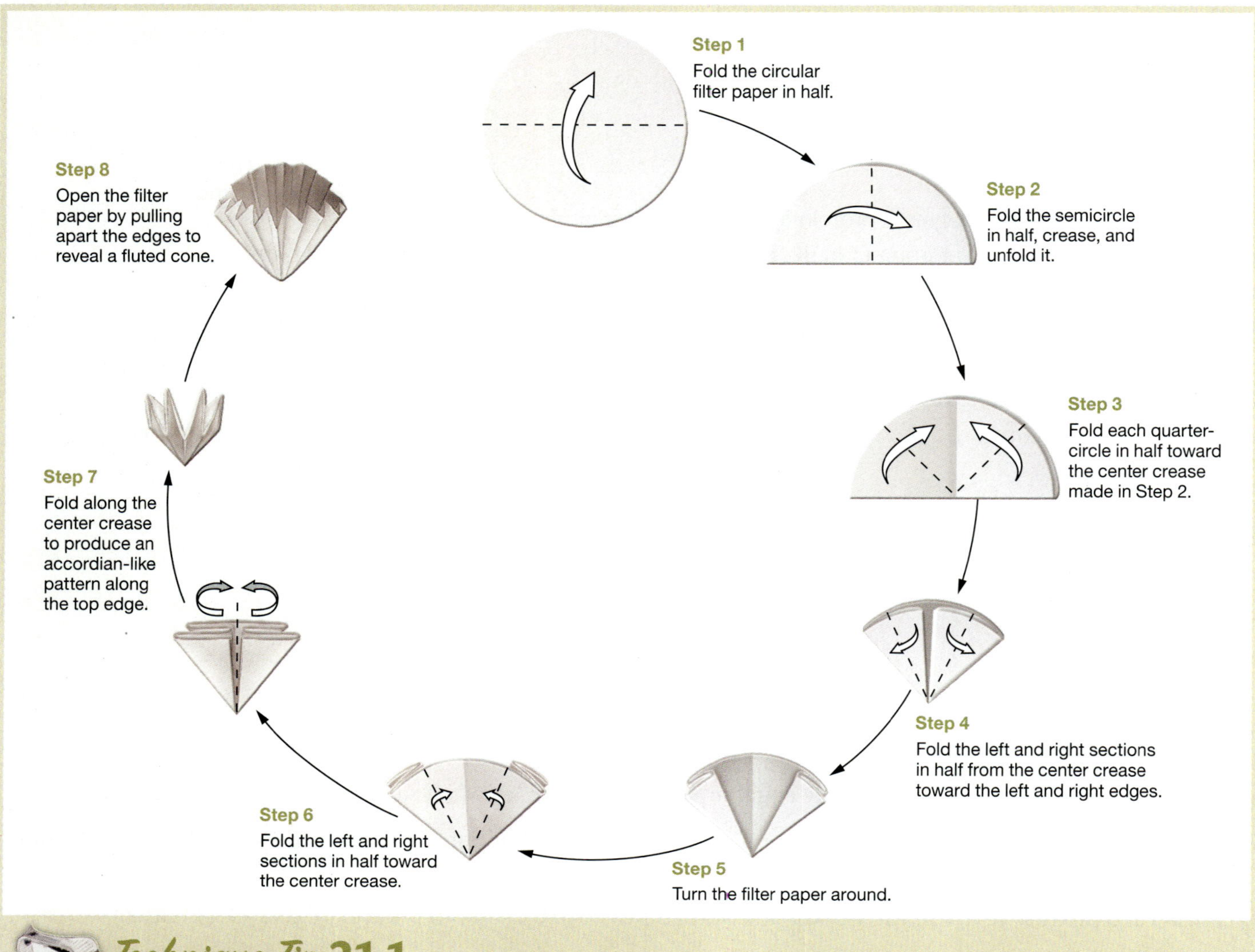

Technique Tip 21.1

Fluted filter paper has more surface area than a simple cone made from flat filter paper. This speeds the rate of filtration by increasing the contact area between the solution and paper. Follow the steps in the illustration to produce a fluted piece of filter paper. Fold along dotted lines in direction indicated by arrows.

Part B Bradford Test for Proteins

1. To each of six wells on a spot plate, add 10 drops of Bradford reagent.
2. To each well add 2 drops of the appropriate solutions as indicated below:
   - Well 1: Glycine solution
   - Well 2: Tyrosine solution
   - Well 3: Gelatin solution
   - Well 4: Albumin solution
   - Well 5: Phycocyanin solution (from Part A)
   - Well 6: Sucrose solution
3. Record your observations on the data sheet, page 317, and indicate whether each sample contains protein.
4. Dispose of the spot plate contents as directed by your instructor.

Part C Phycocyanin Denaturation

1. Use a 250 mL beaker to prepare a boiling-water bath on a hot plate.
2. Label 10 test tubes 1–10. (If fewer than 10 test tubes are available, you may empty test tubes after obtaining your data, rinse them with distilled water, and reuse them.)
3. Add 10 drops of phycocyanin solution (from Part A) to each test tube.
4. Place Test Tube 1 in a boiling-water bath for up to 5 minutes. Record your observations on the data sheet, page 318.
5. To each of the remaining test tubes, add *up* to 20 drops of the appropriate additive as indicated below. Add each substance *dropwise*, counting the drops and observing any changes as you proceed. Stop the addition if a significant change in the sample is noted. Record your observations on the data sheet, *including the number of drops necessary to cause any observed change.*
   - Test Tube 2: Distilled water
   - Test Tube 3: Isopropanol
   - Test Tube 4: Acetone
   - Test Tube 5: Bleach
   - Test Tube 6: 0.1 M NaCl
   - Test Tube 7: 0.1 M $ZnCl_2$
   - Test Tube 8: Detergent
   - Test Tube 9: 1.0 M HCl
   - Test Tube 10: 1.0 M NaOH
6. For any test tube that exhibited a significant change, perform the Bradford test on the solution: In a spot plate well, add 10 drops of Bradford reagent followed by 2 drops of the test-tube mixture. Record your observations on the data sheet.
7. Dispose of the test tube and spot plate contents as directed by your instructor.

21

Name

Lab Partner

Lab Section   Date

Lab 21 Pre-Laboratory Exercise

1 Provide a term that matches each description below.

a Bond that joins two amino acids in a protein.

b Atom that is bound to a carboxylic acid, amine, and variable side chain in an amino acid.

c Level of protein structure associated with the folding of the entire polypeptide chain into a specific overall three-dimensional shape.

d Process of disrupting the secondary, tertiary, and/or quaternary structure of a protein that often leads to loss of its function.

e Part of the amino acid structure that makes each amino acid unique.

f Localized folding of a polypeptide chain into a regular pattern caused by hydrogen bonding between amide groups.

2 What is the purpose of mixing the sand with the *Spirulina* cells during the extraction of phycocyanin? Why is it important to grind the *Spirulina* vigorously in the mortar and pestle?

3 The primary structure of a protein is formed by the condensation of amino acids in a certain sequence. Consider the dipeptide formed by the condensation of glycine and tyrosine in Figure 21.2B.

a Draw the structure of the dipeptide that would be formed if the two amino acids condensed in the opposite sequence.

b How are the structures of the dipeptides in Figure 21.2B and your drawing related to each other? Are they isomers? If so, what kind?

4 Determine the type(s) of interaction and the level of structure that would be affected by treating a protein with each additive or action:

	Additive/Action	Interaction(s) Disrupted	Protein Structure Affected
a	Adding acetic acid		
b	Boiling		
c	Adding thioglycolate (a reducing agent)		

5 Many proteins become insoluble in water upon denaturation. Provide an explanation for this behavior.

Name ______________________________

Lab Partner ______________________________

Lab Section ____________________ Date ______________

Lab 21 DATA SHEET

Data and Observations

Part B Bradford Test for Proteins

Well	Sample	Observations	Protein (+/−)?
1	Glycine		
2	Tyrosine		
3	Gelatin		
4	Albumin		
5	Phycocyanin		
6	Sucrose		

Part C Phycocyanin Denaturation

Test Tube	Additive	Observations	Bradford Test Observations (if necessary)
1	Heat		
2	Distilled water		
3	Isopropanol		
4	Acetone		
5	Bleach		
6	0.1 M NaCl		
7	0.1 M $ZnCl_2$		
8	Detergent		
9	1.0 M HCl		
10	1.0 M NaOH		

Name ______________________

Lab Partner ______________________

Lab Section ______________________ Date ______________

Lab 21 DATA SHEET

(continued)

Reflective Exercises

1 Albumin is the primary constituent of egg whites. Would you say that eggs are high in protein? Use your data to justify your answer.

2 Did all of the additives denature the phycocyanin? Identify the additives that had no effect on the protein and explain why they are different from the others.

3 Among the additives that denatured phycocyanin, were some more effective than others? Use your data to justify your answer.

4 Did your experimental evidence support the assertion that denaturation only disrupts the shape of a protein and not the peptide bonds? Use your data to justify your answer.

5 A student capped a test tube containing 10 drops of phycocyanin solution and shook it vigorously for 60 seconds. Afterward, the solution was colorless. Explain how the shaking caused this result.

6 Some proteins are embedded in the nonpolar phospholipid bilayer of a cell membrane. How would you expect the three-dimensional structures of these proteins to differ from phycocyanin and other water-soluble proteins?

7 Determine whether each statement below is true or false. Explain your reasoning.

a Two different proteins may be generated from the same number and type of amino acids.

True/False

b Denaturation of a protein produces amino acids.

True/False

c No covalent bonds are ever disrupted during protein denaturation.

True/False

▲ Lactase is an enzyme that helps the body process lactose, crystals of which are shown here.

Lab 22

Got Milk Sugar? Lactase and Enzyme Activity

Objectives

After completing this lab you should be able to:

- Describe an enzyme-catalyzed reaction cycle.
- Explain why pH and temperature may affect the catalytic activity of an enzyme.
- Explain how the presence of an inhibitor may affect the catalytic activity of an enzyme.
- Analyze the specificity, pH sensitivity, and temperature sensitivity of the enzyme lactase.
- Identify certain inhibitors of the enzyme lactase.

One of the major functions of proteins is to serve as enzymes. An **enzyme** is a **catalyst** for a biochemical reaction. In general, catalysts are substances that increase the rate of a chemical reaction without undergoing a permanent change during the reaction. As a consequence, an individual catalyst molecule may be reused repeatedly during a chemical transformation.

After participating in one reaction, the catalyst is regenerated and able to participate in a subsequent reaction with fresh reactant. Enzymes are vital for proper biological functioning because they increase the rate of reactions that would otherwise occur far too slowly to sustain life.

Enzymes ordinarily catalyze precise reactions involving very specific reactants, also called **substrates.** The reaction takes place when a substrate interacts with a specialized region, or **active site,** on the protein to form an enzyme-substrate complex. After the reaction is complete, the products are released from the enzyme's active site, which is then available to catalyze another reaction (Fig. 22.1).

The specificity of enzyme-catalyzed reactions is largely a result of the fact that the shape of the active site is complementary to the shape of substrate(s) in a manner similar to the way a hand fits into a glove. Just as your foot would not be a good fit for your glove, compounds other than the substrate do not fit well in the enzyme's active site. As a result, the catalytic activity of an enzyme is significantly depressed or nonexistent for compounds other than the intended substrate.

Any change in the three-dimensional shape of an enzyme can alter or inhibit its reactivity. In general, enzymes exhibit maximum reactivity under certain optimal conditions, such as within a narrow pH or temperature range. Outside of these conditions, the protein shape can change and lead to decreased enzyme activity. Under extreme conditions (e.g., very high heat or very low pH), enzyme activity may be lost completely due to denaturation of the protein.

Inhibitors are compounds that prevent an enzyme from catalyzing a reaction (Fig. 22.2). **Competitive inhibitors** are usually similar in structure to an enzyme's substrate, and they function by competing with the substrate for access to the enzyme active site. Other inhibitors, called **noncompetitive inhibitors,** interact with the enzyme in a location other than the active site. This causes a change in the shape of the enzyme, which can render it less effective or inactive. In all cases, an inhibitor slows the rate of a reaction compared to the observed rate when the substrate and enzyme interact in the absence of the inhibitor.

In this lab, you will study the activity of **lactase**, the enzyme used to hydrolyze the disaccharide lactose into its monosaccharide components (Fig. 22.3). The hydrolysis of lactose and other disaccharides provides monosaccharides used to fuel the body. Lactose intolerance is a condition caused by the absence of lactase in the digestive tract. In individuals with this condition, the undigested lactose may be broken down by intestinal bacteria, leading to abdominal cramping, bloating, and gas.

You will gauge the reactivity of lactase under various conditions by using a common urinalysis test strip to measure the quantity of glucose produced in the enzyme-catalyzed reaction. First, you will examine the specificity of lactase for lactose compared to other potential disaccharide substrates. You then will investigate the effects of pH, temperature, and the presence of potential inhibitors on the lactase activity.

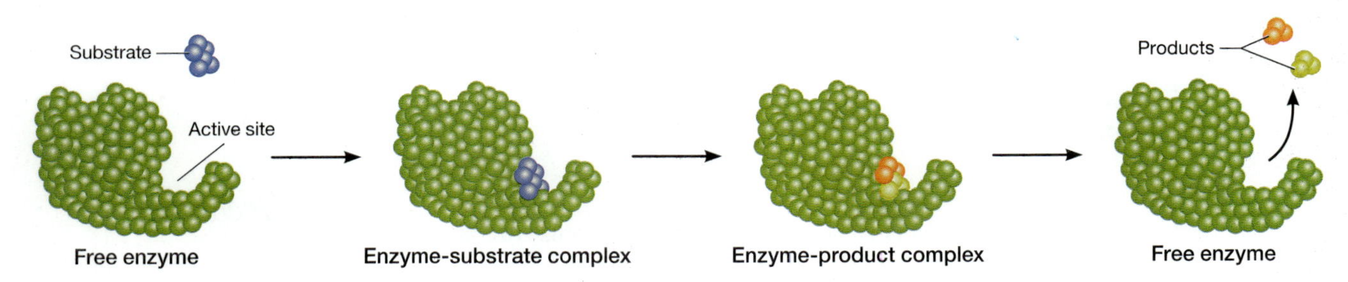

22.1 Enzyme-catalyzed reaction cycle. The substrate interacts with the active site to form an enzyme-substrate complex. The reaction generates new products in the enzyme's active site, after which the products are released and the enzyme is available to catalyze another reaction.

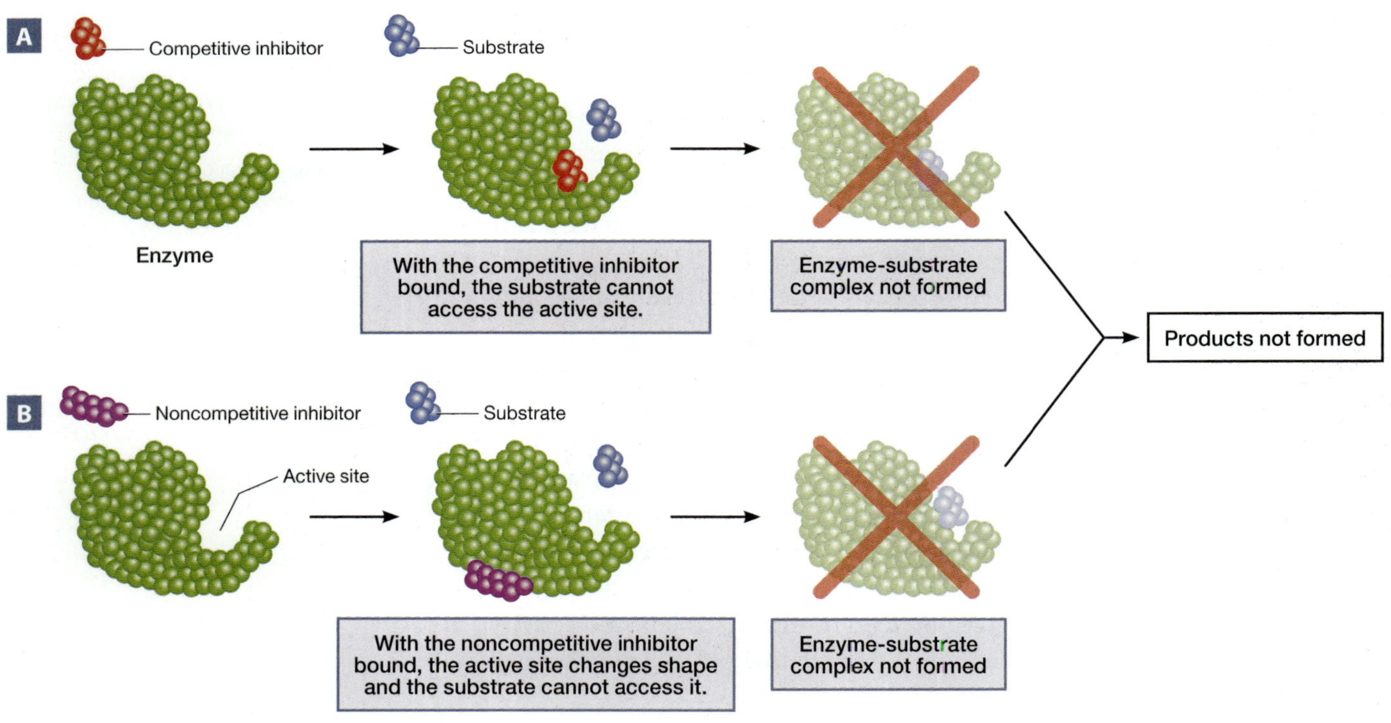

22.2 Action of inhibitors on an enzyme: (**A**) competitive inhibitors effectively block the substrate from gaining access to the active site; (**B**) noncompetitive inhibitors alter the shape of the enzyme's active site by binding to the protein in a remote location.

Lactose + H_2O $\xrightarrow{\text{Lactase}}$ D-galactose + D-glucose

22.3 Hydrolysis of the disaccharide lactose yields the two monosaccharides D-galactose and D-glucose. The transformation is catalyzed by the enzyme lactase, which is shown above the reaction arrow because it is neither a reactant nor a product of the chemical reaction.

Procedure

What Am I Doing?

Part A: Preparing a solution of lactase to use throughout the experiment.

Part B: Measuring the quantity of glucose produced by separately treating lactose and two other disaccharides with lactase.

Part C: Measuring the quantity of glucose produced by treating lactose with lactase at six different pH values.

Part D: Measuring the quantity of glucose produced by treating lactose with lactase at four different temperatures.

Part E: Measuring the quantity of glucose produced by treating lactose with lactase in the presence of three potential inhibitors.

Materials

- ❑ Hot plate
- ❑ 50 mL beaker
- ❑ 400 mL beakers (4)
- ❑ Thermometer
- ❑ Mortar and pestle
- ❑ Glass stirring rod
- ❑ 50 mL Erlenmeyer flask
- ❑ Test tubes (6)
- ❑ Test-tube holder
- ❑ pH meter
- ❑ Glucose test strips (17)
- ❑ Dropper
- ❑ 25 mL graduated cylinder
- ❑ 10 mL graduated cylinder

Part A Preparing a Lactase Solution

1. Use a 400 mL beaker on a hot plate to prepare a warm-water bath with a temperature of approximately 37°C. You will use this bath in Parts B–E.
2. Use a mortar and pestle to grind one commercial lactase pill into a powder.
3. Combine the lactase powder and 25 mL of distilled water in a 50 mL beaker. Stir the solution with a glass stirring rod to dissolve as much of the powder as possible. Do not be concerned if all the powder will not dissolve.
4. Decant the liquid into a clean 50 mL Erlenmeyer flask and discard any remaining solid as directed by your instructor.
5. Label your solution "Lactase Enzyme" and use it for Parts B–E.

Part B Analyzing the Specificity of Lactase

1. Label three test tubes 1–3, and add 3 mL of the following solutions to the corresponding test tubes:
   - Test Tube 1: 2% lactose solution
   - Test Tube 2: 2% maltose solution
   - Test Tube 3: 2% sucrose solution

2. To each test tube, add 1 mL of your Lactase Enzyme.
3. Gently agitate each test tube to ensure that the contents are well mixed (see Technique Tip 22.1).
4. Place the test tubes in the 37°C warm-water bath for 10 minutes.
5. After the heating period, remove the test tubes from the warm-water bath and use a test strip to measure the glucose content in each. Dip the strip into each test tube as directed on the package label, and then compare the results to the color chart on the label.
6. Record your results (in mg/dL) on the data sheet, page 329.
7. Dispose of the test tube contents as directed by your instructor.

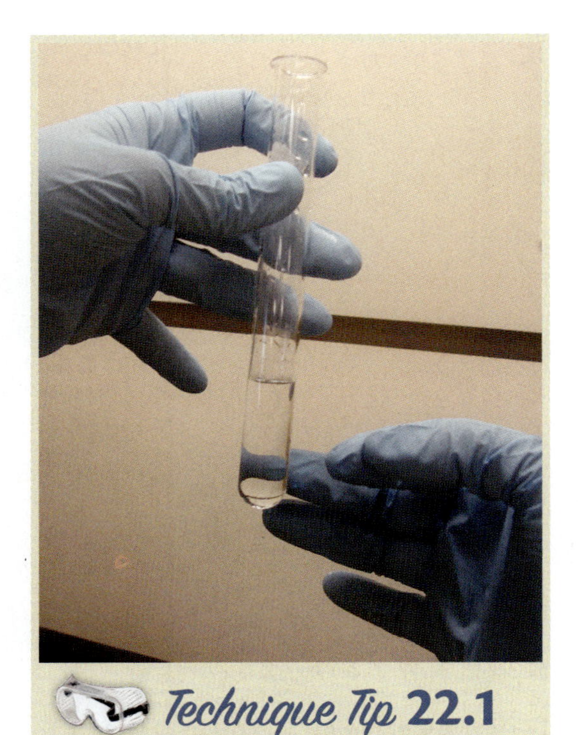

Technique Tip 22.1

To safely shake a test tube, hold it between your thumb and split first two fingers. Flick the test tube using a finger from the opposite hand.

Part C pH Sensitivity of Lactase

1. Label six test tubes with the following pH values: 2, 4, 7, 8, 10, and 12.
2. Add 2 mL of the following solutions to the corresponding test tubes as indicated below:
   - Test Tube 2: pH 2 buffer
   - Test Tube 4: pH 4 buffer
   - Test Tube 7: pH 7 buffer
   - Test Tube 8: pH 8 buffer
   - Test Tube 10: pH 10 buffer
   - Test Tube 12: pH 12 buffer
3. Use a pH meter to measure the actual pH of the solution in each test tube and record these values on the data sheet, page 329.
4. Add 1 mL of your Lactase Enzyme to each test tube.
5. Add 3 mL of a 2% lactose solution to each test tube.
6. Gently agitate each test tube to ensure that the contents are well mixed (see Technique Tip 22.1).
7. Place the test tubes in the 37°C warm-water bath for 10 minutes.
8. After the heating period, remove the test tubes from the warm-water bath and use a test strip to measure the glucose content in each test tube.
9. Record your results (in mg/dL) on the data sheet.
10. Dispose of the test tube contents as directed by your instructor.

Part D Temperature Sensitivity of Lactase

1. In addition to your 37°C warm-water bath, prepare an ice-water bath, room temperature-water bath, and a boiling-water bath. Record the actual temperature of each bath on the data sheet, page 330.
2. Label four test tubes 1–4.
3. To each test tube, add 1 mL of your Lactase Enzyme.

4. Place each test tube in the appropriate water bath for 10 minutes as indicated below:
   - Test Tube 1: Ice-water bath
   - Test Tube 2: Room temperature-water bath
   - Test Tube 3: 37°C warm-water bath
   - Test Tube 4: Boiling-water bath
5. After the test tubes have been in their respective water baths for 10 minutes, add 3 mL of a 2% lactose solution to each test tube.
6. Gently agitate each test tube to ensure that the contents are well mixed.
7. Place the test tubes back in their respective water baths for 10 minutes.
8. After the heating/cooling period, remove the test tubes from the water baths and use a test strip to measure the glucose content in each test tube.
9. Record your results (in mg/dL) on the data sheet, page 330.
10. Dispose of the test tube contents as directed by your instructor.

Part E Effect of an Inhibitor

1. Label four test tubes 1–4, and add 2 mL of the following solutions to the corresponding test tubes:
   - Test Tube 1: Distilled water
   - Test Tube 2: 2% galactose solution
   - Test Tube 3: 2% maltose solution
   - Test Tube 4: 1 M $CaCl_2$ solution
2. Add 2 mL of a 2% lactose solution to each test tube.
3. Add 1 mL of your Lactase Enzyme to each test tube.
4. Gently agitate each test tube to ensure that the contents are well mixed.
5. Place the test tubes in the 37°C warm-water bath for 10 minutes.
6. After the heating period, remove the test tubes from the warm-water bath and use a test strip to measure the glucose content in each test tube.
7. Record your results (in mg/dL) on the data sheet, page 330.
8. Dispose of the test tube contents as directed by your instructor.

Name ______________________________

Lab Partner ______________________________

Lab Section ____________________ Date ____________

Lab 22
Pre-Laboratory Exercise

1 Provide a term that matches each description below.

a Location at which a substrate binds to an enzyme. ____________

b Protein used as a catalyst for a biological reaction. ____________

c Compound that prevents an enzyme from functioning properly by competing with the substrate for the active site. ____________

d Reactant in an enzyme-catalyzed reaction. ____________

e Enzyme responsible for the hydrolysis of lactose. ____________

2 Name the two monosaccharides produced by the hydrolysis of each disaccharide below.

a Lactose ____________ + ____________

b Maltose ____________ + ____________

c Sucrose ____________ + ____________

3 Use your answers to Question 2 to explain why testing for glucose concentration is a good indication of lactase activity on all of those substrates.

4 Convert 37°C to the corresponding value in Fahrenheit. Why do you think most of the experiments in this lab are conducted at 37°C?

5 People who are lactose intolerant often consume lactose-free milk products. How do you think these products are produced from natural milk?

Name ______________________________

Lab Partner ______________________________

Lab Section ____________________ Date ____________

Lab 22 DATA SHEET

Data and Observations

Part B Analyzing the Specificity of Lactase

Lactase: Specificity Analysis

Test Tube	Substrate	Glucose Concentration (mg/dL)
1	Lactose	
2	Maltose	
3	Sucrose	

Part C pH Sensitivity of Lactase

Lactase: pH Sensitivity

Test Tube	Solution pH	Glucose Concentration (mg/dL)
2		
4		
7		
8		
10		
12		

Part D Temperature Sensitivity of Lactase

Lactase: Temperature Sensitivity

Test Tube	Temperature (°C)	Glucose Concentration (mg/dL)
1		
2		
3		
4		

Part E Effect of an Inhibitor

Lactase: Inhibitor Effect

Test Tube	Added Compound	Glucose Concentration (mg/dL)
1		
2		
3		
4		

Name ______________________________

Lab Partner ______________________________

Lab Section ____________________ Date ____________

Lab 22 DATA SHEET

(continued)

Reflective Exercises

1 In the table below, present a claim regarding the specificity of lactase for the substrate lactose. (Does lactase act exclusively or preferentially on lactose?) Describe your evidence and provide a rationale to justify your claim.

Claim	
Evidence	**Rationale**

2 Based on your results from Part C, is the activity of lactase sensitive to pH? If so, at which pH was lactase the most active?

3 Recall that carbohydrate hydrolysis reactions are catalyzed by acid. Should this fact cause you to question any of your results in Part C? Justify your answer.

4 Did you find that lactase displayed very low reactivity at one or more pH values? If so, how might this lack of activity be explained at the molecular level?

5 Based on your results from Part D, is the activity of lactase sensitive to temperature? If so, describe the trend you observed and provide a likely explanation for it.

6 The friend of a lactose-intolerant student suggested adding a powdered lactase supplement directly to dairy foods before eating them. Would you expect this strategy to be very effective if the student's friend consumed ice cream? What about if she consumed a café au lait (hot coffee with milk)? Use your data to help explain your answers.

7 Based on your results from Part E, which (if any) of the added compounds inhibited the activity of lactase? Use your data to justify your answer.

8 Which of the carbohydrate compounds in Part E was the more potent inhibitor of lactase? How might the inhibition caused by this compound provide a mechanism for regulating the amount of lactose hydrolyzed in your digestive tract?

▲ Strawberries have eight sets of chromosomes, making it easy to perform a DNA extraction.

Lab 23

Blueprint of a Strawberry
DNA Extraction

Objectives

After completing this lab you should be able to:

- Describe the chemical components and structure of DNA.
- Explain the chemical processes used to extract DNA from the nucleus of a cell.
- Extract the DNA from a strawberry and use chemical tests to confirm its identity.

Nucleic acids are composed of long chains of covalently bound molecules called **nucleotides** (Fig. 23.1). Each nucleotide includes three structural components: a five-carbon monosaccharide, a phosphate group, and a nitrogenous base (cyclic compound that contains nitrogen).

There are two important biological nucleic acids: **deoxyribonucleic acid (DNA)** and **ribonucleic acid (RNA)**. One major difference between the two is that ribose makes up the sugar portion of the RNA nucleotides, while deoxyribose is the sugar component of DNA nucleotides. The two also differ in the types of nitrogenous bases found in their nucleotide subunits.

In living systems, DNA stores genetic information. When translated, this information provides a chemical blueprint for all of our inherited traits, such as eye color, hair color, growth rate, and metabolic rate. The DNA chain is formed by the condensation of nucleotides, which are linked together through **phosphodiester bonds.** A portion of a DNA molecule is illustrated in Figure 23.2.

Like proteins, nucleic acids adopt complex three-dimensional structures that can be described using a hierarchy that corresponds to increasing scales. Just as the primary structure of a protein is defined by the order of amino acids in the chain, the order of nucleotides in the polymeric chain constitutes the primary structure of DNA.

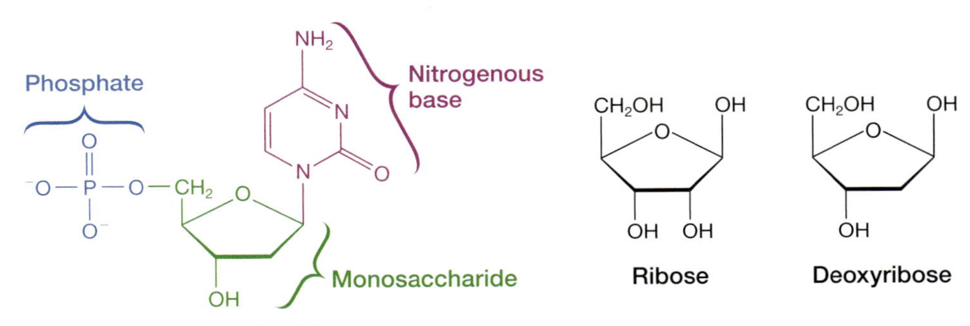

23.1 Overview of nucleotide structure.

The secondary structure of DNA is the famous **double helix** formed by hydrogen bonding between two strands of DNA, while the tertiary structure involves supercoiling of this helix around proteins called **histones**. Individual packages of DNA coiled around histones are called **nucleosomes**. The chain of nucleosomes is further wound into tight fibers called **chromosomes**. These structures are illustrated in Figure 23.3.

In this laboratory exercise, you will isolate DNA from the cells of strawberries. The DNA of higher organisms (eukaryotes) is mostly contained within the nuclei of their cells. In order to isolate the DNA, it is first necessary to disrupt the cell wall and membrane in order to release the cellular components. This will be accomplished by physically grinding the cells and adding an extraction solution, which contains detergent molecules, sodium chloride, and ethylenediaminetetraacetic acid (EDTA).

The detergent interferes with the intermolecular forces that hold the cell membrane together. Along with the EDTA, the detergent also helps denature cellular enzymes that would otherwise degrade the DNA molecules. The sodium chloride increases the solubility of the ionic DNA and helps separate it from the ionic histones.

Once the extraction is complete, addition of cold ethanol causes the precipitation of large molecules in the solution, including DNA, RNA, and various proteins. Unlike the other molecules, the solid DNA is large and fibrous. This enables removal of the DNA from the solution by spooling it onto a rod in the same way you might spool spaghetti onto a fork.

You also will perform a series of chemical tests on your isolated DNA in order to confirm its identity. First the DNA polymer will be hydrolyzed by heating it in aqueous acid. This reaction liberates deoxyribose, which will react with the Dische reagent to produce a blue color in solution. Ribose does not react in this manner. The phosphate in DNA will react with a molybdate stain to produce a blue color upon heating, and nitrogenous bases that belong to the purine family will react with silver ions to generate a white precipitate.

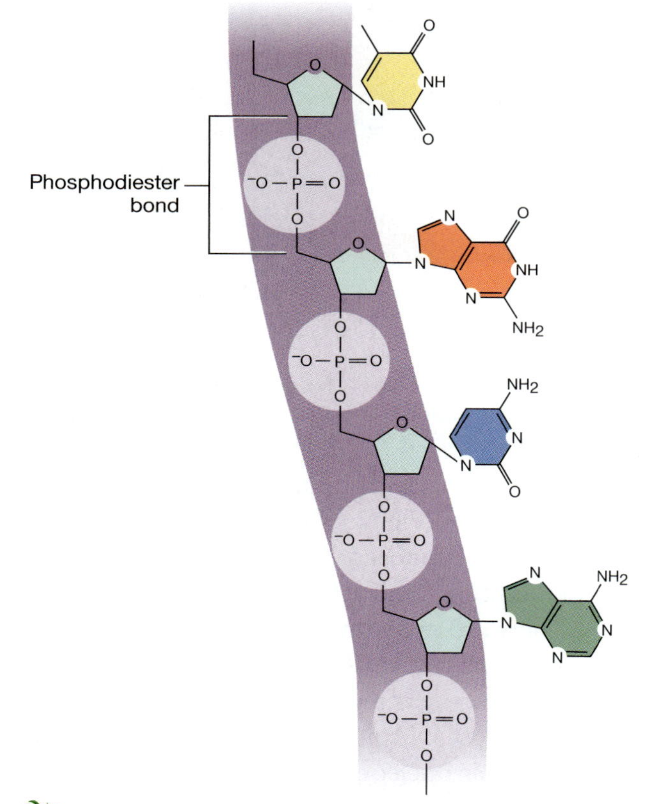

23.2 Portion of a DNA molecule highlighting the primary structure.

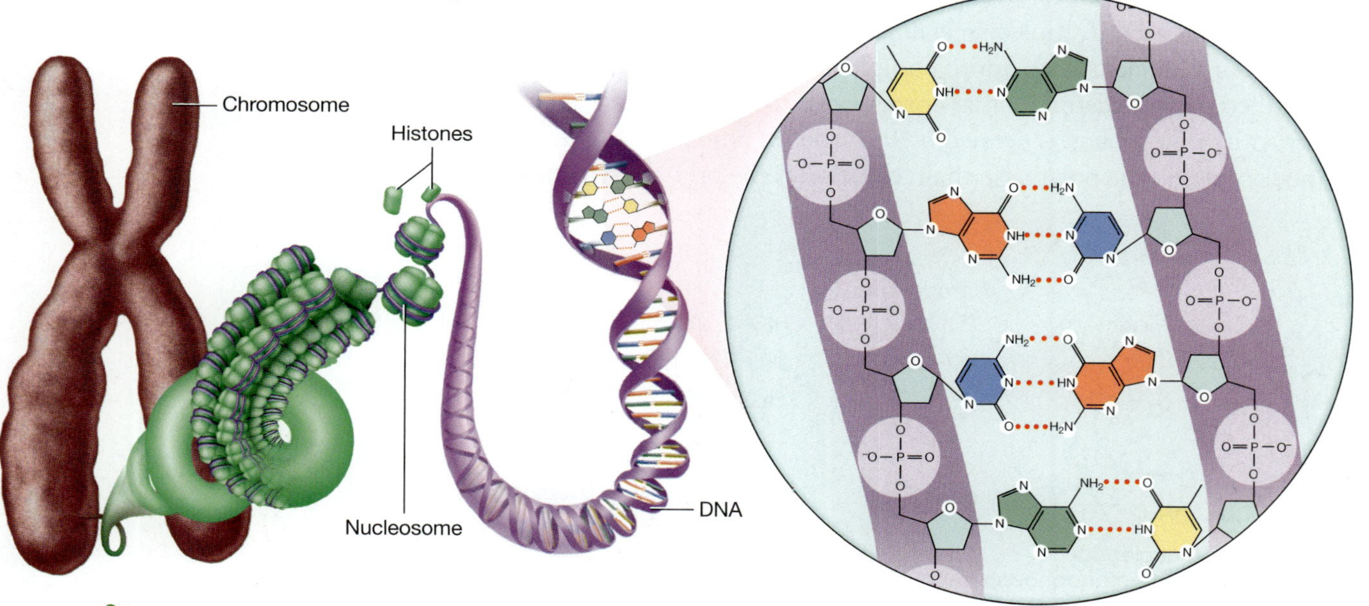

23.3 Secondary and tertiary structures of a DNA molecule combine to form a large structure called a chromosome.

Procedure

What Am I Doing?

Part A: Extracting DNA from a strawberry, precipitating it with cold ethanol, and isolating it by spooling the DNA fibers onto an inoculating loop.

Part B: Hydrolyzing the extracted strawberry DNA and testing the solution for the presence of deoxyribose, phosphate, and purines.

Materials

- ❑ 10 mL, 25 mL, and 50 mL graduated cylinders
- ❑ Glass stirring rod
- ❑ Resealable plastic bag
- ❑ Ring stand
- ❑ Iron ring or clamp
- ❑ Funnel
- ❑ Filter paper
- ❑ Cheesecloth
- ❑ 250 mL and 400 mL beakers
- ❑ Inoculating loop
- ❑ Test tubes (6)
- ❑ Dropper
- ❑ Cellulose paper strip
- ❑ Pasteur pipettes (2)
- ❑ Hair dryer or lab oven
- ❑ Watch glass
- ❑ Red litmus paper
- ❑ Hot plate
- ❑ Forceps

Part A Extracting DNA

1. Add 20 mL of ethanol to a graduated cylinder and chill it in an ice-water bath.
2. Prepare the DNA extraction solution:
   a. In a 50 mL graduated cylinder, mix the following:
      - 1 mL of 10% sodium chloride (NaCl)
      - 1 mL of 10% sodium dodecyl sulfate (SDS)
      - 1 mL of 0.1 M ethylenediaminetetraacetic acid (EDTA)

   b. Add distilled water to the graduated cylinder until the total volume of the solution is 40 mL.

   c. Use a glass stirring rod to ensure that the solution is well mixed.
3. Obtain one strawberry and remove any green sepals (leaves), if present. If a frozen strawberry is used, make sure it is room temperature before proceeding to Step 4.
4. Place the strawberry in a resealable plastic bag. Expel the air from the bag and seal it.
5. Use your hands to thoroughly crush the strawberry inside the bag for about 2 minutes.
6. Pour the extraction solution into the plastic bag with the strawberry. Reseal the bag and mash the strawberry for an additional 1 minute.
7. Secure a funnel onto a ring stand with an iron ring or clamp. Add a double layer of cheesecloth (or a coffee filter) to the inside of the funnel. Place a 250 mL beaker under the funnel.
8. Pour the strawberry mixture from the plastic bag into the filter, allowing the liquid to drain into the beaker.
9. Hold the beaker at an angle and slowly pour the ice-cold ethanol from the graduated cylinder down the side of the beaker. Do this step very carefully so the ethanol forms a layer on top of the strawberry extract rather than mixing with it.
10. Allow the beaker to sit upright undisturbed for at least 2 minutes. Observe the interface and note any changes on the data sheet, page 341.
11. Insert an inoculating loop into the solution just below the interface between the two layers. Rotate the instrument in a single direction to spool the DNA onto the loop (Fig. 23.4).

23.4 To spool the DNA, rotate the inoculating loop in a single direction as if you are winding spaghetti onto a fork.

12. When you have spooled all of the DNA onto the loop, remove the loop from the solution. Place the DNA on a piece of filter paper to dry.
13. On the data sheet, page 341, record your observations about the appearance of the DNA you isolated.

Part B Chemical Identity Tests

1. Prepare a boiling-water bath on a hot plate using a half-filled 400 mL beaker.
2. Label four test tubes 1–4. To each test tube, add the appropriate sample as indicated below:
   - Test Tube 1: 2 mL of 2 M sulfuric acid (H_2SO_4)
   - Test Tube 2: Isolated DNA + 2 mL of 2 M sulfuric acid (H_2SO_4)
   - Test Tube 3: 1 mL of 1% ribose solution + 1 mL of 2 M sulfuric acid (H_2SO_4)
   - Test Tube 4: 1 mL of 1% deoxyribose solution + 1 mL of 2 M sulfuric acid (H_2SO_4)
3. Place the test tubes in the boiling-water bath for at least 15 minutes.
4. After the heating period, remove the test tubes from the bath and allow them to cool for 5 minutes.
5. Label two more test tubes 5 and 6. To Test Tube 5, add half of the cooled DNA/H_2SO_4 solution from Test Tube 2. To Test Tube 6, add 1 mL of 2 M H_2SO_4. **Set aside Test Tubes 5 and 6 for the phosphate and purine tests.**
6. Perform the Dische test on the samples in Test Tubes 1–4 using the following steps:
   a. Add 2 mL of the Dische reagent to each test tube.
   b. Heat the test tubes in the boiling-water bath for up to 15 minutes.
   c. Record your observations on the data sheet, page 341. The appearance of a dark-blue color indicates the presence of deoxyribose in the sample.

SAFETY NOTE

The Dische reagent contains diphenylamine, glacial acetic acid, and concentrated sulfuric acid. Diphenylamine is toxic if ingested, and the acids are highly corrosive! *Wear gloves and handle this reagent with caution!*

7. Perform the phosphate test on the samples in Test Tubes 5 and 6 using the following steps:
   a. Obtain a small strip of cellulose paper. Use a pencil to label two positions for spotting Samples 5 and 6.
   b. Use the narrow end of a Pasteur pipette to spot each sample directly above the corresponding labeled position on the paper: Dip the end of the pipette into the sample and then gently touch it to the surface of the paper, allowing a small sample spot to absorb into the paper.
   c. Using forceps to hold the paper near one edge, dip the entire strip into a molybdate stain solution. Remove it from the solution and place it on a paper towel to absorb any excess liquid.
   d. Heat the paper with a hair dryer or place it on a watch glass and warm it gently on a hot plate or in a lab oven. The development of a blue spot indicates the presence of phosphate. Record your observations on the data sheet.

SAFETY NOTE

The molybdate stain contains sulfuric acid, which is corrosive. *Handle it with care!*

=23

8. Perform the purine test on the samples in Test Tubes 5 and 6 using the following steps:
   a. Add about 2 mL of 2 M ammonium hydroxide (NH_4OH) to each test tube.
   b. Test the pH of the sample by using a glass stirring rod to add 1 drop of the solution to a piece of red litmus paper that has been moistened with distilled water. If the solution is basic (the red paper turns blue), proceed to Step c. If not, add more NH_4OH dropwise until the sample is basic.
   c. Add 10 drops of 0.1 M silver nitrate to each tube.
   d. The appearance of a white precipitate (often as faint cloudiness) indicates the presence of a purine. Record your observations on the data sheet, page 341.
9. Dispose of all waste as directed by your instructor.

Name ______________________

Lab Partner ______________________

Lab Section ______________ Date ______________

Lab 23 Pre-Laboratory Exercise

1 Provide a term that matches each description below.

a Carbohydrate component of DNA. ______________

b Generic term for polymers of nucleotides. ______________

c Secondary structure of DNA. ______________

d Nucleic acid composed of nucleotides that contain ribose. ______________

e Proteins around which DNA coils to produce a tertiary structure. ______________

2 Look at the structures of ribose and deoxyribose in Figure 23.1, page 333. What is the structural difference between these two pentose sugars?

____

____

3 Consider the structure of the nucleotide shown below. What are the structural differences between this nucleotide and deoxycytidine 5′-monophosphate (Fig. 23.1)? Would you expect to find this nucleotide in DNA or RNA? Explain your answer.

NH_2 N N N N $^{-}O-P(=O)(O^{-})-O-CH_2$ O OH OH

Adenosine 5'-monophosphate

____

____

4 Consider the portion of DNA shown in Figure 23.2, page 334. Is DNA neutral or does it bear a charge? If charged, is it positive or negative?

____

____

5 Consider the DNA extraction procedure you will use in this lab. Fill in the boxes on the outline with the appropriate procedural action or expected result.

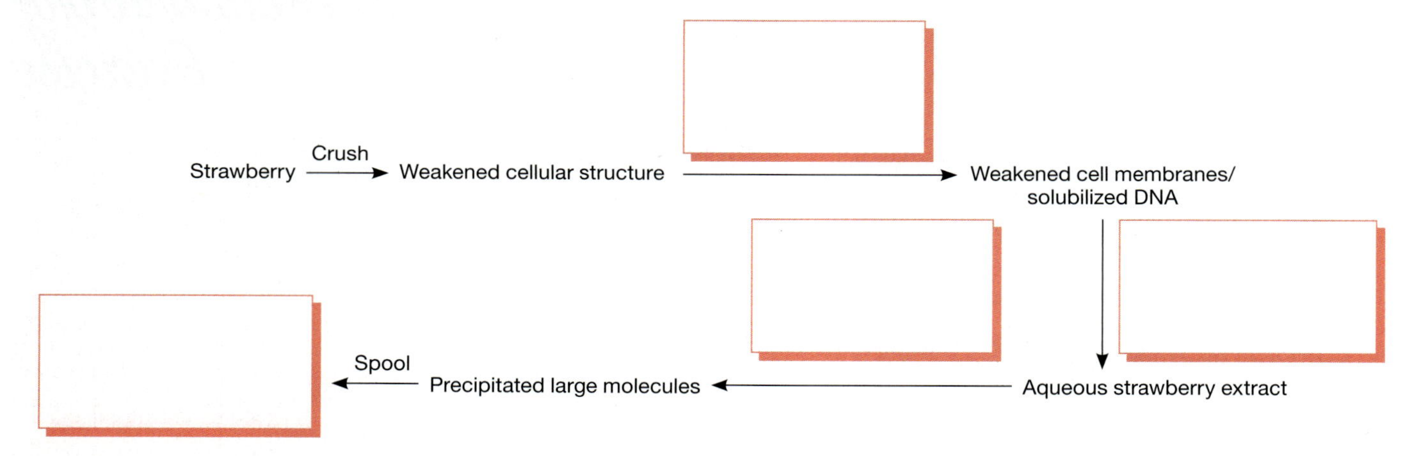

6 Why is it important that the ethanol you use in Step 9 of Part A be ice cold?

Name ______________________________

Lab Partner ______________________________

Lab Section ____________________ Date ____________

Lab 23 DATA SHEET

Data and Observations

Part A Extracting DNA

Procedure References	Description	Observations
Step 10	Strawberry Extract/Ethanol Interface	
Step 13	Extracted DNA	

Part B Chemical Identity Tests

Test Tube	Sample	Dische Test	Phosphate Test	Purine Test
1	H_2SO_4		X	X
2	DNA + H_2SO_4		X	X
3	Ribose + H_2SO_4		X	X
4	Deoxyribose + H_2SO_4		X	X
5	DNA + H_2SO_4	X		
6	H_2SO_4	X		

Reflective Exercises

1 In the table below, make a claim regarding the identity of the material you isolated from the strawberry. Were your results consistent with this material containing DNA? Describe your evidence and provide a rationale for your claim.

Claim	
Evidence	**Rationale**

2 Explain the purpose of mashing the strawberry in the bag.

3 Explain at a molecular level how the detergent in the extraction solution weakened the strawberry cell membranes.

Name ____________________

Lab Partner ____________________

Lab Section ____________________ Date ____________________

Lab 23
DATA SHEET
(continued)

4 The sodium chloride in the extraction solution serves to separate the DNA from the histones. Based on this information and the structure of DNA, what can you infer about the charge of the histone proteins? Use your inference to explain how the sodium chloride helps separate the DNA and the histones.

5 Name some of the classes of compounds that were likely left in the ethanol once you finished spooling the DNA.

6 A single thread of cotton is almost impossible to see with the naked eye, but a rope composed of many cotton threads wound together is clearly visible. Given your knowledge about the size of molecules, do you think the DNA you observed was composed of a single molecular strand or was it many DNA molecules wound together? Explain your answer.

7 What was the purpose of performing the Dische test on the H_2SO_4, ribose, and deoxyribose solutions?

Glossary

– A –

acid compound that produces or donates a hydrogen ion (H^+), usually referred to as a proton, in an aqueous solution

acid-base indicator dye that varies in color depending upon the acidity of the solution, therefore providing a measure of the solution acidity

active site specialized region on an enzyme where the substrate binds and the biochemical reaction is catalyzed

aldose monosaccharide that contains an aldehyde functional group

alkane hydrocarbons that contain only single bonds

alkene hydrocarbons that contain one or more carbon-carbon double bonds

alkyne hydrocarbons that contain one or more carbon-carbon triple bonds

amphipathic compound containing both polar and nonpolar regions

analgesic medicine used to alleviate pain

anion negatively charged ion

anomeric carbon aldehyde or ketone carbon atom in a linear monosaccharide, and the site of monosaccharide ring formation

aqueous solution homogenous mixtures in which water is the solvent

aromatic hydrocarbon hydrocarbon recognizable by the presence of one or more benzene rings, although other bonding patterns also are possible

– B –

Barfoed's reagent chemical reagent often used to distinguish between monosaccharides and disaccharides; like Benedict's test, Barfoed's test relies on the carbohydrate acting as a reducing agent toward Cu^{2+}, and a positive test is visualized by the generation of a copper(I) precipitate; however, Barfoed's reagent reacts much more quickly with monosaccharides than disaccharides, allowing the two types of carbohydrates to be distinguished from each other

base compound that accepts a proton in an acid-base reaction

Benedict's reagent chemical reagent commonly used to test for the presence of reducing sugars; the test relies on the reduction by the sugar of Cu^{2+} to Cu^+, generating a precipitate that is usually red in color

buffer solution that resists large changes in pH upon addition of either an acid or a base

– C –

calorimeter device used to perform calorimetry experiments

calorimetry experimental technique that measures heat transfers in chemical reactions

carbohydrate polyhydroxylated aldehyde, polyhydroxylated ketone, or a substance derived therefrom

casein dominant class of proteins found in milk

catalyst substance that increases the rate of a chemical reaction without undergoing a permanent change during the reaction

cation positively charged ion

cell membrane semipermeable barrier between the inside and outside environments of a living cell; it is composed predominantly of a bilayer of phospholipids

Charles' Law directly proportional relationship between the volume and temperature of a fixed quantity of gas at constant pressure

chemical change process that involves a change in the composition of a substance

chemical properties qualities that are displayed by a substance when it changes composition

chromatography technique used to separate compounds in a mixture based on differences in the way they interact with a stationary medium and a mobile medium

chromosomes structure composed of DNA tightly wound around histone proteins found in the nucleus of a cell

claim statement that summarizes the outcome or implications of an experiment in the form of an answer to a research question, explanation of data, conclusion, or description

colloid two-phase system of any state in which one component is permanently suspended in the other

concentration quantity of a solute dissolved in a given amount of solution

condensation reaction chemical reaction in which two or more small molecules combine to form a larger molecule; a small molecule, such as water, is also produced

conjugate acid substance formed when a base gains a proton

conjugate acid-base pair chemical species that differ from each other by a single proton

conjugate base substance formed when an acid loses a proton

constitutional isomers compounds with the same molecular formula in which the atoms are connected differently

control baseline for comparison when the conditions of an experiment are modified

covalent compound molecule formed by the sharing of electrons between nonmetal atoms

-D-

deoxyribonucleic acid (DNA) nucleic acid used as the genetic material of living cells; it is a polymer composed of deoxyribonucleotides

dialysis process involving the movement of solutes across a semipermeable membrane from the region of higher solute concentration to the region of lower solute concentration

dialyzing solution solution in which a dialysis tube is immersed

disaccharide type of carbohydrate composed of two monosaccharide subunits linked through a glycosidic bond

double helix shape describing the secondary structure of DNA formed by hydrogen bonding between base pairs of two strands

-E-

empirical formula smallest whole number ratio of a compound's components

emulsifier substance that causes polar and nonpolar compounds to be suspended in the same mixture

emulsion mixture of two normally immiscible liquids

enzyme catalyst for a biochemical reaction

evidence measurements or observations gathered over the course of an experiment or scientific investigation

extraction process of selectively transferring one component of a mixture into a different phase

-F-

formula expression of a compound's composition that uses element symbols and subscripts to denote the identity and quantity of each atom or ion present in a single unit of the substance

functional group set of atoms that are bound in a characteristic pattern and exhibit a predictable chemical behavior

furanose five-membered ring form of a monosaccharide

-G-

glycosidic bond bond between the anomeric carbon of one sugar and an alcohol group of another compound

group column on the periodic table

-H-

heat flow of energy between a warmer substance to a cooler one

histone protein around which DNA coils

homogenization mechanical treatment process that breaks fats into smaller globules that remain suspended in the aqueous layer for extended periods of time

hydrocarbon molecule composed only of carbon and hydrogen atoms

hydrolysis chemical transformation that involves the breaking of a large molecule into smaller subunits by reaction with water

hydrophilic literally "water-loving"; refers to polar compounds or portions of compounds that readily dissolve in or are attracted to water

hydrophobic literally "water-fearing"; refers to nonpolar compounds or part of compounds that do not readily dissolve in and/or repel water

hypertonic having a concentration higher than the surrounding or adjacent solution

hypotonic having a concentration lower than the surrounding or adjacent solution

-I-

immiscible term describing liquids that do not mix with each other to produce a homogeneous solution

insoluble inability of a solid to fully dissolve in a liquid

ion chemical species that bears an electrical charge

ionic compound neutral compound composed of ions

ionize process of an atom or molecule gaining or losing electrons to form an ion; also, the process whereby a species dissociates into ions when dissolved

isoelectric point pH at which the net charge of a charged species is zero

isomer (*see* constitutional isomer) compounds that share a molecular formula but have different structures

isotonic having a concentration equal to the surrounding or adjacent solution

ketose monosaccharide that contains a ketone functional group

-L-

lactase enzyme that hydrolyzes the disaccharide lactose into its monosaccharide components

Lewis structure visual representation used to describe the arrangement of atoms and electrons in covalent compounds, where the chemical symbol for elements are surrounded by dots, representing nonbonding electrons, and dashes, which represent bonds

-M-

meniscus curve exhibited by certain liquids when adhered to a glass cylinder at the surface

mineral element that is essential for life and must be obtained from the diet

miscible term describing liquids that mix to form a homogeneous solution

mobile phase medium in a chromatographic system that flows over or through the stationary phase

molar mass mass of one mole of a substance

molecular formula chemical formula of a covalent compound that represents the actual number of each type of atom in a single molecule

monosaccharide literally "single sugar"; a carbohydrate that cannot be hydrolyzed to a simpler carbohydrate; monosaccharides generally contain three to seven carbon atoms and possess multiple hydroxyl groups and either an aldehyde or a ketone functional group

-N-

neutralization reaction chemical reaction involving the generation of neutral products from the combination of an acid and a base

nucleic acid polymer composed of nucleotides

nucleotide molecule composed of a nitrogenous base, a monosaccharide, and a phosphate group; nucleotides are the monomer subunits of nucleic acids

nutrient chemical components of food that are required for growth and proper biological function

-O-

octet rule tendency of most atoms to react to achieve a configuration with eight electrons in their outer shell

origin starting line of a thin layer chromatography experiment where the sample was first spotted

osmosis process of solvent molecules (typically water) passing across a semipermeable membrane in a direction that would equalize the concentrations on either side

-P-

partition redistribution of molecules between two solvents

percent yield quantity, expressed as a percentage, of product actually isolated relative to the theoretical yield of a chemical reaction

period row in the periodic table

periodic law principle that the physical and chemical properties of the elements recur in a repeating pattern when they are ordered by increasing atomic number

periodic table table in which elements are arranged into characteristic horizontal rows and vertical columns by increasing atomic number

pH quantitative expression of the hydronium ion concentration in a solution: $pH = -\log[H_3O^+]$

phosphodiester bond type of bond that links nucleotides in a nucleic acid

physical change transformation, such as a change in physical state, that does not alter the composition of a substance

physical properties qualities of a substance that are displayed without changing the identity of the substance; examples include smell, color, physical state, melting point, boiling point, density, and solubility

polyatomic ion ion composed of more than one atom

polysaccharide carbohydrate polymer composed of many monosaccharide subunits; polysaccharides can be hydrolyzed into small carbohydrates and/or monosaccharides

precipitate insoluble solid formed in solution during a chemical reaction

pyranose six-membered ring form of a monosaccharide

-Q-

qualitative analysis experiment or test designed to indicate identity

quantitative analysis experiment or test designed to indicate quantity

-R-

rationale an explanation of how or why the evidence gathered during an experiment or scientific investigation supports or justifies a claim

recrystallization process in which a solid is purified by dissolving it in a hot solvent and then allowing the solvent to cool slowly, causing the material to resolidify in a purified form

reducing sugar carbohydrate that can be oxidized by a mild oxidizing agent; reducing sugars generally possess the ability to generate a free aldehyde group in solution and give a positive Benedict's test

retention factor in a thin layer chromatography experiment, the ratio of the distance up the plate traveled by a compound to the distance up the plate traveled by the mobile phase

ribonucleic acid (RNA) biological nucleic acid composed of ribonucleotides

saponification chemical reaction that uses hydroxide ions to hydrolyze ester-containing molecules into alcohols and carboxylate salts; saponification is used to produce soaps by treating animal fats or plant oils with lye (NaOH) or potash (KOH)

saturated hydrocarbon hydrocarbons in which each carbon atom has the maximum number of possible bonds to hydrogen atoms; saturated hydrocarbons contain only single bonds

Seliwanoff's reagent chemical reagent often used to distinguish between aldoses and ketoses

semipermeable membrane membrane that allows only certain solutes to move freely across it while the movement of others is inhibited

soap ionic compound formed by the interaction of a small cation, generally sodium or potassium, with the carboxylate anion of a long-chain fatty acid

solubility measure of the ability of a solute to dissolve in a particular solvent

soluble ability of a solid to fully dissolve in a liquid

solute component of a solution present in a lesser amount

solvent component of a solution present in the greatest amount

solvent front in a thin layer chromatography experiment, the line demarking the farthest distance travelled by the mobile phase from the origin

specific heat capacity (*also* specific heat) intrinsic ability of the substance to absorb thermal energy; specifically, the amount of heat required to raise the temperature of 1 g of substance by 1°C

stationary phase medium that does not move in a chromatography system

substituent atom, or group of atoms, that is not a part of the main skeleton in an organic compound but is attached to it

substrate organic compound that binds to an enzyme and is the reactant in the enzyme-catalyzed reaction

-T-

theoretical yield maximum quantity of product that can potentially be produced during a chemical reaction

thin layer chromatography technique used to separate small quantities of compounds in a mixture. The mixture is applied to a plate coated with a very thin layer of stationary medium and a solvent (mobile phase) is then allow to travel up the plate through the stationary phase; compounds in the mixture travel different distances up the plate based on their relative affinities for the two phases

triacylglycerol molecule composed of three long-chain fatty acids components attached to a single glycerol backbone through ester bonds; triacylglycerols, also called triglycerides, are the chemical species best known as fats and oils

-U-

unsaturated hydrocarbon hydrocarbons in which one or more carbon atoms do not contain the maximum number of bonds to hydrogen; unsaturated hydrocarbons contain one or more double or triple bonds

-V-

valence electron electrons found in the outermost energy level of the atom

-W-

whey protein non-casein, water-soluble proteins found in milk

Photo Credits

All photos are courtesy of William G. O'Neal unless noted here.

- **Chapter 1 opener**
  Wladimir Bulgar/Science Source
- **Chapter 2 opener**
  Will & Deni McIntyre/Science Source
- **Chapter 3 opener**
  Richard Beech Photography/Science Source
- **Chapter 4 opener**
  Manfred Kage/Science Source
- **Figure 4.2B**
  Patrick Llewelyn-Davies/Science Source
- **Chapter 5 opener**
  Charles D. Winters/Science Source
- **Chapter 6 opener**
  Andrew Lambert Photography/Science Source
- **Figure 6.1**
  Andrew Lambert Photography/Science Source
- **Chapter 6 Reflective Questions 6, 7**
  Andrew Lambert Photography/Science Source
- **Chapter 7 opener**
  Amy Stevens
- **Chapter 8 opener**
  Martyn F. Chillmaid/Science Source
- **Chapter 9 opener**
  Gary Retherford/Science Source
- **Figure 9.1A**
  Amy Stevens

- **Chapter 10 opener**
  Charles D. Winters/Science Source
- **Chapter 11 opener**
  Philippe Garo/Science Source
- **Chapter 12 opener**
  Sam Pierson/Science Source
- **Chapter 13 opener**
  SCIMAT/Science Source
- **Chapter 14 opener**
  Gusto Productions/Science Source
- **Chapter 15 opener**
  Andy Crawford/Dorling Kindersley/Science Source
- **Chapter 16 opener**
  Charles D. Winters/Science Source
- **Chapter 17 opener**
  John Crawley/*Exploring Biology in the Laboratory*, 2nd edition, © 2014 Morton Publishing
- **Chapter 18 opener**
  Frédérique Bidault/Science Source
- **Chapter 19 opener**
  Jean-Paul Chassenet/Science Source
- **Chapter 20 opener**
  Ted Kinsman/Science Source
- **Chapter 21 opener**
  James King-Holmes/Science Source
- **Chapter 22 opener**
  Antonio Romero/Science Source
- **Chapter 23 opener**
  Amy Stevens

Index

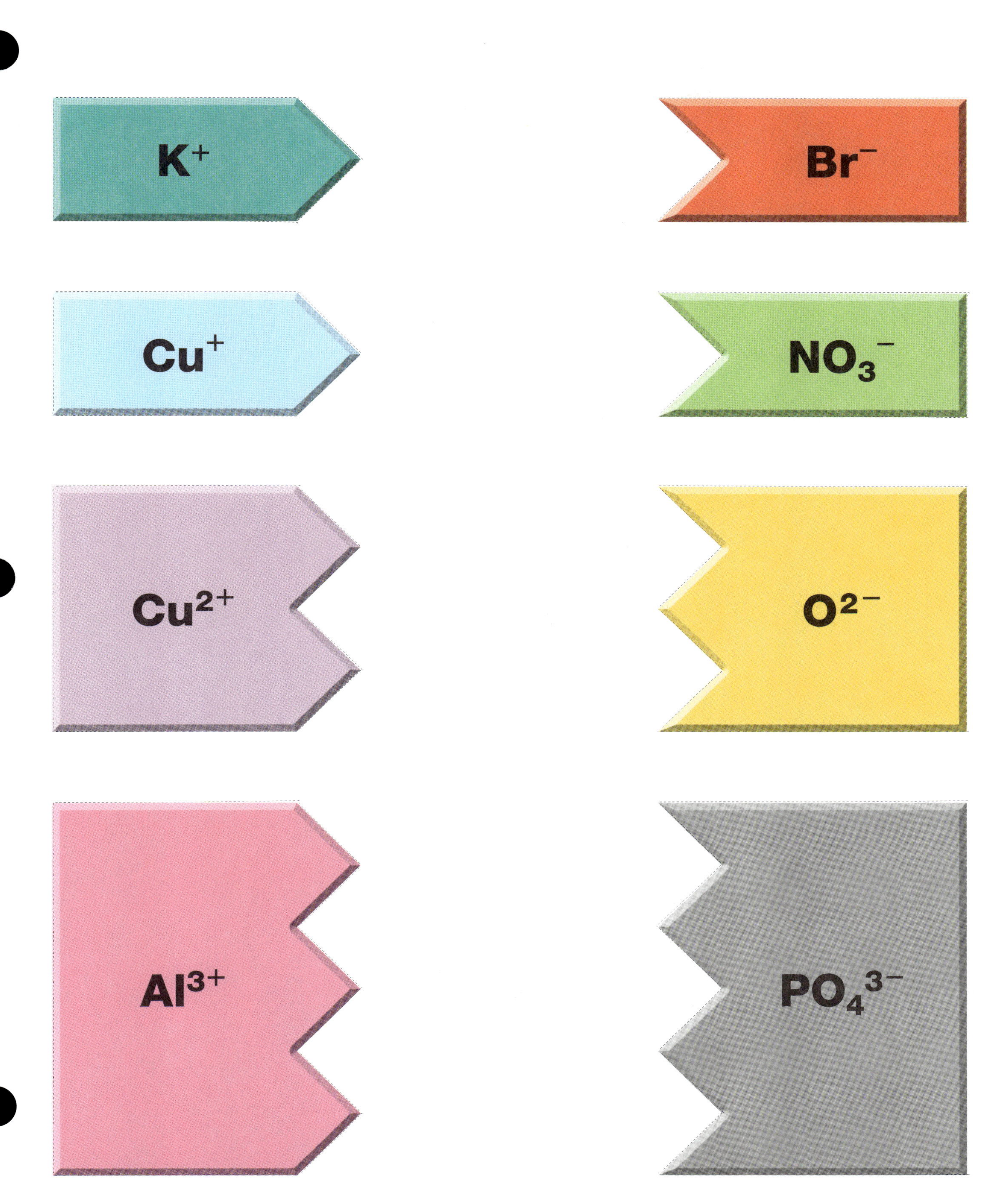
K^+
Br^-
Cu^+
NO_3^-
Cu^{2+}
O^{2-}
Al^{3+}
PO_4^{3-}

K^+

Br^-

Cu^+

NO_3^-

Cu^{2+}

O^{2-}

Al^{3+}

PO_4^{3-}

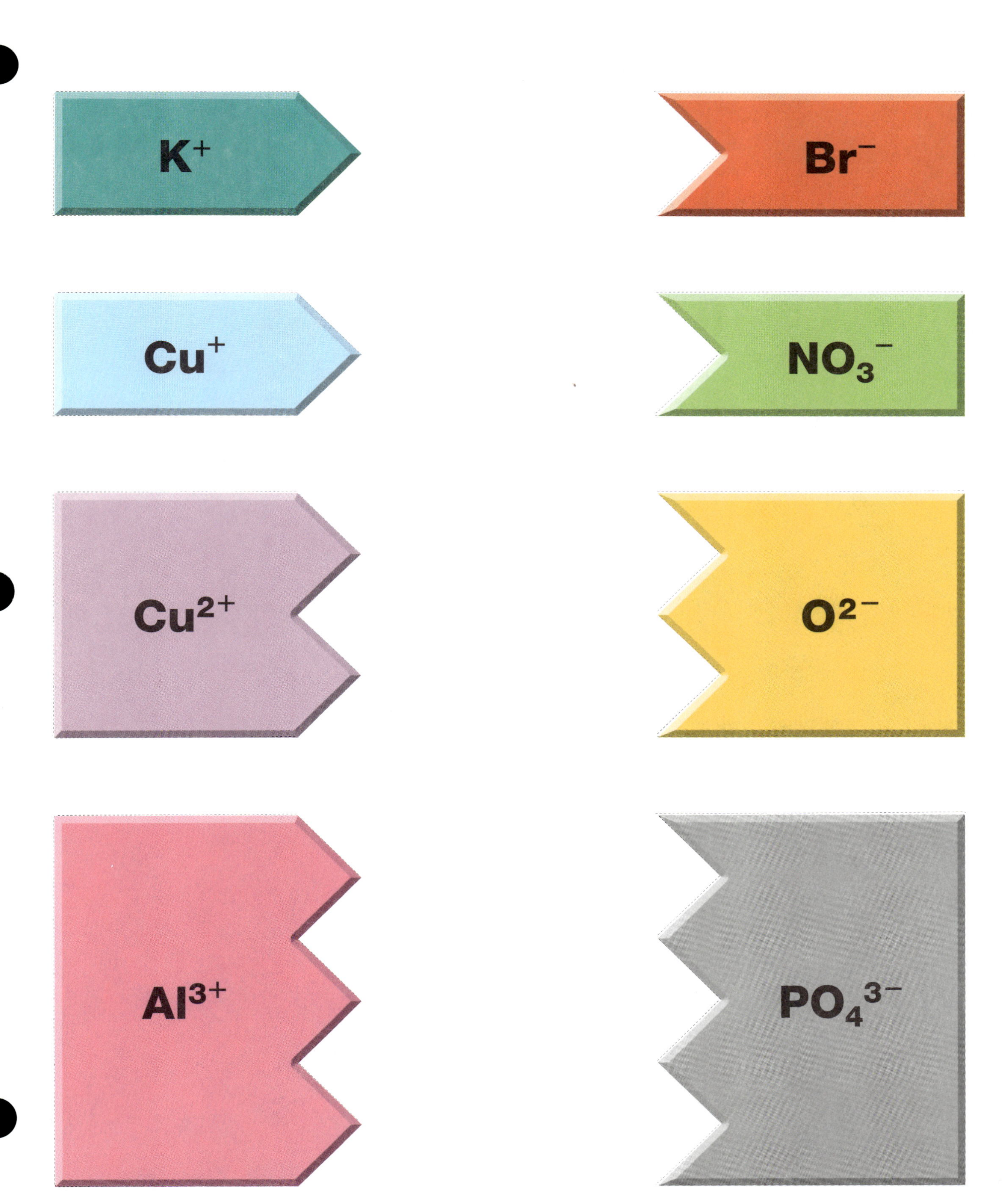
K^+
Br^-
Cu^+
NO_3^-
Cu^{2+}
O^{2-}
Al^{3+}
PO_4^{3-}

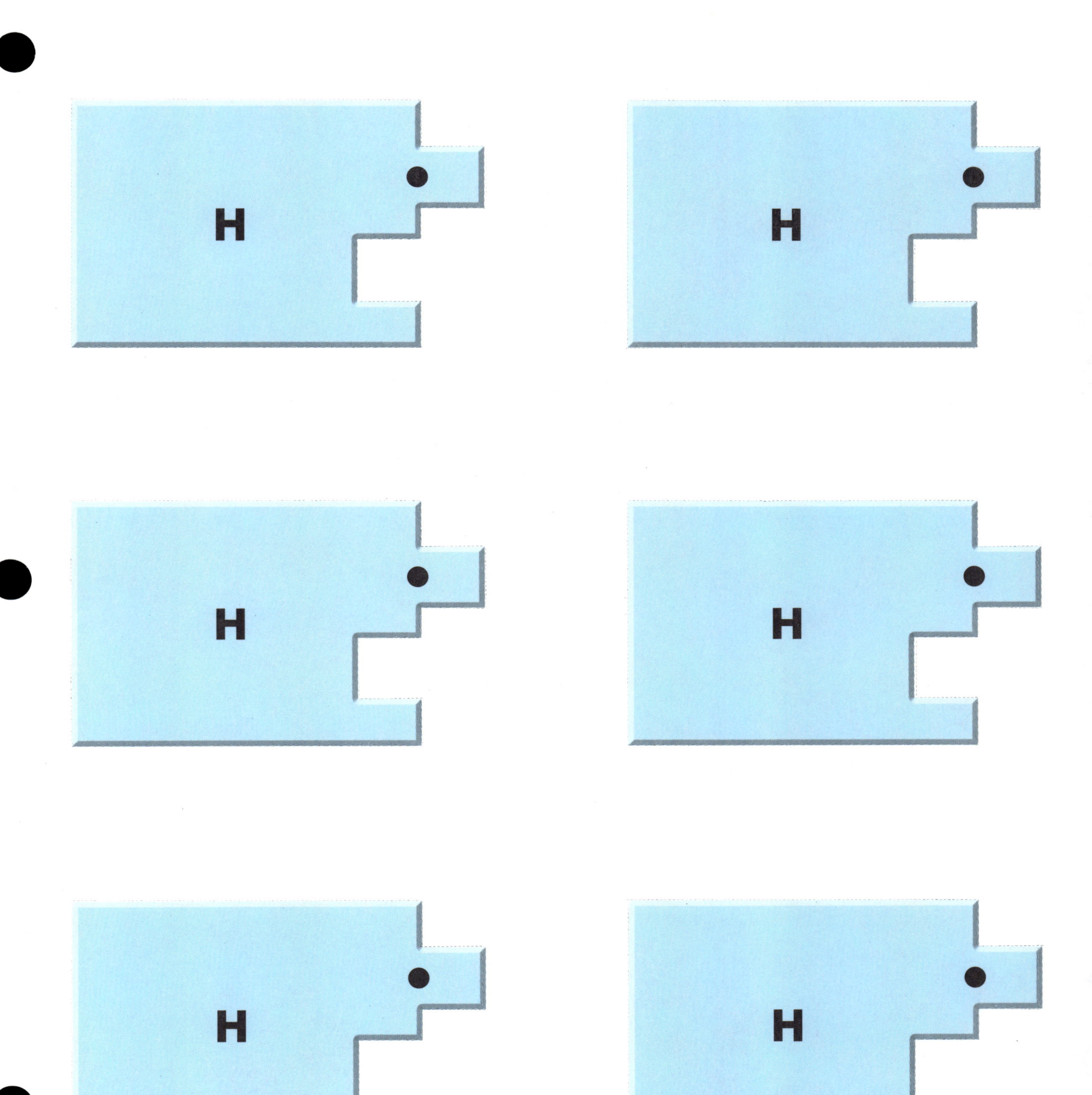
H
H
H
H
H
H

C

C

C

C

C

C

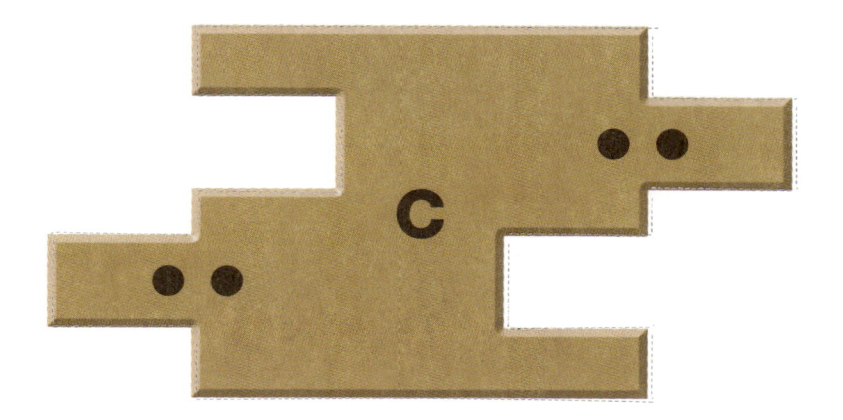
C

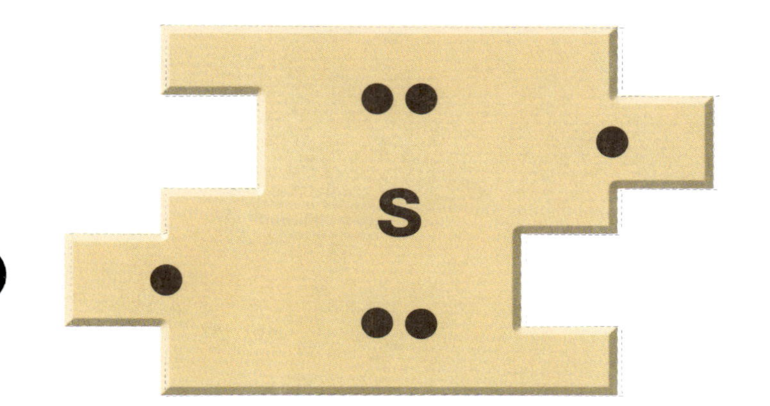
S

N

N

P

N

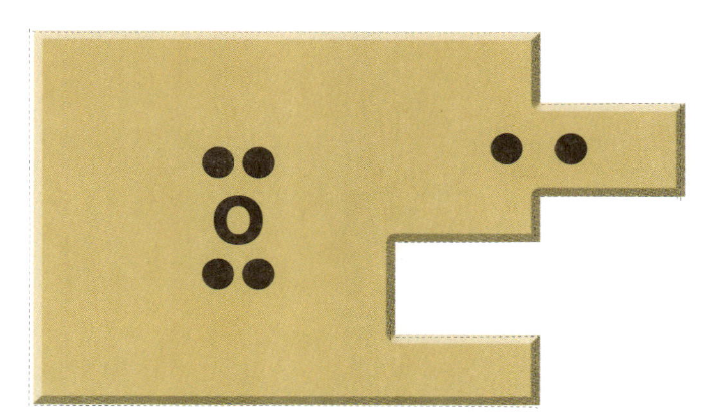
O

N

N

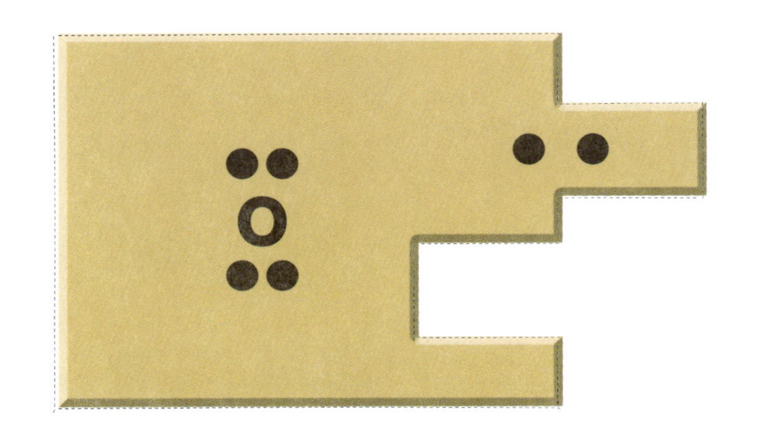
O

N

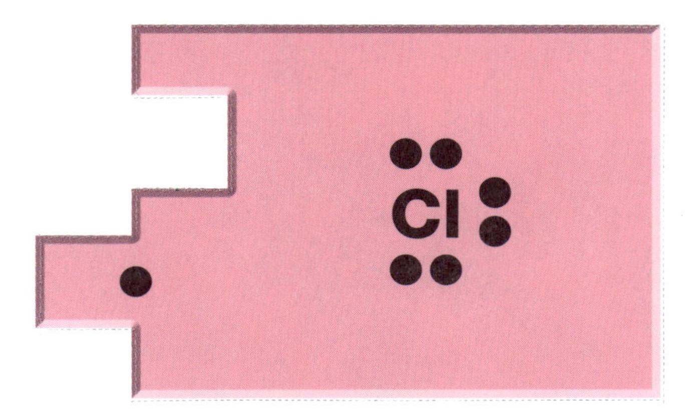
Cl

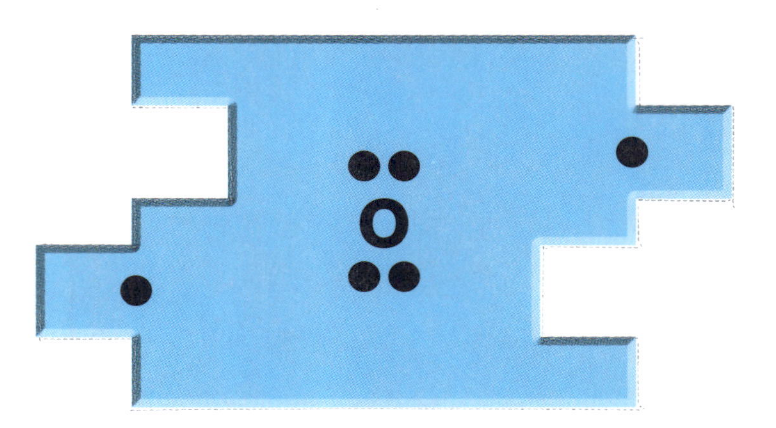
O

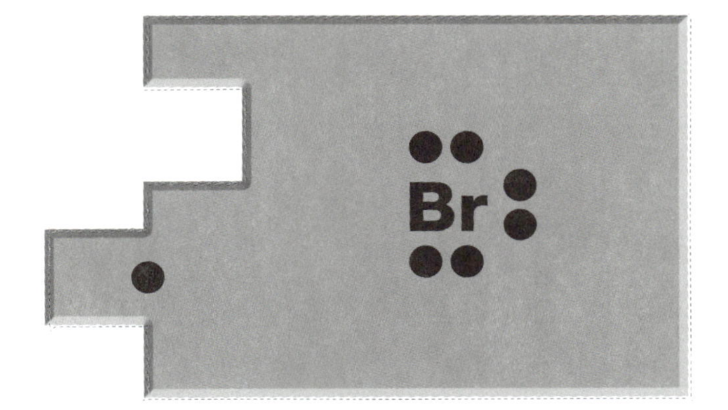
Br

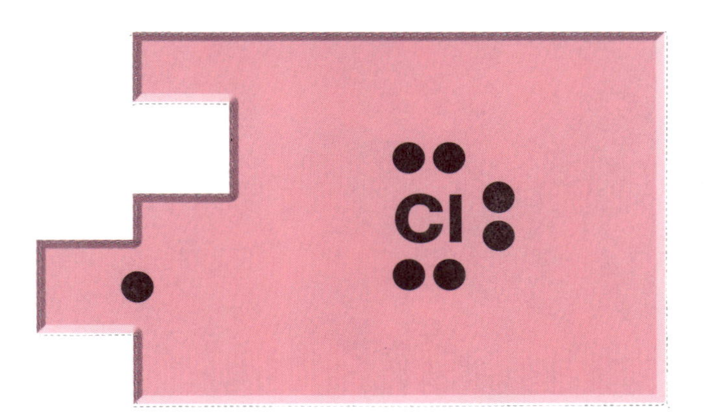
Cl